Computer-Mediated Learning for Workforce Development

Dominic Mentor
Teachers College, Columbia University, USA

A volume in the Advances in Educational Technologies and Instructional Design (AETID) Book Series

IGI Global
DISSEMINATOR OF KNOWLEDGE

Published in the United States of America by
IGI Global
Information Science Reference (an imprint of IGI Global)
701 E. Chocolate Avenue
Hershey PA, USA 17033
Tel: 717-533-8845
Fax: 717-533-8661
E-mail: cust@igi-global.com
Web site: http://www.igi-global.com

Library of Congress Cataloging-in-Publication Data

Names: Mentor, Dominic, 1968- editor.
Title: Computer-mediated learning for workforce development / Dominic Mentor, editor.
Description: Hershey, PA : Information Science Reference, 2018.
Identifiers: LCCN 2017028585| ISBN 9781522541110 (hardcover) | ISBN 9781522541127 (ebook)
Subjects: LCSH: Computer-assisted instruction. | Educational technology. | Internet in education. | Mobile communication systems in education. | Academic-industrial collaboration.
Classification: LCC LB1028.5 .C5535 2018 | DDC 371.33--dc23 LC record available at https://lccn.loc.gov/2017028585

This book is published in the IGI Global book series Advances in Educational Technologies and Instructional Design (AE-TID) (ISSN: 2326-8905; eISSN: 2326-8913)

British Cataloguing in Publication Data
A Cataloguing in Publication record for this book is available from the British Library.

Advances in Educational Technologies and Instructional Design (AETID) Book Series

Lawrence A. Tomei
Robert Morris University, USA

ISSN:2326-8905
EISSN:2326-8913

Mission

Education has undergone, and continues to undergo, immense changes in the way it is enacted and distributed to both child and adult learners. From distance education, Massive-Open-Online-Courses (MOOCs), and electronic tablets in the classroom, technology is now an integral part of the educational experience and is also affecting the way educators communicate information to students.

The **Advances in Educational Technologies & Instructional Design (AETID) Book Series** is a resource where researchers, students, administrators, and educators alike can find the most updated research and theories regarding technology's integration within education and its effect on teaching as a practice.

Coverage

- E-Learning
- Hybrid Learning
- Instructional Design
- Adaptive Learning
- Collaboration Tools
- Curriculum Development
- Online Media in Classrooms
- Web 2.0 and Education
- Educational Telecommunications
- Virtual School Environments

IGI Global is currently accepting manuscripts for publication within this series. To submit a proposal for a volume in this series, please contact our Acquisition Editors at Acquisitions@igi-global.com or visit: http://www.igi-global.com/publish/.

The Advances in Educational Technologies and Instructional Design (AETID) Book Series (ISSN 2326-8905) is published by IGI Global, 701 E. Chocolate Avenue, Hershey, PA 17033-1240, USA, www.igi-global.com. This series is composed of titles available for purchase individually; each title is edited to be contextually exclusive from any other title within the series. For pricing and ordering information please visit http://www.igi-global.com/book-series/advances-educational-technologies-instructional-design/73678. Postmaster: Send all address changes to above address.

Titles in this Series

For a list of additional titles in this series, please visit: www.igi-global.com/book-series

Assessing Social Support and Stress in Autism-Focused Virtual Communities Emerging Research and Opportunities
Amit Saha (Wake Forest Baptist Medical Center, USA) and Nitin Agarwal (University of Arkansas at Little Rock, USA)
Information Science Reference • copyright 2018 • 131pp • H/C (ISBN: 9781522540205) • US $135.00 (our price)

Learner Experience and Usability in Online Education
Imed Bouchrika (University of Souk Ahras, Algeria) Nouzha Harrati (University of Souk Ahras, Algeria) and Phu Vu (University of Nebraska at Kearney, USA)
Information Science Reference • copyright 2018 • 316pp • H/C (ISBN: 9781522542063) • US $195.00 (our price)

Pedagogy Development for Teaching Online Music
Carol Johnson (University of Melbourne, Australia) and Virginia Christy Lamothe (Belmont University, USA)
Information Science Reference • copyright 2018 • 368pp • H/C (ISBN: 9781522551096) • US $195.00 (our price)

Enhancing Education Through Open Degree Programs and Prior Learning Assessment
Carolyn N. Stevenson (Kaplan University, USA)
Information Science Reference • copyright 2018 • 306pp • H/C (ISBN: 9781522552550) • US $185.00 (our price)

Innovative Applications of Online Pedagogy and Course Design
Ramesh C. Sharma (Wawasan Open University, Malaysia)
Information Science Reference • copyright 2018 • 451pp • H/C (ISBN: 9781522554660) • US $195.00 (our price)

Enhancing Social Presence in Online Learning Environments
Michael Marmon (University of North Texas, USA)
Information Science Reference • copyright 2018 • 317pp • H/C (ISBN: 9781522532293) • US $185.00 (our price)

Digital Technologies and Instructional Design for Personalized Learning
Robert Zheng (University of Utah, USA)
Information Science Reference • copyright 2018 • 391pp • H/C (ISBN: 9781522539407) • US $195.00 (our price)

Curriculum Development for Gifted Education Programs
Jessica Cannaday (Azusa Pacific University, USA)
Information Science Reference • copyright 2018 • 284pp • H/C (ISBN: 9781522530411) • US $175.00 (our price)

IGI Global
DISSEMINATOR OF KNOWLEDGE

701 East Chocolate Avenue, Hershey, PA 17033, USA
Tel: 717-533-8845 x100 • Fax: 717-533-8661
E-Mail: cust@igi-global.com • www.igi-global.com

Table of Contents

Detailed Table of Contents

In this chapter, the authors argue for designing the context, not just the technology, of e-learning simulations (eSims). They examine the limitations of "pure" eSims (i.e., not involving human-delivered feedback) in light of design challenges for informal learning, and learning transfer, in today's knowledge-intensive work environment when solutions call for creative responses to messy, non-structured tasks in different organizational contexts. Authentic learning on realistic tasks is needed in the design of eSims in such situations. Neaman describes ways that directed performance support (DPS), an approach he pioneered, helps to overcome such limitations. He identifies a core set of DPS guidelines that others can use to adapt and utilize DPS designs in their eSimS. Using a system dynamics framework, Marsick discusses organizational barriers and supports that also need to be considered in eSim learning design for effective engagement and learning transfer.

In recent years, there has been a worldwide commitment to integrating technology into classrooms to train students in the computer skills they will need to be workforce ready. This chapter provides an overview of the Secret Agent Society – Whole of Class Project: a randomized controlled trial of a gaming-based social skills training curriculum that was conducted in Grade 5 classrooms throughout Australia. The chapter explores the content, structure, and delivery format of the social skills program, together with the design and implementation of an online teacher training course to accompany it. Teacher feedback on the online teacher training course is presented, together with recommendations for enhancing the design, implementation, and evaluation of future online professional development courses for school staff.

Technological advancements of the twenty-first century have yielded a greater need for systematic aptitude and preparedness for the current U.S. and global workforce. To meet the needs of the twenty-first century workforce requires teachers to alter the manner in which they teach because student needs call for a new type of learning. The twenty-first century teacher has to assist students develop their technological identities for a new culture of learning shaped by technological advances and globalization. This new culture of learning is less teacher focused and more student centered. As more employers demand graduates who have the requisite twenty-first century skills, higher education must also continue to develop students who are problem solvers, effective communicators, critical thinkers, and effective collaborators. This chapter focuses upon the existing digital divide in historically marginalized communities.

In the aftermath of the Arab Spring, Occupy Wall Street, and Black Lives Matter movements, protest has become the default response to social problems. As students and youth become more involved in political upheaval, they turn to the technology that surrounds them. This chapter focuses on computer-mediated youth civic action and interaction. It examines past trends in youth activism and how social media skills acquired through activism could help these same youth activists transition to the workforce.

This chapter reports on efforts to create a sustainable model to increase engagement, success, and retention in a MOOC for learning computer programming, for a U.S.-based national vocational program. In 2014, the training organization was one a few national and regional organizations who were awarded scholarships by a telecommunication's company to participate in a MOOC whose curriculum was informed and designed by multinational corporations to try and address the dearth of young computer programming talent. The vocational training program aimed to convert MOOC registrations into active and supported participation, with a view to increase completion. Theoretical frameworks were employed to ramp up knowledge of an unknown subject area and skill. Social connectedness methods were used to create teaching and learning communities (TLC) of support. Key results allowed the organization's trainees to outperform all other participating organizations. Resulting in the organization being awarded 500 more scholarships for computer programming that could be used over a three-year period.

Mentoring, the tutelage of one person by a more experienced one, is well documented to have a multitude of benefits. Mentoring, which can be traced back to Greek mythology and beyond, has seen its format evolve, especially with the advancement of technology. Traditional models of mentoring include in-person mentoring or more recently, online mentoring. For some underrepresented groups, mentors provide models for success. In 2014, capitalizing on technological advancements and the need for in-person dialogue with a larger constituency, Weill Cornell Medicine launched a science, technology, engineering, and math (STEM) blended learning curriculum during its summer program that prepares undergraduates for careers as physician-scientists. This initiative fuses the positive aspects of in-person mentoring along with the opportunities provided by virtual mentoring by a larger group of peer mentors. Blended learning, together with in-person and virtual mentoring, offers a newly charted multi-dimensional approach to fulfilling academic and career goals during a STEM summer program.

This chapter focused on cultivating mobile activism mobile journalism with middle and high schoolers of a town in Long Island. The youth film production effort was in response to a hate crime. An immigrant was attacked and killed by a group of young males after a suspected spree of other attacks that same night. After the murderous incident, immigrant parent and students of the local schools feared for their lives. Working towards the goals, the organizers set out to teach students how to use mobile and computer technologies for filmmaking. Using themes of human rights, they also focused on responding to hate crimes and immigration issues. This chapter offers key discoveries and lessons. The short intensive program provided academic and workforce development skills as well as how to use computer technology for digitizing personal narratives. The program also offered informal academic purposes, along with observations, opportunities, and recommendations from the findings for other K-12 digital video filmmaking endeavors.

Selecting and implementing the platforms and tools to support online learning effectively in higher education is currently a challenge for which there is not a one-size-fits-all solution. Finding the right balance between supporting strong pedagogy, offering training and support, providing data security and privacy, ensuring ease of use, among other factors, shape the decisions that leaders in higher education make as they develop and implement online learning environments. This chapter explores the results of a study conducted during the Fall of 2016 to better understand the efforts higher education experts undergo to develop online learning environments. The study considers the sometimes-competing pedagogical and pragmatic needs such efforts entail and seeks to identify trends and best practices.

Currently, technology is used for various purposes in every aspect of life. One of these areas is the field of education, which has an important role in advancing the technologies. Educators often desire to concentrate on their own courses or fields. However, the changing perception of the students and the technologies they are interested in are necessitating educators to follow technological innovations and developments in instructional technologies. The objective of this chapter is to attract attention to some of the global virtual, augmented, and mobile applications that are currently utilized within the medical learning arena. The authors also address technologies that aim to train human resources in health and other fields. The chapter illuminates the use of virtual, augmented, and mobile learning environments via their use in health education.

This chapter introduces how mobile devices can be used on a personal and macro level for professional development. The chapter also covers theories posited for practical applications from pedagogy and andragogy perspectives. There are multiple layers of considerations in terms of context, content, and collaboration factors to optimize mobile learning. There are more mobile devices in the world than people, and more than three quarters of the world's population already has some type of mobile phone, making it the most wide-spread technology and most common electronic device in people's hands. Un-tapping this ubiquitous technology creates a wide array of educational possibilities. Hence, a mobile first learning design is crucial in organizational leadership and professional development, to help bridge the gap between personal lives, schools, colleges, and the workplace. Furthering the concept of learning is everywhere as a natural segue for ownership of learning and engagement.

With perceptual capabilities, computers can intelligently function as a part of our everyday lives, helping us to make sense of what is happening and joining in with us as we experience and navigate through many different types of situations. In this way, computing systems can be situated when they combine machine perception with background knowledge to observe, explore, and interpret human and environmental activity in a way that supports decision making. Many of the current, prominent situated computing solutions are consumer-focused in nature. But these systems will, in time, change the way we work as well as how we learn. Many disciplines are adapting and changing in the process, and many opportunities remain, including the use of situated computing systems for workforce education. There are many opportunities to innovate and improve workforce education by leveraging the power and affordances of mobile-device technology to turn everyday situations into everyday learning opportunities.

Personal response systems (PRSs) are prevalent across a range of educational settings, and this increasing importance has prompted many researchers to examine their various aspects. Their effects on students' learning performance have three main categories of antecedents: the learners' characteristics, the instructors' characteristics, and other contextual factors. A theoretical model is developed on the effects of PRSs on students' learning performance. This chapter describes the characteristics of PRSs, reviews their advantages and disadvantages, and proposes a theoretical model of their antecedents on students' learning performance. It concludes by exploring research implications and directions for future PRS research.

This chapter aims to equip readers with a conceptual understanding to help them leverage experience-based learning in electronic (e) and mobile (m) learning environments. We are in times where learning goals needs to cater to increasingly complex scenarios that require non-didactic methods. Experiential learning emerges as a promising way to deliver such outcomes. David Kolb's experiential learning model emerges as a popular model to conceptualize such learning. E-learning and m-learning cater to the needs of an increasingly mobile learner who seeks situated and personalized learning. There is a need to incorporate experiential features in e/m-learning in a workplace informed manner. The authors weave the learnings from a pilot research and from real world examples to conceptualize a model of experiential learning for e/m-learning environments. The purpose of the models is to help learners critically evaluate other learning applications in digital environments or even design their own.

Foreword

Despite technology's pervasive reach into nearly every aspect of our lives, classrooms and training centers often, shockingly, remain disconnected. There are many reasons for this, including poorly-thought-out technology strategies leading to closets of unopened smartboards because trainers haven't been taught how to use them, or mandatory "web-enhancement" programs that result in artificial student or adult learners' discussions on stilted message boards. For whatever reason, there remains a clear need to discuss, evaluate, and propose novel ways of using computers to integrate instruction into learners' real-world lives, improve outcomes, and make learning more efficient.

Consequently, I was thrilled to discover that Dr. Dominic Mentor was curating this collection on computer-mediated learning for workforce development, and I'm delighted to offer this introduction to Dr. Mentor and his work. Over the years, I have had the chance to work with academics in multiple contexts, from my time serving as the principal investigator on government-funded research projects at the University of Maryland to most recently in my role at Chief Education Officer at Voxy, one of the world's fastest-growing educational technology companies focused on offering web- and mobile-based English instruction to learners around the world. It is rare to find scholars who can easily conduct high-quality research while at the same time apply their findings in the world outside of academia, and Dr. Mentor is an excellent example of such a person. His work and thought leadership contribute to both current and future scholarship as well as external education programs, giving him a unique perspective on how best to apply the theoretical findings of empirical research in an accessible, practical, actionable way. He builds bridges between academia, non-profits, and commercial enterprises, connecting researchers with practitioners and making evidence-based best practices available and accessible, improving instruction for learners in a wide variety of settings.

This makes him the perfect person to have put together this thought-provoking volume on using computers for workforce development. In addition to his experience as an educator and researcher, he's spent over twenty years developing e-learning for both formal and informal academic settings. He has served as the Senior National Director of a unique workforce development organization, collaborating with academic directors and staff to convert a program from a predominantly paper-based model to an engaging multimodal and mobile digital environment, with notable business partners including American Express, the Bill & Melinda Gates Foundation, Brown University, Coca Cola Company, Facebook, Google, Harvard University, Microsoft, New York University, and Viacom. In addition, he co-initiated mlearning infused workforce development programs for the New York Mayor's Office of Adult Education, where he advised on, among other initiatives, a program to help previously incarcerated women obtain employment. Dominic also served as a consultant and trainer on a mobile activism film project with young immigrant children in response to a hate crime. Dr. Mentor operates from the philosophy that learning is everywhere and works best when it is integrated into students' lives, leveraging technology, social connections, and personalized instruction. His strong understanding and application of how and what is required to create a thriving ecosystem of learning is the essence of why he is the editor and contributing author of this book on Computer Mediated Learning for Workforce Development.

Dr. Mentor proposed, compiled, and edited this collection to report on innovative research such as integrating learning into workplace specific contexts, virtual mentoring, game-based social skills training, and the use of personal response systems as well as to foster meaningful conversations on computer-mediated civic action, the debilitating impact of technological disparities on the development of urban youth, and mobile journalism. In addition, the volume offers instructive chapters on establishing teaching and learning communities in MOOCs, best practices for selecting online learning tools, guidelines for creating mobile training environments for health educations, and a roadmap for developing computer-mediated solutions for workforce development.

At the heart of these diverse contributions is the central theme that computer technology should be integrated to enhance dynamic opportunities for learning, making them useful and accessible. As these chapters illustrate, there is ample opportunity to use technology to improve workforce training programs at every level, from offering much-needed skills development to youth in under-resourced urban environments, to providing agile, flexible, and accessible in-service professional development for busy professionals. It is my hope that the takeaways from this book include not only offer a comprehensive overview of technology-mediated learning, but also an awareness of how technology can be leveraged to help solve the real problems faced by the training programs, schools, colleges and workplaces tasked with developing our current and future workforce.

Katharine B. Nielson
Voxy, USA

Preface

Over the past decade, computing power and bandwidth development has grown exponentially. Accompanying this rapid growth is an increased expectancy and dependence on accessibility to information and communication technologies. While children and adults alike interact with this technology in dynamic and intuitive ways in their personal lives, many do not have the same experience when learning in school, at college or on the job. In addition, workplace and learning organizations do not readily utilize these technologies in teaching and training and miss out not only on key operational purposes, but also on gaining strategic advantages for themselves, their communities and for their stakeholders.

There is a need to rethink replicating mono-directional and linear educational engagement, like the pushing of content in the form of uploading PDF's or extensively long video lectures. This mono-directional approach does not make use of the digital age, and how people interact with computer media. In a hypermedia and social media rich world, celebrating non-linear training engagements will also aid the bridging of the informal with formal learning environments. While the ubiquity of mobile devices aids microlearning opportunities and in conjunction with the informal to formal bridging, cater for the transfer of knowledge and experience from one discipline to another area or context (Starr, Ganco, & Campbell, 2017). Our mobile and hypermedia infused world also caters for "working learners [who] places more value on learning the right thing at the right time as opposed to taking in all the content at once" (Von Ins, 2016). Furthermore, the learning and development does not need to be over after the completion of the learning experience, but rather should constantly be accessible for revisits and repetition as part of a continuous learning framework.

Computer mediated learning is rife with opportunities for pulling content and getting learners to be active participants and/or co-authors of their learning journey. Opening an organically curated digital grassroots approach, provides space for listening to learners' training needs. Which can now be responded to dynamically and offers opportunities to leverage their own talents and experiential knowledge, through the fruitfulness of hosting multiple perspectives to add to the richness of the learning community's educational engagement as they critically review and vote up the legitimacy and relevance of the content with the help of the facilitator. And embracing powerful online search functionalities becomes less of a competitor of learning, but rather a powerful partner that embraces digital disruptions the way other industries have done (Niles-Hofmann, 2016). While context is still key, collaboration crucial, content is now accessible nearly everywhere via online search. Using the power of those algorithmically informed search processes can aid individuals' online exploration paths through enhanced automation of similar matching search results.

In an age of increasing automation through digitization, there is a need to move from digital literacy to digital fluency, including the use of computer mediated learning to develop soft skills in conjunction with technical knowledge practice. For schools and organizations to provide and sustain professional careers in the coming century, entry level talent and more seasoned workers, as well as college graduates from all levels, will need to:

- Be proficient with current technology, but also comfortable with the ever-advancing nature of technology and its changing role in the workplace
- Be able to work successfully in teams using technology, including teams that span digital geographies
- Understand and make-use of electronic performance monitoring tools, for managing their own work, but also the work of others
- Have effective facilitated communication across the full range of in-person, distance, and technology-driven tools as well as professional workplace norms present in virtual training engagements.
- Be life-long learners, taking individual responsibility for continuous professional development, training and using technology to create space for "just in time" learning opportunities
- Create transformative opportunities by leveraging e-learning and m-learning tools by integrating these digital methods into teaching and learning engagements
- Increase learning outcomes by acting on cyclical, evidence based, digital learning analytics, invest in, and building capacity in electronic and mobile engagement will also help with translating market demand at a faster pace while establishing and sustaining supportive, as well as collaborative digital learning community networks (Global Communities, 2016)
- Be ready to adapt and adopt to new digital systems with these career skills strongly identified, educational and training practices are still lagging in many

K–12 and Higher Education contexts, with no mobile accessibility nor mobile first design principles driving these curriculums (de Aquino, Robertson, Allen, & Withey, 2017). There seems to also be a need to change the rules around banning devices, and more of an approach of embracing mobile devices and their use in formal learning contexts. The latter would also stand a higher chance of aiding the teaching of responsible use and etiquette. It is evident that there is a continuous need for the K-12, secondary and post-high school outcomes of children and young adults to come under review (Dumford & Miller, 2017). Making Cents International's Demand-Driven Training for Youth Employment Toolkit (Minic & Olenik, 2017) indicate a stronger need to enhance programs for youth and to provide entry level talent with more of the workplace needed skills. The Aspen Prize for Community College Excellence lists labor market success as a fourth criterion for its prize and Wyner (2014) expounds on some of the methods used by those colleges that have won the Aspen Prize in his book *What Excellent Community Colleges Do – Preparing All students for Success*. To increase its employment outcomes, Thomas Friedman (2014) from the *New York Times* indicates in his piece "It Takes a Mentor" that everyone wants talented employees out of college or technical schools, who are as ready to plug and play as possible to become or remain competitive in the digitally integrated labor market.

Talent training is of utmost importance to remain competitive in the fast-paced knowledge economy. This training impacts every area, be it improved employee engagement, skills development, accountability, staff retention, or succession planning. Unfortunately, it can be a quagmire of minefields, varying from one-dimensional engagement, mono-directional lecturing, out of date learning management systems, untrained or unskilled staff placed in training and learning roles, limited social and collaborative opportunities with peers or managers, as well as a lack of opportunities to offer and host user generated content. Oftentimes, without being informed by instructional design, compliance training becomes a boring click through exercise. The driver then becomes completion rather than engagement and sits squarely opposed to the objectives of the training as well as the mission and expectations of the organization. The #MeToo movement, sexual harassment and all training deserves better than that. Engaging people in training deserves and requires a prudent approach, especially when demanding people's time and attention, but even more so if a company is truly dedicated to informing, growing and/or changing a culture. Being able to show that actual learning engagement occurred, versus simply completing a slide deck, is the type of learning accountability that needs to be instilled in many organizational learning ethos and departments.

In many cases, Human Resource departments are tasked with the Learning and Development (L&D) of their organizations, yet have not been trained in constructing virtual, blended, mobile or eLearning engagement. The 2015 Global Human Capital Trends report by Deloitte indicates that learning technology broadens corporate Learning and Development (L&D) capabilities (Bersin, Agarwal, Pelster, & Schwartz, n.d.). And while addressing the skills gaps is fundamentally crucial, the 2015 research in Deloitte's HR Trends reports that less than 25% of companies feel comfortable with today's digital learning environment, even though companies are facing a radical shifting context for the workforce and the workplace (Schwartz, Collins, Stockton, Wagner, & Walsh, 2017). The world of work and school is changing the way we do business and the way we learn. This is not only because of rapid technological advances, but also because of a faster pace of data availability and with information more readily available at our finger tips. "Learning how to use data to make better and faster instructional design decisions that have impact" will aid faster adaptation and adoption (Niles-Hofman, 2016). The use of data could help the adaptiveness of colleges and corporations to adjust the glacial speed at which some of them move in response to the cultural technology shifts and labor market.

Some noteworthy cases exist, like the Aspen prize-winning colleges, where the skills gap is being purposefully targeted and rewarded to help students graduate with the technical and professional knowledge to have the requisite skills necessary to enter and succeed in the local and global job market for a chosen career track (Wyner, 2014). The focus on supporting student, trainee, entry level or veteran learning, both inside and outside of traditional instructor led, or online training sessions, with a primary focus on micro or bite size learning and engagement across screens, including mobile devices, will aid engagement and motivation, yielding higher results. There is a great need for mentoring or coaching with immersive and engaging onboarding. A dire need to incorporate digital tools and social media that would stimulate and build a social connectedness, cultivating a sense of belonging to aid onboarding, quality of engagement, productivity, and retention (Mentor, 2016, 2017). The return on investment for including collaborative and virtual learning (Battou, 2017) as part of a continuum through internships, peer-to-peer welcome systems, or shadowing, will further benefit the individuals and workplace environments. Combined with learner curated content, all the above could cater for responsive learning designs that create experiences, not just passive participation in courses, further cultivating workforce readiness. A workforce ready talent pipeline is substantial both in economic benefit to any company, but also the impact for society.

THE BOOK

The computer mediated elements and needs mentioned above are essential items that would further enhance the learning capabilities, increase retention and transferability of skills, as well as embrace flexible, blended learning in a world dominated by digital engagement and multimodal learning platforms. This book can be used as a guide to aid and benefit HR training teams, chief learning officers, senior learning strategists, schools, after-school programs, professional developers, professors, colleges, and non-collegiate workforce development training programs.

This book will offer case studies and examples of how this can be achieved through a multi-pronged approach that would resonate with multi-generational workforces. In today's knowledge economy, the investment in, and training of talent will help organizations, schools, colleges, businesses and institutions, become or remain competitive to their own goals, standards and peers. Today's knowledge economy is also moving and growing at a rapid speed due to fast-paced technology innovations. This book will present various cases that will elucidate why some talent development programs have become stagnated, stalled, stumbled, course corrected and succeeded. This book will break down and bridge the silos between schools, colleges and businesses, and offer valuable examples of innovative talent development. The use of mobile learning, eLearning, blended virtual, distance as well as Massively Open Online Courses (Egloffstein & Ifenthaler, 2017) and other learning technologies will be at the center of all the cases and offer valuable insights and lessons.

The concepts, cases and chapters in this book can help prioritize learner contexts and instructional inputs. The case studies will offer some succinct research examples and short narratives that can be used to inform learning and development designs. The cases will also offer how to inform learning design ecologies with digital processes for improved talent and product development. The book also offers methods on how to increase and impart skills, abilities, engender transferable knowledge capabilities, productivity, virtual mentoring, talent retention, as well as increase job placement rates across a network of labor markets.

Case examples presented in this book cover in detail these SMARTer, more Agile and Scrum goals, operating in sustainable timely sprints, as part of preparing for, and running of a project marathon in staggered deliverable fashion:

- Establishing standards for instructional quality in F2F or virtual settings
- Empowering all learning through formative assessments
- Aligning learning to a user's needs
- Offering and updating multimodal methods to deliver market-relevant core curriculum
- Synchronous and continuous performance transparency through immediate feedback
- Enhance Personal Learning Environment through automated (Maycock & Keating, 2017) and informal learning contexts

The book will also aid the reader on how to inform and evaluate any learning design. Will aid the development, implementation, change management as well as the monitoring and evaluation of computer mediated training and learning initiatives.

Overview of Chapters

The book is organized into 13 chapters. A brief description of each chapter follows:

Chapter 1 sets the scene for discussions on computer mediated learning for workforce development as presented by authors of this book. The chapter identifies the pressing issues regarding the integration of learning into businesses processes and practices. Looking at the related challenges of designing the context for learning, and not just focusing on the technology. It also identifies the importance of managing the learning policies to support the processes and practices that could aid the structure and success of learning engagement plans.

Chapter 2 establishes the need for upskilling teachers in the delivery of a game-based social skills youth program with a focus on students on the autism spectrum as well as presents some samples that were designed and used in the program. The authors of this chapter contend that by investing in the development of upskilling teachers, will benefit not only the educational organization, its community of children and parents, but can also benefit the larger global society.

Chapter 3 looks at the struggle that America is grappling with on the disparity of urban education and technological opportunities. The authors contend that a digital divide is still impacting the United States of America with educational resources still inequitably distributed. They offer philosophical and practical examples of how debilitating the impact of not being able to adequately prepare urban students for 21st century employment affects American society as a whole.

Chapter 4 reviews the ongoing rise of worldwide protests that aim to bring attention to how computer, mobile and social media organized protest interaction can aid workforce readiness. The author pays attention to how mobile devices have been converted into mobile activism tools, offering youth opportunities to be plugged into the pulse of others who are in the same predicament. The chapter highlights the need to consider technological 'solutions' in their broader context and assess the impact of social and political factors on the business environments and organizational processes.

Chapter 5 reviews the opportunity of mass educational methods to address the massive gap between computer technology labor markets and the upcoming labor force. Information and computer technology companies have not only identified the need to address the shortage of computer programmers, but have invested in massively open online courses (MOOCS) as a means to address the issue in bulk. The author looks at increasing engagement and lowering attrition rates of these MOOCS. The author also offers a model that was designed for a workforce development program to participate in computer programming MOOCS. The chapter sets out to classify the required instructional design in terms of cultivating a social connectedness community, as well as virtual and on-the-ground support structures to enhance engagement and completion of the MOOC computer programming courses. While also offering the methods to replicate, duplicate, and organically grow the digital adoption of this mass engagement methodology successfully for geographically dispersed locations.

Chapter 6 presents an analysis of issues in managing a virtual mentoring community in a medical school. The chapter looks at enhancing the way we mentor in a blended learning environment. The chapter also offers recommendations for those who wish to create, maintain or enhance their digitally infused mentoring programs.

Chapter 7 looks at creating space for the generating of digital film narratives by immigrant youth. The chapter covers the methods used to cultivate mobile activism and mobile journalism skills in response to a hate crime. The process uncovers how parental and community involvement were of paramount importance and impact.

Chapter 8 analyses and compares recent approaches for selecting online learning platforms. The author offers insight into finding the right balance between supporting strong pedagogy, offering training and support, providing data security and privacy, while ensuring ease of use as a top priority. The chapter systematically reviews the gaps and problems in many of the tried approaches. Other research is cited that refers to the importance of creating a criteria matrix for selecting online learning platforms. Furthermore, a systematic position for future research and practices are established.

Chapter 9 reviews virtual and mobile learning environments usage for health education as well as the issues surrounding a need for faster adoption of these methods for a more digitally proficient medical health workforce. The authors provide numerous examples showcasing the possibilities and advantages of virtual, augmented and mobile eLearning environments and how it can facilitate better medical health education. The authors present an array of technical tools and techniques that help in achieving realistic simulations, with tactile engagement.

Chapter 10 introduces how mobile devices can be used on an individual and macro level for professional development. The author uses the rise of mobile apps and mobile learning concepts to focus and showcase the lessons from the USA's first graduate mobile learning course. The author contends that a mobile first learning design is crucial in organizational leadership and professional development, to help bridge the gap between personal lives, schools, colleges, and the workplace, as a natural segue for ownership of target audiences' learning.

Chapter 11 identifies the role mobile device technology can play in improving workforce development by presenting a roadmap for developing computer-mediated solutions for situated workforce learning with a hint at artificial intelligence and the algorithms that helps us better navigate our daily lives and learning experiences. Citing sources, the author contends that computing systems can be situated when they combine machine perception with background knowledge to observe, explore, and interpret human and environmental activity in a way that supports decision-making. The author highlights the huge potential to leverage these technologies for education and training, offering opportunities to bridge informal, experiential and situated learning through perceptual computing.

Chapter 12 looks at the impact and relevance of personal response systems (PRS) on students' learning where it is still being utilized. The authors offer a theoretical model that was developed on the effects of PRS on students' learning performance. The chapter also reviews advantages and disadvantages of PRS. The authors conclude by exploring implications and relevance for PRS' future in a ubiquitous mobile device world.

Chapter 13 presents concepts, theories, as well as principles necessary for Conceptualizing Experiential Learning for Computer Mediated Engagement in a digitally rich world. The chapter aims to equip readers with a conceptual understanding to help leverage experience-based learning in electronic and mobile learning environments. The authors assert that there is still a dearth with incorporating experiential features in e/m-learning, and more so in a workplace informed manner. Utilizing learning design cycles, the authors offer two cases that offers pragmatic, formal and technical examples. The purpose of the models proposed is to equip readers with the ability to critically evaluate other digital learning applications.

REFERENCES

Battou, A. (2017). Designing an Adaptive Learning System Based on a Balanced Combination of Agile Learner Design and Learner Centered Approach. *American Scientific Research Journal for Engineering, Technology, and Sciences*, *37*(1), 178–186.

Bersin, J., Agarwal, D., Pelster, B., & Schwartz, J. (n.d.). *Global Human Capital Trends 2015*. Retrieved from https://www2.deloitte.com/insights/us/en/focus/human-capital-trends/2015.html

de Aquino, C. T. E., Robertson, R. W., Allen, P., & Withey, P. (2017). A global learning-centered approach to higher education: Workplace development in the 21st century. *Revista Tecnología Ciência & Educação (Bauru)*, 6.

Dumford, A. D., & Miller, A. L. (2017). Assessing alumni success: Income is NOT the only outcome! *Assessment & Evaluation in Higher Education*, *42*(2), 195–207.

Egloffstein, M., & Ifenthaler, D. (2017). Employee perspectives on MOOCs for workplace learning. *TechTrends*, *61*(1), 65–70.

Friedman, T. L. (2014). It takes a mentor. *New York Times*, p. A12.

Global Communities. (2016). *Partners for good, adapting and sustaining demand-driven training programs: A case study from the Middle East*. Retrieved from https://www.globalcommunities.org/publications/2016-MENA-YES.pdf

Maycock, K. W., & Keating, J. G. (2017). The impact of an automated learning component against a traditional lecturing environment. *Journal of Computer Assisted Learning*, *33*(6), 597–605.

Mentor, D. (2016). EMxC3= e&mLearning Cultivating Connected Communities: Sustainable Workforce Talent Development. In Handbook of Research on Mobile Learning in Contemporary Classrooms (pp. 240-259). Hershey, PA: IGI Global.

Mentor, D. (2017, April 27). Cultivating Digital TLC Teaching and Learning Communities. In *ModSim 2017 International Congress on Modelling and Simulation - Modeling and Simulation in the Age of Data*. Retrieved December 24, 2017, from http://modsimworld.org/papers/2017/Cultivating_Digital_TLC_Teaching_and_Learning_Communities.pdf

Minic, B., & Olenik, C. (2017, December 27). *A Demand-Driven Training Toolkit for Youth Employment*. Retrieved from http://www.makingcents.com/single-post/2017/12/27/A-Demand-Driven-Training-Toolkit-for-Youth-Employment

Niles-Hofmann, L. (2016, July 6). *Data-Driven Learning Design: How to Decode Learner Digital Body Language*. Retrieved January 24, 2018, from http://www.loriniles.com/ebook

Schwartz, J., Collins, L., Stockton, H., Wagner, D., & Walsh, B. (2017). *Rewriting the Rules for the Digital Age*. 2017 Deloitte Global Human Capital Trends.

Starr, E., Ganco, M., & Campbell, B. (2017). *Strategic Human Capital Management in the Context of Cross-Industry and Within-Industry Mobility Frictions*. Robert H. Smith School Research Paper No. RHS 2753720. http://dx.doi.org/10/2139/ssrn.2753720

Von Ins, C. (2016). *A Working Learner Is a Modern Learner*. Retrieved March 31, 2018, from https://www.td.org/insights/a-working-learner-is-a-modern-learner?_ga=2.139790707.1582201096.1522499837-1585515252.1522499837

Wyner, J. (2014). *What excellent community colleges do*. Cambridge, MA: Harvard Education.

Chapter 1
Integrating Learning Into Work: Design the Context, Not Just the Technology

Adam Neaman
Citadel Global Equities, USA

Victoria J. Marsick
Teachers College, Columbia University, USA

ABSTRACT

In this chapter, the authors argue for designing the context, not just the technology, of e-learning simulations (eSims). They examine the limitations of "pure" eSims (i.e., not involving human-delivered feedback) in light of design challenges for informal learning, and learning transfer, in today's knowledge-intensive work environment when solutions call for creative responses to messy, non-structured tasks in different organizational contexts. Authentic learning on realistic tasks is needed in the design of eSims in such situations. Neaman describes ways that directed performance support (DPS), an approach he pioneered, helps to overcome such limitations. He identifies a core set of DPS guidelines that others can use to adapt and utilize DPS designs in their eSimS. Using a system dynamics framework, Marsick discusses organizational barriers and supports that also need to be considered in eSim learning design for effective engagement and learning transfer.

INTRODUCTION

The 70-20-10 framework of Lombardo and Eichinger (1996) claims that the way people learn to do their jobs is 70% through challenging assignments, 20% through developmental relationships, and 10% through formal training. While this breakdown has never been precisely measured because of the inherent difficulties of doing so, there is a great deal of common-sense evidence that 70-20-10 is directionally correct. In other words, people build the majority of their professional capabilities on the job—informally, incidentally, and socially. This poses a challenge for those seeking to develop a workforce. How does one design informal, incidental, and social learning given that it occurs anywhere, any time, and any place?

DOI: 10.4018/978-1-5225-4111-0.ch001

Informal learning is central to today's knowledge-intensive work environment where success often depends on individuals' abilities to creatively respond to messy, non-structured tasks in different organizational contexts. This chapter characterizes informal and incidental learning and the paradox of designing for these kinds of learning. It then examines a successful example of a design solution, positioning it within the context of its precursors and their limitations in engaging people. The first designs that Neaman discusses are eLearning simulations (eSims). Despite eSims' potential effectiveness in some domains, they take a formal training approach to workplace learning that can limit learner engagement – especially in busy workplaces. To get past eSims' limitations, Neaman moved away from simulations altogether. Instead, he used technology to integrate learning more naturally into the messiness of people's day-to-day work. He describes Directed Performance Support (DPS), an approach he developed to achieve this kind of integration. He identifies core DPS design guidelines that others can use to develop their own DPS solutions. Finally, in the last section, the authors discuss the often-overlooked organizational system dynamics conditions that are essential for effective learning transfer.

SIMULATIONS: A FORMAL APPROACH TO LEARNING BY DOING

Simulations are useful for developing many skills that are best learned in the workplace. One reason is that simulations embody implicit guidelines and rules that emerge through action. Tacit knowing (Polyani, 1966), which is fundamental to expert performance, remains below the level of consciousness and verbal expression in the course of actual practice—whether that practice occurs in "real" situations or simulated ones.

Marsick and Watkins (1990) defined informal and incidental learning in contrast to classroom-based, structured training. Control of informal learning choices rests primarily in the hands of the individual (Watkins & Marsick, 2014). It is stimulated and guided by intentions or interests, even though learners might not set formal goals or follow planned activities. Incidental learning, by contrast, is an unintentional byproduct of some other activity. It is organic, and acquired through action, often in interaction with others at work (Eraut, 2004; Marsick & Watkins, 2014). This recast perspective supports examination of ways that technology-supported learning has changed over time, and how it is being woven into the context of people's work in ways that drive engagement.

Many theories posit that people learn by doing in cycles of action and reflection (Dewey, 1916, 1938; Kolb, 1983; Piaget, as cited in Gruber & Voneche 1977; Schön 1983, 1989). The authors argue that people naturally learn to do their jobs by doing them, not in a classroom. Bear, Tompson, Morrison, Vickers, Paradise et al. (2008) "estimated that informal learning accounts for up to 75% of learning in organizations" (as cited in Noe, Clarke, & Klein, 2014, p. 147). The learning by doing that naturally occurs in organizations is therefore an excellent guide for impactful learning design in organizations. Want people to learn accounting? Have them do the accounting for an organization. Want them to learn to facilitate meetings? Have them do that.

Widespread recognition that people learn on the job is consistent with the fact that job postings for senior roles typically seek candidates with some minimum number of years of experience. However, time on the job, even with challenging work assignments, does not necessarily result in new learning or skill (Ericsson, Krampe, & Tesch-Romer, 1993). The problem with simply using stretch assignments (assigning people real work—not training —that exceeds their current capabilities) as an approach to professional development is that the learning process can be hard to predict, control, or facilitate. And, of course, employers do not really want a brand-new trainee learning meeting facilitation by running real meetings with key clients. In contrast to challenging work assignments, simulations allow learning designers to carefully design and control learning by doing. The authors argue here that simulations make it easier to keep learners focused on the particular challenges they must learn to handle, minimize the time they spend practicing skills they either already have or can learn with minimal effort, and make sure their mistakes (fundamental to the process of learning by doing) are not catastrophic and are fully realized for their potential as learning opportunities.

In the 1990's and early 2000's, workplace learning technology was primarily focused on formal training. Gloria Gery (1991) drew attention to the value of computer-mediated Electronic Performance Support Systems (EPSS), an alternative that helps employees learn just enough to perform a task competently without comprehensive training prior to executing a task. However, partly due to technological limitations, early EPSS work focused largely on tasks involving relatively straightforward rules, scripts, and facts and so EPSS use grew slowly.[1] In recognition of the primacy of learning by doing, simulations became the state of the art in computer-mediated workforce learning (Schank, 1997; Aldrich, 2005; Quinn, 2005). "Simulation" here means simulating real experiences, but not necessarily in the 3D-virtual-worlds sense.

Simulations, as defined here, have people learn to do something by actually doing it, pursuing realistic goals in simulated versions of the contexts and roles in which those skills will ultimately be employed (Schank & Neaman, 2001). In workplace learning simulation design, absolute real-job similitude is not necessary—as long as the designer replicates the most salient features of the goals, roles, tasks, and context of the skills being learned. For example, a person learning accounting needs a realistic experience of using spreadsheets to compute cash flows; what s/he does *not* need is a 3D view of a simulated accounting office, complete with plants and conference tables. The virtual reality that the term "simulation" often evokes is best saved for tasks where the physical environment is salient, like architectural design or military tactics.

Here, then, are a few examples of workplace eLearning Simulations ("eSims" for short) from Neaman's past work:

- In a simulation designed to teach corporate finance basics, designers had new hires evaluate a fictional company's financial documents to determine how effectively the company was creating value so they could then make decisions about future direction.
- In a simulation designed to teach people how to use the Myers-Briggs Type Indicator (a popular personality framework also referred to as MBTI), learning designers had their real colleagues observe videos of conflicts between collaborating fictional colleagues. Based on the scenarios, the learners first had to figure out each fictional colleague's MBTI preferences. Then, they had to use that information to determine strategies that would be likely to improve the collaboration.

The Uses and Limits of eSimulations

To discuss eSims' capabilities, the authors narrow their focus to only those that are highly scalable. This means limiting the discussion to "pure" eSims – i.e., those that do not have inherently-hard-to-scale human instructors reviewing and providing feedback on students' work. To illustrate the significance of this distinction, in an eSim that is *not* pure, students might write an essay, submit it online, and, a day later, get rich qualitative feedback from an instructor. The range of skills that can be taught this way is orders-of-magnitude greater than with pure eSims, because students receive substantial, meaningful feedback on work that calls for creative judgment, e.g., advertising strategies for new markets, or facilitating change in large organizations. The cost of this increased range of skill development, however, is vastly decreased scalability primarily due to the costs of human labor.

By contrast, in pure eSims, all feedback must be generated by the software. Despite advances in AI, current software cannot independently provide meaningful feedback on work that is not well characterized or evaluated with straightforward rules and computational logic. Outside of highly logical/mathematical domains then, pure eSims' effectiveness drops off precipitously.

Providing meaningful feedback—at scale, with no instructor intervention—has been a recurring challenge in the history of pure eLearning—whether via eSims or through other forms of eLearning such as Massive Open Online Courses (MOOCs) like those offered by Khan Academy and Coursera[2]. People who have been in the business of building eLearning to teach relatively mechanical skills, like basic physics or computer programming, have been optimistic about the power of pure eLearning because it can deliver reasonably insightful personalized feedback in those domains. When those same optimists try to build pure eSims focused on creative skills, they often fail. Except in limited situations, described below, the software is not up to the task of providing relevant feedback.

There *are* effective pure eSims that cover skills in creative domains, but they tend to focus on developing only the most mechanical, rule-bound skills within those domains. For example, Neaman and colleagues developed a pure eSim to successfully teach meeting facilitation to people in a professional services firm. Across a population of hundreds of students, average "value-for-time-spent" ratings were significantly higher than those for a popular live training session of comparable length on the same topic. In the eSim, employees are asked to help structure fictional meetings and then use a formal facilitation framework to critique fictional colleagues as they run those meetings. The objectives of the eSim focus on enabling people to straightforwardly employ a particular, formal framework. What it doesn't teach, however, are the more fluid, creative aspects of facilitation like coming up with solutions for handling complicated moments in a conversation or designing new activities tailored to engage a particular group of participants more effectively.

Learning to do anything hard takes work—practicing what one is learning to do. Moreover, people's expectations about feedback quality seem to inform their willingness to do that work. Years of informal observation by Neaman and his colleagues suggest that the amount of *simulated* work people willing do is strongly correlated with both the quality of personalized feedback they expect to get, and the likelihood that someone whose opinion they value might review their work. For example, in an eSim to teach business writing, students might be unwilling to write substantial business letters if they did not expect to receive meaningful, expert feedback on their particular letters.

The Structure of eSimulations: Why Performance Support Is Not Enough

As discussed earlier, developing simulations is, in part, a process of designing fictional learning tasks that approximate what people ultimately need to do in the world. Around those tasks, designers build in resources, tools, and feedback systems to help their audiences figure out how to succeed and how to make sense of successes and mistakes. The participants' activity in the simulation "tells" the system what participants are doing at any given moment, and therefore, what supports would be most relevant to offer.

Simulations, and especially pure eSims, are costly and time consuming to build and update because of the level of complexity and internal consistency required for them to operate without ongoing instructor intervention. They are also costly to run because they take people away from their real jobs. This raises

Figure 1. The structure of eSimulations

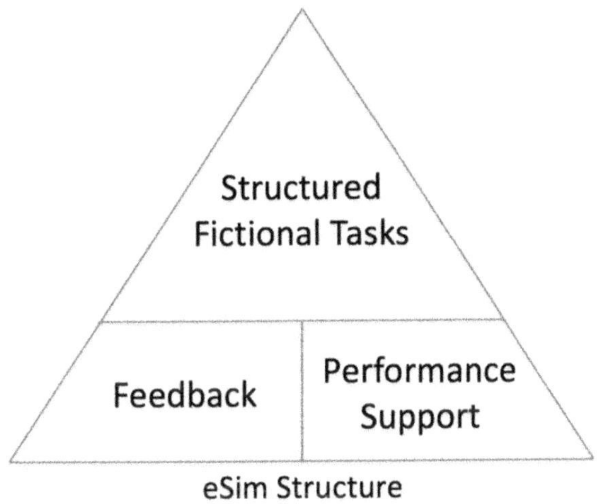

a question: Why not just skip the simulated activities and have people learn on the job with the help of performance support? The primary reason is that simulations give designers a great deal more control over the learning process. This control enables them to do four important things:

1. **Make Sure People Encounter the Right Challenge:** Opportunities to learn from life experiences are unpredictable; this leaves learning to happenstance. For example, it could be dangerous for a customer service representative to learn, on the job, how to handle a physically aggressive customer. And, it might take a long time before one was encountered. With simulations, designers can pre-plan what challenges their audiences will face and thus what skills they will build, driving them to the targeted learning consistently and providing them with opportunities to practice safely before they find themselves in real, non-fictional hot water. Further, simulations standardize training so that everyone practices the same set of skills not only efficiently and safely, but also in ways that make it easy to consistently assess and compare individuals' capabilities.
2. **Focus Only on What Is Hard and Necessary to Learn:** Simulations can be tailored to focus on just the hard parts of real tasks—the places where novices typically fail. That way, learners aren't wasting time covering things they already know how to do, that they can learn effortlessly, or that get covered in traditional training as what one "should" know.
3. **Provide Personalized, Just-in-Time Feedback:** Simulations "know" what their users are working on; and can thus provide highly relevant feedback just-in-time—a key ingredient in making a learning experience memorable, engaging, and not frustrating (Askew, 2000).
4. **Drive the Process of Learning Through Both Immediate Needs and Accountability:** Performance support systems are easy to ignore if the skills they address, however crucial, only arise rarely in the normal course of work. But assigned simulations create "pull" for the necessary skill development through work that cannot be done without the requisite expertise. Simulations also provide learners with a sense of progress—drawing attention to the tasks they are completing and goals they are achieving. And, management can support completing simulations in a way they cannot for ad hoc use of performance support.

The fact that eSims are limited to developing rule-bound skills is a problematic constraint in that most of the skills that differentiate experts from novices are anything but rule-bound (Ericsson, Krampe, & Tesch-Romer, 1993). They typically involve nuance, creativity, and shifts in mindset—reframing not only the immediate problems people are tackling, but also their broader perspective on their work world and their roles and goals in it.

Whether eSims are pure or involve human feedback, they inevitably have several serious limitations beyond those already discussed. Simulations are only as good as their authenticity—their ability to approximate the real work someone needs to learn how to do. Good simulations are more authentic than sitting in classrooms, but still not as authentic as learning on the job. In addition to taking people away from work while their inboxes fill, simulations necessarily focus on helping learners overcome well-established problems that others have already solved. In the racing knowledge economy, where information is ubiquitous and even relatively new knowledge rapidly becomes a commodity, it is critical to also give people practice solving *novel* problems—generating fresh ideas and inventing new solutions. Getting past such limitations requires a better way to support learning by doing—one that draws on the strengths of simulations without all their shortcomings.

Directed Performance Support: *Designed* On-The-Job Learning

Neaman sought a new approach to supporting informal learning by doing when he was asked to develop an eLearning solution for senior leaders in a professional services firm. Senior leaders often have very little time or patience for traditional eLearning formats, including simulations. Looking for other ways to engage and support them in online learning-by-doing, Neaman developed an approach he calls "Directed Performance Support" (DPS). DPS is like simulations in that it involves carefully structured tasks, surrounded by performance support and mechanisms for (in this case, human-delivered) feedback. The key difference is that the tasks are not fictional. Participants are assigned to perform tasks that are part of their actual job—just in a particularly structured way.

Figure 2. Directed performance support replaces eSims' fiction with real tasks

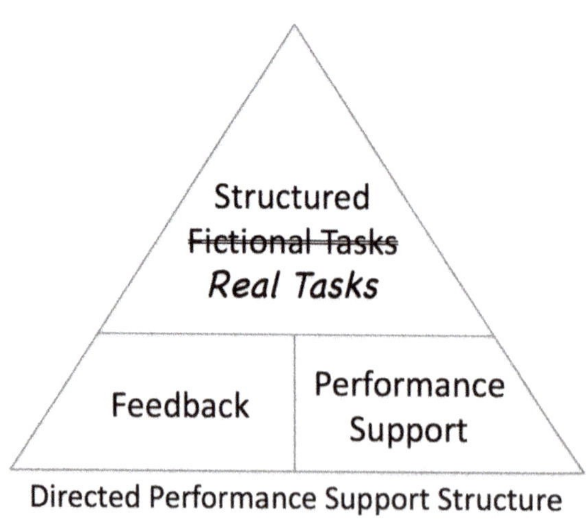

DPS Replaces the Fictional Tasks of Simulations With Structured Real Tasks

Directed performance support evolved from eSims. This section uses examples from Neaman's prior work, starting with the New Horizons Program at Worldwide Diverse Law[3], to illustrate how directed performance support differs from eSims in its approach to tasks, performance support, and feedback.

WD Law, for short, is a large firm with a broad range of capabilities. The goal of New Horizons is to prepare lawyers who are partners at WD Law to have specific types of conversations with their clients—conversations that are outside those lawyers' areas of expertise. An example of a problem New Horizons was created to solve is:

Jane is a senior partner with substantial expertise in patent law. Her client, Acme, is very well served with respect to patents, but desperately needs help with union contract negotiations, a field of law outside Jane's expertise. If only Jane could have a relatively informed conversation on union contracting with her Acme client, it would open the door to bringing in real union contracting experts from Jane's firm and everyone would benefit.

Traditional approaches to tackling this goal, like classroom training and/or reference libraries containing videos and brochures about the capabilities of each part of the law practice, had failed in the past. Very few partners showed up for classes and very few accessed the reference libraries. Past experience made it clear to Neaman that eLearning simulations would similarly fall flat with this audience. He developed the concept of DPS as an alternative. In DPS, learning is facilitated by assigning people to do specific tasks that are part of their regular jobs (work they already need to get done). DPS uses the structure imposed in those assignments to enable the DPS to "know" what people are working on so that the system can point to relevant support materials and opportunities to get feedback when most useful.

To illustrate how DPS solutions differ from typical eSims, the DPS components in the New Horizons Program at WD Law—tasks, performance support, and feedback—are described below.

Tasks in DPS Solutions

Partners were asked to pick a client with whom they wanted to engage and then fill out a "conversation planner." The planner is a structured set of questions about the client's situation, presented in an online form. If the partner fills out the form, answering the questions, s/he will have created all the notes needed to inform the intended client conversation:

When Jane, the aforementioned partner, filled in her online conversation planner to prepare to discuss her firm's other capabilities with her client at Acme, she was not answering generic questions like "What is WD Law's approach to union contract negotiations?" Rather, she answered critical client-specific questions like "Which aspects of WD Law's unique approach are most salient to Acme's current union contracting challenges and how will they deliver effective impact?" and "Where is the Acme team most likely to have concerns about WD's approach?"

Performance Support in DPS Solutions

To support the work of conversation planning, one of the teams Neaman was leading worked with WD Law's employment law experts to compile documents, video libraries of expert stories, preparatory fieldwork activities, and other resources on WD Law's approach to union contract negotiations and its broader perspective on employment law. Rather than assign reading and videos week by week, each question in the conversation planner linked back to specific online resources, which the partners could use at their discretion. Only the task of answering the questions—not the consumption of resources—is assigned work. The message is: "The questions in the planner are designed to help you prepare for a conversation with your client. Here are specific resources WD Law's employment law experts think will be helpful in answering each of them. Use the resources and reach out to your employment law colleagues only as you see fit. The assignment is not to use the resources, it's to come up with answers that leave you prepared for a compelling, insightful conversation with your client."

Feedback—and More Performance Support—in DPS Solutions

Peer feedback is central to the DPS approach, but the lines are blurred between feedback and performance support. Partners share work with peers and seek their input. They do get feedback, but much of it is also performance support in the forms of insight sharing and problem solving.

Collaborative problem solving was thus a central feature of the New Horizons program. When a partner had filled out a conversation planner and was preparing to go talk to that client, she would typically run some of her thinking past peers. In doing so, she might call out problem areas where, for example, she wanted help figuring out her approach or conceiving of new ways to align WD's capabilities with client needs. This embodies a shift from peers just delivering feedback toward peers also working as collaborators. The shift is also a pivot from learning what is already known to learning in the form of knowledge creation or novel problem solving.

As collaborators solve novel problems, they generate new knowledge. In the high-speed ephemeral knowledge economy, this is arguably the most important form of learning. In addition, this process produces valuable meta-learning—understanding how better to learn. These collaborations tend, incidentally, to strengthen the collaborators' problem solving and collaboration skills and thus their ability to generate new knowledge, as well as build their networks (Watkins & Marsick, 1993).

Observation provides insight into a secondary benefit of collaborating with peers to solve real, personally meaningful problems: engagement. People often like to help one other; they find the collaborations motivating (Eraut, 2004). Neaman has found engagement in online social learning activities is higher when discussions focus on individuals' real-world challenges. Research on this phenomenon is limited. However, an experimental study (Ge, Planas, & Er, 2010) of scaffolding for students' web-based problem-based learning (PBL) provides support for hypothesizing these benefits from peer engagement and questioning. PBL involves collaborative problem solving of typical, realistic cases with no defined solutions. The PBL study involved structured support, rather than the social learning environment described here.

Nonetheless, this phenomenon was borne out in analyses and experiments that Neaman later conducted at Everwise, a company whose online learning platform emphasizes mentoring and peer-to-peer learning. Neaman found that people's willingness to help one other in online discussion forums increased substantially if the person seeking help was struggling with a specific, pressing concern versus asking a more generic question. When someone posted a reasonably specific question—e.g., *My boss keeps contradicting me in front of my direct reports*—there were often 10-15 responses, many of which were thoughtful and detailed. In contrast, when someone posted a general question—e.g., *What are the top ten things to do to impress your new manager?*—they might receive 3-4 relatively brief responses that were not particularly insightful.

Neaman replicated these results by enlisting some Everwise colleagues who were on the platform to post their own questions in generic, impersonal terms and enlisting others to post more clearly personal questions about specific situations. As predicted, across roughly 30 questions posted, with questions split more or less evenly between personal and impersonal, the specific, personal questions consistently garnered more responses that were often longer and more insightful than responses to impersonal questions.

In hindsight, people's seeming inclination to help each other with specific problems was likely part of Neaman's success with DPS at WD Law. Partners, who seldom spent time posting in the firm's online forums, were surprisingly active in the New Horizons forums when they saw opportunities to help their colleagues solve real, messy, pressing problems.

DPS Design Guidelines

Research was not identified that accurately documented e-learning participation rates among partners in professional service firms. However, anecdotal information gained through conversations with Chief Learning Officers and other learning leaders suggests that engagement levels in the New Horizons program were unusually high for a senior professional services audience. Over roughly two months, a substantial majority of all partners spent several hours voluntarily completing their planners for specific client conversations. Though data on subsequent activity is not publicly available, Neaman expects the partners went on to have the actual client conversations they planned. Partners are sufficiently driven and focused that they would not have engaged if they did not believe it would deliver value to their clients.

Based on his success at WD Law, Neaman developed core DPS design guidelines:

- First and foremost, only assign tasks that are the real work of people's actual jobs. (Motivation and engagement are driven by task accomplishment. If your audience feels like they are primarily *learning how* to do their jobs, or *doing simulated versions* of their jobs, it's not DPS,)
- Start the learning design process by digging deeply into the learning need with three questions (which should, in fact, be the starting point for all learning projects):
  - What do you want them to be able to do?
  - What's hard about doing it?
  - *(i.e., where do they make mistakes or underperform)*
  - Why is it hard?
  - *(i.e., what beliefs, knowledge gaps, and contextual factors are the root causes of their mistakes?)*
- Structure the assignments so that people are asked to do the hard parts of the tasks in the challenging contexts you've identified.
- For each task, point the audience to supporting resources or activities you believe would be helpful, with a particular focus on addressing the root causes of their mistakes.
- Never make supporting resources or just-for-practice activities required assignments (unless using the resources is inherently part of doing the job).

- Build in opportunities not only to get feedback from peers and/or experts, but also to enlist their support in problem solving.

Integrating Learning Into Work Is Not Enough

Energized by his initial success at WD Law, Neaman used the same principles to develop DPS solutions for professionals using the Everwise online learning platform—a diverse audience of thousands employed by a broad range of leading-edge corporate clients. While there was considerable enthusiasm from people *who used* Everwise's initial DPS offerings, that group comprised less than 2% of Everwise's highly engaged audience. Roughly 15 user interviews conducted to explore this phenomenon revealed a simple and not-exactly-shocking truth: people are busy. Demands on time from day-to-day jobs created an inertial force that discouraged users not only from stepping away from desks to go to formal training, but also from even making an effort to diverge from "good enough" standard approaches to work in order to improve it. DPS primarily asks people to perform tasks directly in service of their real jobs. It also demands that they question and change the way they work. A typical response in user interviews was something along the lines of the following: *I know I could be better at setting goals for my team and running quarterly reviews of those goals. It probably would make the team more effective, but I just don't have the time to try out a new way of doing things. I'd love to work on it—just not right now.*

Contrasting the success of New Horizons with initial outcomes at Everwise, it is clear that centering the work of learning on getting real work done (which is the essence of the DPS) was a critical factor in New Horizons' success—something also reflected in the enthusiasm of those who actually did use the Everwise systems. However, it was clearly not the only reason, given their limited success at Everwise. What, then, is needed to ensure fuller participation?

CREATING THE ORGANIZATIONAL CONDITIONS FOR ADOPTION

Neaman conjectures that New Horizons' DPSs engaged a much larger portion of their audience because they were aligned with critical, well-incentivized shifts in the direction of New Horizons; and this was not so at Everwise. Among other things, targeted new behaviors at New Horizons were shown—with real data—to be revenue generating. They were also incorporated into assessment criteria in annual performance reviews, with implications for compensation.

Considering the Learning Context

Neaman hypothesized at least three other important factors in New Horizons' success:

- Internal marketing efforts promoting the program, with clear, compelling messaging from the CEO and other top leaders
- Peer collaboration teams, each with a senior leader in the role of "captain"
- Gamification mechanisms, such as showing individuals' progress vs. peers, that created accountability to peers, leadership, and even oneself

After months of observing usage and interviewing users (and non-users) at Everwise, Neaman concluded that the Everwise DPS systems were unlikely to succeed without the kinds of supports put in place by New Horizons. This led to expansion of the Everwise conception of, and approach to, the ways it serves its customers. Ultimately, the Everwise experience points to a broader principle: without the right organizational supports, better learning designs will not be sufficient to drive learning engagement.

To generate learning engagement with any degree of reliability, appeals to individual motivations are not enough; organizational motivations have to be integral to the solution. Effective incentivization requires a system dynamics approach to understanding human motivation. This is the kind of thinking that goes into change management: ensuring the organization is signaling that it values the new practices and supports the target behaviors; giving people the resources and time they need to change; and providing *credible* reassurances that there is room for safe trial-and-error as people learn.

Deci, Koestner, and Ryan (2001) described ways in which external rewards can undermine intrinsic motivation. While simple "carrot and stick" strategies like compensation and performance reviews have an important role to play in motivating change, they must be used judiciously. Individual motivation is also powerfully influenced by work environment. Changing the environment can permit, or even incentivize, effective application of newly acquired capabilities or practices. While learning designers might not be able to effect such change, they will be more effective if they develop their own skills in accounting for organizational constraints that support or inhibit desired learning processes and outcomes.

Real Work Learning: Example of Specific Leverage Points Using System Dynamics Framework

This section introduces a system dynamics framework (Gephart & Marsick, 2016) that provides a lens through which to identify and illustrate ways to work around barriers and draw upon supports in order to motivate learners, and institutionalize learning gains. It can help designers ensure that what is learned can be successfully used on the job during and after that experienc—often called "transfer of learning" (Holton III & Baldwin, 2003). As will be illustrated, authentic learning on real work tasks benefits when key conditions in the organization's context that affect incentivization and learning transfer are considered in designing DPSs.

Marsick starts with an example from research that Gephart and Marsick conducted with a division of a global manufacturing company (hereafter called "Global") on transfer of learning in a high potential management development (HiPo) program. Like the earlier DPS examples, the HiPo program did *not* use simulations; real work was central in its design. The point of this example is to shed light on how finding the right leverage points for organizational support might lead to improvement in both the design of simulations and directed performance support, as well as transfer of learning on the job. The case example is limited in that at the time of the research (between the years 2000 and roughly 2005), technology was just being introduced to support communication. A knowledge management system had recently been launched, but was not yet universally used. On the other hand, virtual communication via telephone, tele-conferencing, and email were prevalent.

The example is based on two years of survey and observational research that provided baseline assessments relevant to the HiPo program, and that compared and contrasted responses of two consecutive HiPo cohorts. Results were based on a 65% response rate of 84 associates in the first year cohort, and 75% response rate of 88 associates in the subsequent year cohort.

The umbrella company in which Global was situated had set a strategic goal of service excellence. Global promoted behaviors to reach this goal, for example: high-quality customer response and working as if it were an extension of the customer's staff. A key value was to succeed or fail as a team. Global was highly decentralized and locally managed. Technologies to support social networking and collaboration were adopted to work seamlessly together across countries and functions; help employees share knowledge and solutions; and build on what others know rather than starting each time from scratch.

The HiPo program was designed to catalyze and support networking, teamwork and collaboration. Action Learning (AL) brought these goals into the development program. In AL, peers work in small teams on real work problems (that have, as yet, not been well or fully solved) in order to learn by doing (O'Neil & Marsick, 2007). Action learning designers assume that skills needed to address complex, situated challenges can best be developed when people face urgent needs in actual circumstances surrounding learning transfer. AL combines real work with learning support (e.g., learning coaches, reflection and learning reviews, reframing of assumptions) for trying out new behaviors and learning from errors and successes. Its aim is to use situated, peer-supported problem solving for improving skills that typically emerge in and through action.

Action learning groups in the HiPo program met during three, week-long development sessions spaced over six months. HiPo team members communicated virtually, and sometimes face-to-face (especially regionally). Interactions with peers were also computer-mediated. Networking across boundaries was encouraged, but associates surveyed in two consecutive HiPo cohorts reported that it could be hard to get information or advice they needed from others who did not understand enough about their local context. Engineers, for example, needed to verify other engineers' qualifications before consulting with them. At the time of the study, such verification could be difficult to ascertain without some face-to-face interaction. Local country preferences, language differences, regulations, and work practices could make communication across boundaries challenging because of differences in experience and geographical contexts.

The program was organized in tracks based on function, e.g., supply chain, sales, or engineering. Business leaders worked with HR to design, teach, and support learning. Each participant set a realizable goal for an Action Learning project with his/her manager for measurable change to occur through development of new practices. The CEO held participants accountable for results. In the program's first half, team members investigated barriers and alternative solutions to achieving goals. In the second half, each implemented, assessed, and adjusted new practices.

Gephart and Marsick researched barriers and supports to the ways in which participants developed and leveraged their networks in projects and on the job. The study used the framework that Gephart and Marsick (2016) had developed and used over time, a system dynamics framework: *Strategic Leverage through Learning*©. See Figure 3. This model is described more fully below, following the HiPo example.

Interviews—coupled with HiPo training observations—first probed barriers and supports with an elite sample of 12 experienced networkers representing different functions, backgrounds, cultures and languages. Surveys were then developed and administered to two successive cohorts from four functional clusters (engineering, sales, supply chain, and administration that span functions. The surveys identified—among other things—networking practices in which participants engaged, as well as supports and barriers that learners experienced when they returned to their jobs and tried to use skills developed through projects.

Surveys helped the company understand when networking, teamwork, and collaboration were successfully transferred back to the job. The program aimed to improve collaboration across autonomous

Figure 3. System-level factors (Burke and Litwin 1992) and newly defined learning and performance outcomes in Strategic Leverage through Learning©
Published with permission of Gephart and Marsick.

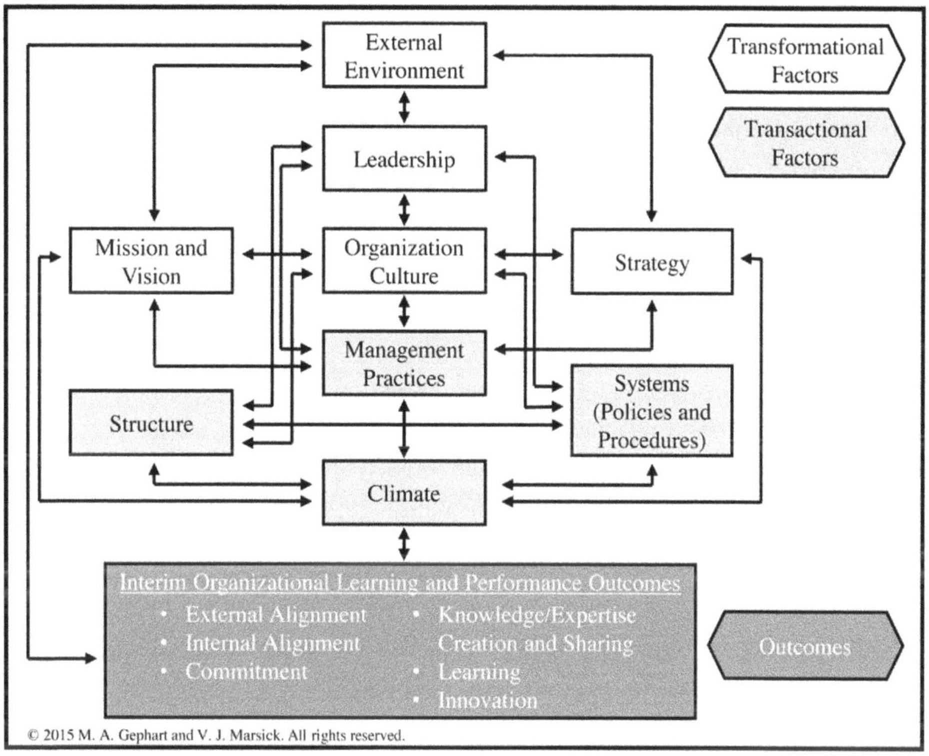

local offices. Global efficiencies were gained by putting mechanisms in place to enable or enhance communication and cooperation across geographic boundaries These included:

- New e-resources for knowledge sharing based on feedback from HiPo participants to network in projects
- Networking opportunities realized through Training &Development, meetings, task forces
- Regionally-appointed thought leaders who connected people through face-to-face and/or technology-mediated modalities
- Regularly scheduled meetings, often by teleconference or WebEx, for sharing information and problem solving

People came together, who would otherwise not meet, to solve challenges. For example, through facilitated data gathering, sharing, and discussion, employees discovered new alternative strategies. Offices set up communities of interest (partially technology-mediated) and monthly videoconferences for knowledge sharing related to metrics, performance, or project team work. Using a facilitated protocol for improved decision making, participants gathered and examined evidence with key players to make on-the-spot decisions.

Leadership and managerial commitment influenced whether, and for whom, factors operated as barriers or supports to networking, learning, and/or performance. Survey results from the second-year cohort suggested that managerial support shaped networking. Supervisors and managers provided resources and contacts, encouragement and authorization. They helped develop networks and networking skills; modeled cooperative values; and went the extra mile to change the way they, themselves, worked in order to meet challenging goals. Senior leaders discussed value they gained from networking. Managerial support of this kind was central to shaping a culture that encouraged networking through expectation setting and career advancement.

There were barriers, some of which were resolved by bringing in outside perspectives that led to new ways of thinking. For example, the existing teams lacked deep, specialized knowledge of equipment, people, and regulations that had to be considered to meet local needs. Experienced specialists were asked, as part of their job, to consult on such projects and act as "go-betweens" to put employees working on challenges in contact with specialists across the globe they would otherwise not know or access. Through formal exchanges, they could discuss technical requirements, design needs, past problems, previous market analyses and how to best translate findings into new perspectives and plans. Some barriers had to be accepted and worked around. For example, structural challenges requiring managerial support, new perspectives, and questioning assumptions included: competition within or across markets, geo-cultural differences that affected communication mechanisms and patterns, and limited resources.

In a summary list: supports that affected networking included:

1. Support from participants' direct managers
2. Networking opportunities that supported getting information and advice
3. "Structure" and mechanisms that supported information sharing and keeping people informed)
4. Organizational climate and culture that supported collaboration through rewards, meaningful incentives, and benefits for one's career; barriers included
5. Structural challenges posed by the job or organization that required "work-arounds."

This extended example illustrates organizational factors important to solutions when learning through real work. New capabilities cannot transfer without leveraging the right contextual system dynamics. Hence, Gephart and Marsick (2016) proposed a system dynamics framework, based on Figure 3, to identify key influences and leverage points. Some system dynamics can easily be altered by individuals or identifiable work units; others call for system-wide change requiring attention by leaders and top management. The next section explains the system dynamics framework, followed by a discussion of how designers can use it.

System Dynamics Framework

Gephart and Marsick (2016) recognized that: "For learning and performance gains to be institutionalized, organizational changes are often required. The organizational system is a critical determinant of whether and how individual and group learning enhance innovation and organizational performance" (p. 7). Assessment frameworks often focus on either learning or performance, but not both together. Gephart and Marsick decided to look at both and how they interact, given barriers and supports in any group or organizational system. They sought to understand how system-level dynamics affect progress toward strategic goals.

They started by using system-level factors identified by Burke and Litwin (1992). Based on these, they "define a set of learning and performance outcomes that represent critical capabilities needed for effective performance in a rapidly changing or uncertain environment: external alignment, internal alignment, commitment, innovation, group and organizational learning, and knowledge/expertise creation and sharing" (p. 8). These are the interim results of actions or steps taken, demonstrated behaviors or initiatives, or other observable phenomena. A review of research described in Gephart and Marsick (2016) suggested that these interim outcomes serve as strategic drivers of longer-term organizational performance; i.e., they can be linked to generation of revenue, savings, reduction in errors, increase in market sector, productivity gains, or other metrics representing tangible results. Gephart and Marsick also identified organizational critical factors that serve as supports for success because they affect learning and performance gains.

Organizational critical success factors can be transformational or transactional (Burke & Litwin, 1992). Transformational dynamics represent fundamental, enduring factors that shape how an organization functions. They include: external environment, mission/vision, strategy, leadership, and organizational culture. Transactional dynamics describe daily operating factors. They include: structure, management practices, systems (including policies, procedures, mechanisms), and work and business climates.

The Global HiPo example above showed how transformational factors shaped transfer of networking, teaming, and collaboration skills. No country office could produce and distribute its signature products without international collaboration. Leaders modeled effective collaboration in their own work and supported managers in offices who put beneficial practices, processes, mechanisms and conditions in place—without which results would be difficult, if not impossible.

Transactional practices feed into effecting transformational ideals and goals. These two levels interact to make it easier to take full advantage of new behaviors and mindsets. This is especially critical when changes in organizational incentives, rewards, accountability mechanisms, alignment, communication, climate and culture, are necessary to support learning and performance gains.

Learning designers cannot fundamentally change these dynamics, but they can incorporate ways of managing known barriers and leveraging supports. They can incorporate contextual information to more effectively and reliably build skills and practices. Efforts to identify this information may reveal situations where systemic or organizational changes will be needed before any approach to individual learning can succeed. Mechanisms can be designed, too, to support transfer of learning on the job. And in some cases, the organization might change some of its structures, practices, policies, rewards and expectations to better leverage the learning gained by individuals in ways that improve progress toward strategic goals.

Researching What Else in Organizations Has to Change

The title of this chapter underscores its thrust: that learning designers must address the context as well as the technology. A key to addressing barriers and supports in the learning transfer context is to find the right situated factors that make a critical difference to success or failure for learners working on particular kinds of challenges in each situation.

The eSims and DPS systems discussed in this chapter are situated in *particular* contexts and organizations, each of which calls for the *right* leverage points to support and motivate individual learning that contributes to bottom-line results. For example, Neaman's New Horizons DPS example above spoke to various organizational system dynamics that drove engagement and real behavioral change, such as alignment with business strategy, core values, and relevant rewards.

Learning designers can use the system dynamics framework in Figure 3 as a map for data gathering to help them identify a prioritized list of key influences on learning engagement and successful transfer of learning. They can elicit prototypical experiences that are framed in examples and language that learners will recognize when back on the job as guides to adapting what was learned to each unique context. They can help learners to look for trouble spots and recognize danger signals; build in ways to contact peers willing to share relevant experiences; and incorporate tools, case studies, tips, and other resources to figure out how to succeed.

Critical success factors are best customized for access and application. Why is context so important when deciding on which factors in the organization to attend?

First, each organization has a different configuration and strengths profile that makes it work. Each organization's culture is uniquely shaped due to founding values, geography, industry, senior leaders, degree to which its focus is innovation, and so forth. To illustrate, in the global manufacturing company, leaders shaped the culture by introducing and modeling new values that demonstrated cooperative behavior, even though competition was also part of its sales culture. Global was locally managed and needed to support customized solutions tailored to different country contexts. Its culture honored decentralization, so Global put mechanisms and practices in place to help employees network across boundaries. Other cultures build on national and organizational cultures favoring consensus—as was the case in a high technology company, Ericsson, which Gephart and Marsick (2016) also researched. Ericsson did not adopt new structures or create new roles and processes to work across competitive boundaries. It relied, instead, on a natural preference for consensus decision making. People and social processes played a role that was similar to that of Global's reliance on mechanisms and practices.

Thus, learning designers have to take into consideration the situated organizational context in which skills are applied in deciding what to build into DPS or eSims to support successful learning transfer. It is overwhelming, and very likely ineffective, to attempt to address a long list of decontextualized best practices that might sometimes be important, but vary with the situation. Designers can research context-specific organizational preferences, traps, and work-arounds that need addressing in their designs.

Effective simulations of real life challenges can, and often do, build on commonly understood "messy reality" factors which, when the determining factors are identified, often turn out to be common root causes of many performance problems. For example, in a simulation (designed by Roger Schank) to train community-based workers, learners were invited to begin their first day of work by confronting real-life challenges they would likely encounter when early in their role in a new job: e.g., telephone calls by community members, correspondence with clients, and "office politics." With little time to get a grasp of issues, the learner is asked to attend meetings to solve problems that engender competing priorities and conflicting views—some with logical cases built on and supported by evidence, others the result of erroneous views that need to be examined.

When a learning designer knows the context, he/she can research turning points, critical junctures, and common challenges and populate support systems with tools and resources that very specifically strengthen the learner's work. Learners can face the simulated real world consequences of making wrong decisions, and learn why their approach would not be effective in that situation. Additional strategies can be incorporated for adjustment and fine-tuning when learning is transferred, for example, fostering peer learning communities that enable discussion and problem solving in the midst of, or after, taking action.

Eraut (2004) assumes, as do this chapter's authors, that the "relative significance and the ways in which these factors interact will differ greatly from one context to another" (p. 268). He, nonetheless, also summarizes learning and context factors that are likely to impact learning transfer. These include:

"challenge and value of the work," "feedback and support," and "confidence and commitment." Context factors include "Allocation and structuring of work," encounters and relationships with people at work," and "expectations of each person's role, performance, and progress" (Ibid., p. 269).

In summary, research-based frameworks provide good starting points for learning designers who want to identify the unique mix of success factors that bear most on the target skills and practices that should be prioritized in a given learning intervention. Designers can then talk with key informants about the work context that surrounds these skills to identify system dynamics that pose likely barriers and supports. These interviews can provide input into the design in order to provide context-based lessons that enable learning from others' experiences in lieu of live on-the-job peer learning. Finding a way to also track experiences of learners as they apply what they know and encounter specific situations can enable formative assessment for improving simulation and DPS-based learning modules as well as addressing any need for change in organizational barriers and supports.

FUTURE RESEARCH

This chapter reflects a natural tension between the research methodologies utilized by its two authors. While both conduct their research in real business contexts, one is an academic researcher and the other a business practitioner. A gross caricature of the differences might be that academic research generates small precise claims with high degrees of confidence seeking generalizable models and principles. Business research generates large actionable claims with low degrees of confidence (or high degrees of undue confidence) that are pragmatically tested step-by-step within context. While workplaces are obviously the best places to study such learning, businesses often do not have interest in allocating the budgets or time for the rigorous assessments required for well-founded academic research.

One approach that Neaman illustrated in refining the DPS model is action research: that is, a form of research for improved action that builds data collection, analysis, interpretation, and practice alterations into work routines. This research mindset—aided by theory-based rules of thumb—yields evidence-based decision-making that can power context-specific improvements.

Marsick used systematic procedures and measures for the applied research illustrated in this chapter, but its purpose was also to assist organizations in improving their practice. Secondarily, and over time, Gephart and Marsick (2016) have developed and tested an extended framework and research instruments that contribute to system dynamics for group and organizational learning. While it is often extremely difficult to authoritatively attribute performance outcomes to learning interventions, proxy measures can be found. Gephart and Marsick have conducted applied research by using a theory-of-change approach (Chen & Rossi, 1980). Theory-of-change captures the causal link pathway underlying any intervention. The contrasting theory-of-action is what actually occurs. It is charted by collecting metrics during interventions, shows whether or not interim outcomes are achieved, and suggests leverage points for course adjustment to achieve desired goals. Gephart and Marsick's interim learning and performance outcomes could guide research to map organizational factors critical to learning transfer when developing DPS and track outcomes expected from learning design that takes these factors into account.

A potentially fruitful line of future research would be to find a middle ground between the two authors' methodological extremes by adding more academic rigor to practitioner research. To arrive at this middle ground the authors suggest starting with two central foci of this chapter: learner engagement and the development of transferable knowledge.

Learning design can heighten engagement and increase transfer by narrowing the disparities between learning and working. On its own, the work typically considered within the scope of learning design is often an insufficient stimulus to engage people, and thus is unreliable in developing meaningful transferable knowledge. To succeed, learning designers must expand their thinking to consider and build in relevant organizational context.

While it is often extremely difficult to authoritatively attribute performance outcomes to learning interventions, it is reasonably easy to measure learning engagement, especially when interventions involve learning by doing. Learning opportunities are among those job resources that Bakker and Demerouti (2008) consider drivers of work engagement. There are reliable measures, as well, of work engagement. Moreover, "a few quantitative studies have shown that work engagement is positively related to job performance" (Ibid., 214). Identifying ways that learning engagement contributes to work engagement is thus a meaningful, if imperfect, predictor of impact. The authors hypothesize that studies in this vein would be meaningful in organizations. People learn more when they pay attention. Inattention can be an indication that a learning audience has tuned out after determining that what is being taught is incompatible with the culture, processes, beliefs, and reward systems in their organization.

To test this last hypothesis, the authors suggest identifying the theory of change each DPS seeks to address. Research could look for relationships between learning engagement and work engagement. Barriers and supports identified by using the system dynamics framework and addressed in DPS or other learning interventions could be researched. Data on system dynamics could reveal relationships of system dynamics factors to learning engagement and work engagement, as well as highlight other outcomes believed to be influenced by the intervention.

The ideas put forth in this chapter, then, could be more substantially validated by business-academic partnerships focused on measuring engagement across two sets of conditions—learning interventions that do and that don't follow DPS design guidelines—in sum, that are deployed with and without the organizational conditions that prompt engagement discussed above.

CONCLUSION

The authors conjecture that the scrap heap of learning initiatives that have failed due to lack of engagement is several orders of magnitude larger than the collection that have succeeded. More than one learning designer has thrown up his/her hands and said "Why bother? Nobody wants what I'm making." Learner engagement is both a constant struggle and a holy grail for workplace learning. So too is alignment across system dynamics in ways that support effective performance. This chapter has explored ways to address what the authors see as the root cause of the problem: the rare co-occurrence of three pre-conditions for engagement:

- **Practice:** Workplaces must have people learn by doing
- **Commitment:** They must provide time and support for them to do so
- **Alignment:** They must be truly conducive to the behaviors they ask people to learn

A quotation often attributed to computer science pioneer Fred Brooks is relevant here: *Good judgment comes from experience. Experience comes from bad judgment.* The preconditions for successful adoption of learning interventions and for transfer of learning are embodied in the trial-and-error stories told in this chapter: classroom training that did not engage people highlighted the critical importance of *practice*; simulations that did not engage busy people who could get by on "good enough" highlighted the critical importance of *commitment* and DPS systems that did not engage people when the target behaviors, or even the activity of learning itself, were at odds with workplace incentives highlighted the critical importance of *alignment*.

If any of these three preconditions—practice, commitment, and alignment—is not satisfied, then many will likely not engage. It is therefore incumbent upon learning designers, managers, and change leaders, working individually or in concert, to take a systemic approach to design, addressing all three, both when they are creating new learning interventions and when they are trying to ensure engagement with existing ones.

A system dynamics approach is therefore needed to support learning through real work for improved performance. Real work can be "messy." This framework can assist learning designers in identifying critical leverage points in the organization that affect people's ability to apply what they have learned to their real work. Designers can use these leverage points to help them determine which resources will be most helpful to their audiences in achieving the best results. Further, research into system dynamics can build knowledge within the organization as to which dynamic factors currently in operation should be supported, and which should be modified by leaders to better reach strategic goals.

Research into DPS and into system dynamics relevant to their effective use, then, can contribute to designing the context—not just the technology—that enables eSims to move from the classroom into the knowledge-intensive workplace and that enables DPS systems to truly engage their time-starved audiences. In so doing, steps can be taken to foster sustainable learning on the job that is naturally motivated by goal accomplishment.

REFERENCES

Akanegbu, A. (2013, April 30). Does the Khan Academy pass the MOOC "Duck Test"? *Ed Tech*. Retrieved from https://edtechmagazine.com/higher/article/2013/04/does-khan-academy-pass-mooc-duck-test

Aldrich, C. (2005). *Learning by doing*. San Francisco, CA: Pfeiffer.

Askew, S. (Ed.). (2000). *Feedback for learning*. London: RoutledgeFalmer.

Bakker, A. B., & Demerouti, E. (2008). Towards a model of work engagement. *Career Development International*, *13*(3), 209–223. doi:10.1108/13620430810870476

Burke, W., & Litwin, G. H. (1992). A causal model of organizational performance and change. *Journal of Management*, *18*(3), 532–545. doi:10.1177/014920639201800306

Chang, C.-C. (2007). A study of the quantitative analysis of the development and implementation of Electronic Performance Support Systems. *International Journal of Instructional Media*, *34*(3), 275–284.

Chen, H.-T., & Rossi, P. H. (1980). The multi-goal, theory-driven approach to evaluation: A model linking basic and applied social science. *Social Forces*, *59*(1), 106–122. doi:10.1093f/59.1.106

Deci, E. L., Koestner, R., & Ryan, R. M. (2001). Extrinsic rewards and intrinsic motivation in education: Reconsidered once again. *Review of Educational Research*, *71*(1), 1–27. doi:10.3102/00346543071001001

Dewey, J. (1916). *Democracy and education: An introduction to the philosophy of education*. New York: Macmillan.

Dewey, J. (1938). *Experience and education*. New York, NY: Macmillan/Collier.

Eraut, M. (2004). Informal learning in the workplace. *Studies in Continuing Education, 26*(2), 247–273. doi:10.1080/158037042000225245

Ericsson, K. A., Krampe, R. T., & Tesch-Romer, C. (1993). The role of deliberate practice in the acquisition of expert performance. *Psychological Review, 100*(3), 363–406. doi:10.1037/0033-295X.100.3.363

Ge, X., Planas, L. G., & Er, N. (2010). A cognitive support system to scaffold students' problem-based learning in a web-based learning environment. *The Interdisciplinary Journal of Problem-Based Learning, 4*(1), 30–56. doi:10.7771/1541-5015.1093

Gephart, M. A., & Marsick, V. J. (2016). *Strategy organizational learning: Using system dynamics for innovation and sustained performance*. Springer-Verlag. doi:10.1007/978-3-662-48642-9

Gery, G. J. (1991). *Electronic performance support systems*. Gery Associates.

Gruber, H., & Voneche, J. (1977). *The essential Piaget: An interpretive reference and guide*. New York, NY: Basic Books.

Holton, E. III, & Baldwin, T. T. (Eds.). (2003). *Improving learning transfer in organizations*. San Francisco: Jossey-Bass.

Kolb, D. (1983). *Experiential Learning: Experience as the Source of Learning and Development*. Upper Saddle River, NJ: Prentice Hall.

Lombardo, M. M., & Eichinger, R. W. (1996). *The career architect development planner*. Minneapolis, MN: Lominger Limited.

Marsick, V. J., & Neaman, A. (2017). Adult informal learning. In Informelles lernen. Eine Standortbestimmung. Wiesbaden: Springer VS.

Marsick, V. J., & Watkins, K. E. (1990). *Informal and incidental learning in the workplace*. London, UK: Routledge.

Marsick, V. J., Watkins, K. E., Scully-Russ, E., & Nicolaides, A. (2017). *Rethinking informal and incidental learning in terms of complexity and the social context. Journal of Adult Learning, Knowledge and Innovation*. doi:10.1556/2059.01.2016.003

Noe, R. A., Clarke, A. D. M., & Klein, H. J. (2014). Learning in the twenty-first-century workplace. *Annual Review of Organizational Psychology and Organizational Behavior, 1*(1), 245–275. doi:10.1146/annurev-orgpsych-031413-091321

O'Neil, J., & Marsick, V. J. (2007). *Understanding action learning*. New York: AMACOM.

Perna, L. W., Ruby, A., Baruch, R. F., Wang, N., Scull, J., Ahmad, S., & Evans, C. (2014). *Educational Researcher*. DOI 10.3102/001389X14562423

Polanyi, M. (1966). *The tacit dimension*. London, UK: Routledge & Kegan Paul.

Quinn, C. (2005). *Engaging Learning: Designing e-Learning Simulation Games*. San Francisco, CA: Pfeiffer.

Schank, R. (1997). *Virtual Learning: A Revolutionary Approach to Building a Highly Skilled Workforce*. New York, NY: McGraw Hill.

Schank, R., & Neaman, A. (2001). Motivation and failure in educational simulation design. In K. D. Forbus & P. Feltovich (Eds.), *Smart machines in education: The coming revolution in educational technology* (pp. 37–69). Cambridge, MA: The MIT Press.

Schön, D. (1983). *The reflective practitioner: How professionals think in action*. New York, NY: Basic Books.

Schön, D. (1989). *Educating the reflective practitioner: Toward a new design for teaching and learning*. San Francisco, CA: Jossey-Bass.

Watkins, K., & Marsick, V. (1993). *Sculpting the learning organization: Lessons in the art and science of systemic change*. San Francisco, CA: Jossey-Bass.

Watkins, K., & O'Neil, J. (2013). The Dimensions of the Learning Organization Questionnaire: A Nontechnical Manual. *Advances in Developing Human Resources*, *15*(2), 133–147. doi:10.1177/1523422313475854

ENDNOTES

1 Gery (1991) and Chang (2007) also identified barriers to effective deployment of EPSS that are relevant to the arguments made in this chapter, such as weak organizational links that made coordination across boundaries for design and implementation challenging; learning measures that were not aligned with performance support goals; and incompatible rewards and incentives.

2 MOOCs "can accommodate large numbers of users from around the globe ('massive'), are available at no charge … with minimal registration requirements ('open'), and are delivered 'online'" (Perna et. al. 2014, 1). While MOOCs scale knowledge dissemination, personalized, quality feedback by experts is not easily provided. The Khan Academy solves this problem by customizing its MOOCs, adding the "on-demand capabilities" of more than 4,000 videos, interactive challenges, and assessment elements (Akanegbu, 2013).

3 The program name, as well as the organization and industry for which it was developed, have been changed for the purposes of client confidentiality.

Chapter 2
The Secret Agent Society:
Upskilling Teachers in the Delivery of a Game-Based Social Skills Youth Program

Renae Beaumont
Weill Cornell Medicine, New York Presbyterian Hospital, USA

Sharon Hinton
University of Queensland, Australia

Kate Sofronoff
University of Queensland, Australia

ABSTRACT

In recent years, there has been a worldwide commitment to integrating technology into classrooms to train students in the computer skills they will need to be workforce ready. This chapter provides an overview of the Secret Agent Society – Whole of Class Project: a randomized controlled trial of a gaming-based social skills training curriculum that was conducted in Grade 5 classrooms throughout Australia. The chapter explores the content, structure, and delivery format of the social skills program, together with the design and implementation of an online teacher training course to accompany it. Teacher feedback on the online teacher training course is presented, together with recommendations for enhancing the design, implementation, and evaluation of future online professional development courses for school staff.

INTRODUCTION

In recent years, billions of dollars have been spent in classrooms across the globe to provide students with access to technology. This has been done to improve students' learning outcomes and to train them in the computer skills they will need to be competitive in tomorrow's workforce. A 2009 survey by the National Centre for Education Statistics (NCES) showed that in the US, 99% of teachers had computers accessible in their classroom every day and 95% of computers had Internet access (Gray, Thomas & Lewis, 2010). Despite this widespread availability of technology in modern-day classrooms, however, survey findings showed that only 40% of K-12 teachers reported using technology "often" to support students' learning (Gray et al., 2010).

DOI: 10.4018/978-1-5225-4111-0.ch002

Research suggests that the task of integrating technology into classroom instruction in a meaningful and innovative way remains challenging for several reasons. Beyond having the necessary hardware and software, school leaders play a pivotal role in determining teacher attitudes and utilization of technology within classrooms. As described by Weng and Tang (2014), in today's modern schools, school leaders' responsibilities include ensuring that teachers have the skills and confidence to use new technology successfully, establishing mechanisms to provide school staff with real-time technical support, providing staff with access to relevant technological resources and setting up mechanisms for them to collaborate with each other to develop their information and communication technology (ICT) skills.

Pittman and Gaines (2015) highlight the time required to create and implement lesson plans that integrate technology as one of the greatest barriers to primary school teachers using it in the classroom. Pivotal to addressing this issue is access to high quality professional development (PD) to improve teachers' ability to design and implement tech-enhanced curricula quickly and effectively. Common criticisms of technology-related PD for teachers are that they are too general and not subject specific, present too much information within a short period of time for teachers to digest, and introduce technologies that teachers do not have access to (An and Reigeluth, 2012).

Tweed (2013) recommends that teacher tech-PD should focus on enhancing teacher attitudes and beliefs relating to technology and should provide teachers with experiences that enable them to see 'technology in action': that is, opportunities for them to see technology being successfully integrated into classrooms in a fast, practical and effective way. Desimone (2009) also speaks to the importance of effective teacher PD involving active learning and collective participation. Without the integration of these recommendations, research suggests that PD focused on technology integration does not translate into improved adoption of technology by teachers in their classrooms (Pittman & Gaines, 2015).

To improve the convenience, availability and cost-effectiveness of teacher PD, there has been a recent increase in the number of online teacher PD programs available (Bates, Phalen & Moran, 2016; Edinger, 2017). Online courses allow teachers to access PD material in a manner that is optimally convenient for them. Hybrid PD programs offer the advantages of online PD programs and have the added bonus of in-person learning opportunities. Such programs enable teachers to review online material, and to receive support from peers and/or experts in answering questions and creatively applying the content to the students in their classrooms. This collaborative learning framework holds great promise in increasing teachers' adoption of tech-inspired curricula after participating in a content-specific PD event (Bates et al., 2016). With these recommendations in place, recent studies have shown online and face-to-face teacher PD to be equivalent in their rates of curriculum implementation post-training (Fisher, Schumaker, Culbertson & Deshler, 2010; Fishman et al., 2013; Masters, de Kramer, O'Dwyer, Dash & Russell, 2012).

Research supports the effectiveness of technologically innovative classroom resources and programs in improving students' eating habits (Chamberland et al., 2017), students' abilities to conduct scientific experiments (Wang et al., 2014) and their mathematical knowledge and understanding (Bicer & Capraro, 2017). While there is also recognition of the value of computer-assisted interventions in improving school mental health and students' social-emotional skills (e.g. Livet et al., 2017), evaluations have largely focused on individual- or small-group interventions delivered by specialist school staff to students with significant behavioral or emotional issues. Examples include the Stressbusters program for teenagers with depression (Abeles et al., 2009), the Coping Cat program for youth with anxiety (Khana & Kendall, 2008) and the Secret Agent Society small group program for children on the Autism Spectrum (Beaumont, Rotolone & Sofronoff, 2015; Einfeld et al., 2017).

Given the potential benefits of technology in improving students' engagement, capacity for collaborative learning with peers and ability to creatively apply their knowledge and skills in daily life, the current chapter focuses on the development and implementation of a whole-of-class adaptation of the Secret Agent Society multi-media social skills program (SAS-WOC) and an associated online teacher training course. The SAS-WOC Program was implemented and evaluated in Grade 5 mainstream classrooms in schools across three Australian states as part of a randomized controlled trial. The objectives of the program were to improve social-emotional learning (SEL) and educational outcomes for all students, especially those who have SEL challenges (e.g. pupils on the Autism Spectrum). Specifically, the program aimed to improve students' abilities to regulate emotions of anxiety and anger, solve social problems, form and maintain relationships with others and to prevent and cope with bullying and teasing: essential skills for them to become workforce-ready for the future.

This chapter provides an insight into the pros and cons of the online professional training model that was used and an overview of some of the practical considerations for school staff delivering computer-mediated curricula within a classroom setting. The chapter concludes with recommendations for how online training courses can be optimally configured and presented to enhance teacher engagement, learning and application of new teaching practices within classrooms.

BACKGROUND AND THEORETICAL FRAMEWORK

The Secret Agent Society: Whole of Class (SAS-WOC) Program

SAS-WOC was designed to be an engaging, user-friendly social-emotional skills training program that Grade 5 teachers can deliver to their whole class. Over an eight-month period, group face-to-face and online meetings were conducted with school leaders, guidance officers and teachers at eight Queensland schools in Australia to gather information on what would be the optimal program materials, format and content for students, parents and school staff. The schools were from metropolitan, regional and rural locations, and varied greatly in size and school type (Department of Education, Catholic Education Office and Private). Based on school staff feedback, the SAS-WOC Program was structured to consist of an initial one-hour school briefing session for interested school staff and parents, followed by nine weekly 45-minute classroom lessons delivered to Grade 5 students over one school term. These lessons were typically delivered by children's classroom teachers, who in some cases co-facilitated the lessons to offer opportunities for peer support and mentorship. Two 45-minute booster lessons were also included in the program for teachers to schedule at 'high-risk' times for students (e.g. before a school camp or test periods).

The SAS-WOC Program was designed to be brief, feature fun interactive activities for students, cater for a variety of student learning styles (i.e. visual, verbal and kinesthetic) and actively involve school leaders and students' parents in an effort to optimize students' SEL outcomes. These objectives were developed in response to school staff feedback that many existing whole-of-class SEL programs are too long and intensive to deliver in the context of competing academic curriculum priorities, feature boring, outdated content, involve too much lesson preparation time in terms of photocopying or printing student handouts and do not actively engage school leaders or parents to promote skill generalization across the school community.

The primary objective of the SAS-WOC Program was to fulfill the Australian Curriculum General Capability of personal and social capability. However, the program also included activities to upskill students in other Australian Curriculum general capabilities, including critical and creative thinking, information and communication technology (ICT), ethical understanding and intercultural understanding. Competence in all of these skill domains helps to prepare students to meaningfully participate in and contribute to society during their working years as an adult. A summary of the objectives of each lesson of the cognitive behavioral program is shown in Table 1.

Weekly lessons included a mix of activities that involved playing designated sections of the Secret Agent Society Computer Game (individually, in small groups or as a whole class), small group games that incorporated drawing, describing or acting out core concepts introduced in the computer game (e.g. relaxation gadgets or strategies to help people feel happier, calmer and braver and to make smart decisions) and individual activities to help students apply concepts introduced in the computer game to their own lived experience (e.g. attaching stickers to pocket-sized devices that showed the body clues and situations where they felt low, medium and high levels of anxiety and anger and the relaxation gadgets they intended to use in these situations to calm down).

The program content and resources borrowed heavily from the Secret Agent Society small group program curriculum. SAS-WOC relied on a social constructivist model of learning, whereby students worked together to discuss and illustrate their learning of core program concepts through varied personal

Table 1. Objectives for each lesson of the SAS-WOC program

Lesson	Objectives
1	Establish club rules Explain friendship surveillance and reward system Students identify goals and skills/traits needed to achieve them
2	Students develop an understanding that we all have strengths and weaknesses Education about the strengths and challenges faced by individuals on the Autism Spectrum Students learn how to identify emotions from facial expression and body posture/ movement clues Set up SAS peer buddy/partner system
3	Students learn how to identify emotions from voice tone clues Students learn how to identify their own emotions from body sensations and thoughts
4	Students learn how to identify their own emotions from body sensations and thoughts Students learn relaxation skills
5	Students learn relaxation skills
6	Students learn relaxation skills Students learn a step-by-step problem-solving formula - D.E.C.O.D.E.R. (Define problem, Explore solutions, Consider consequences, Organize a plan, Do it, Evaluate, Rewards)
7	Students learn the steps for starting, continuing and ending conversations – The Conversation Code
8	Students learn the steps for playing cooperatively with others – the Play Code
9	Students learn the steps for preventing and managing bullying and teasing - The Bully-Guard Body Armor Students plan how they can continue to use their social-emotional skills
Booster Lessons 1 & 2	Students review and apply their social-emotional skills to new challenges

examples. Teachers were also encouraged to create opportunities for students to review and consolidate their knowledge of the social-emotional skills taught in SAS-WOC through group projects such as creating posters, plays, e-books, songs, documentaries, filmed interviews, blog posts, games and/or puzzles to share with the broader school community.

The Secret Agent Society Computer Game

In the multi-level Secret Agent Society Computer Game, the user assumes the role of a rookie agent, training to be a specialist in controlled remote viewing (i.e. mindreading). In Level 1, students played a range of timed games and activities that taught them how to recognize emotions in other people from facial expression-, voice tone- and body posture clues. For example, in The Line Up game, they watched a series of filmed spy-themed vignettes and clicked on the emotion that best described how the 'suspect' felt based on their body posture and movement. If the user responded incorrectly to an item on the first attempt, they were given a clue. If they responded incorrectly on their second attempt, the correct answer was explained to them by Agent Ardon, their mentor special agent. A screenshot from The Line Up is shown in Figure 1. At the end of each level of the game, the player received a ranking promotion (e.g. from rookie to detective).

In Level 2 of the computer game, students engaged in a series of activities that taught them how to identify emotions within themselves from internal body clues and thoughts and how to piece together face-, voice-, body- and situational clues to detect how someone is feeling. Consistent with a behavioral learning paradigm, when students successfully reached a target score for each activity in Levels 1 and 2

Figure 1. Screenshot from The Line Up activity in the Secret Agent Society (SAS) computer game
Reproduced with permission from the author and publisher

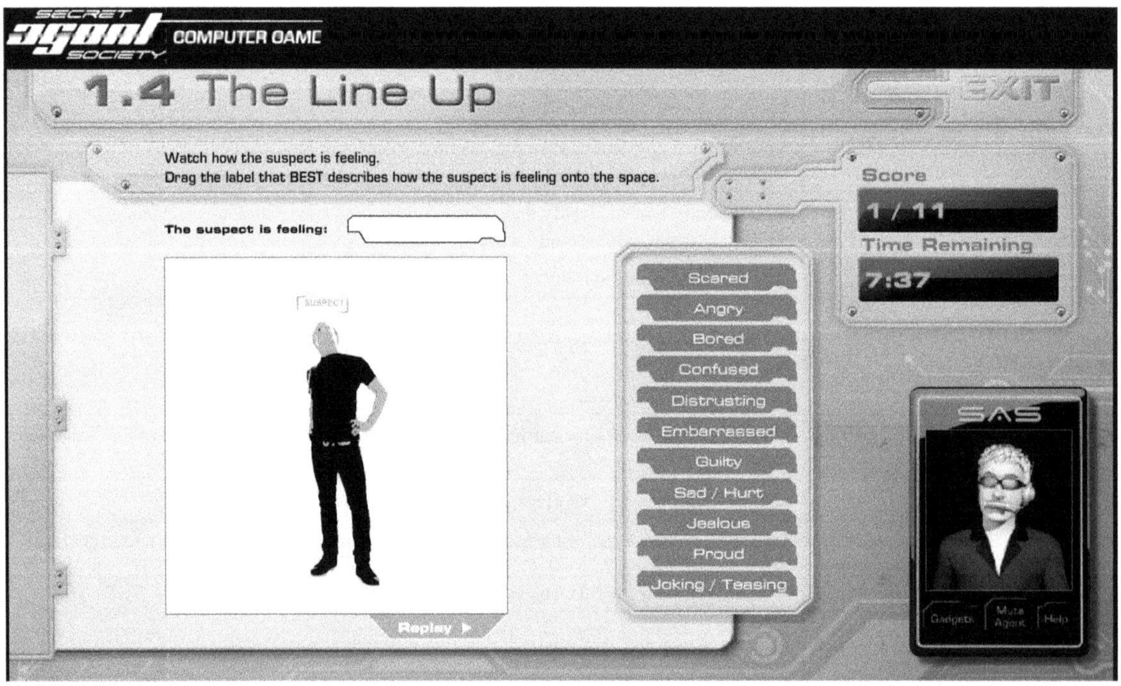

of the computer game, or attempted each activity at least twice, they were reinforced with a secret agent gadget reward (e.g. invisible ink reader, voice frequency amplifier). Players were able to use these gadget rewards to navigate through a series of virtual reality missions in Level 3 of the game, which depicted "choose your own adventure" style interactive scenarios that children often find socially challenging (e.g. trying new things, group work, managing bullying, losing at a game or competition). Students discovered the consequences for their avatar of responding in different ways to these challenges (e.g. breathing slowly and thinking helpful thoughts versus screaming and shouting). A screenshot from Level 3 of the computer game is shown in Figure 2.

In both the SAS-WOC online teacher training course and Class Teacher Manual, teachers were given detailed instructions and trouble-shooting ideas on the variety of ways in which the computer game activities could be played in weekly SAS-WOC lessons by their students, including individually, collaboratively in pairs or small groups, or as a whole of class activity. Decisions regarding the optimal format for computer-game play largely depended on the number of computers available to students and teachers' beliefs about how they would best engage with the game.

Additional Program Materials

In addition to the computer game, the SAS-WOC Program materials included full-color School Leaders Guides and digital Tip Sheets that briefly updated parents and school support staff (e.g. teacher aides, specialist teachers) on what children learned in the program each week and how they could support children in applying these skills at school and at home. Program materials also included student Cadet

Figure 2. Screenshot from a level 3 virtual reality mission in the SAS computer game
Reproduced with permission from the author and publisher

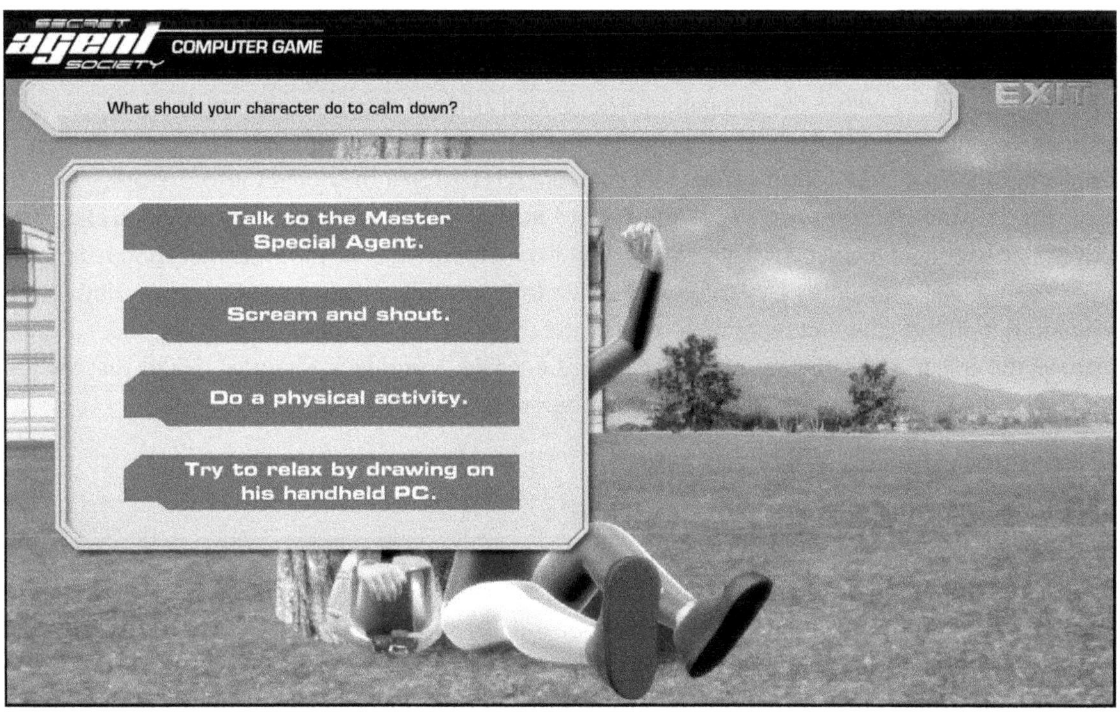

Handbooks which featured graphic illustrations of the skills taught in SAS-WOC and classroom skill posters. A Class Teacher Manual was also given to SAS-WOC facilitators, which briefly summarized the program content, provided weekly lesson plans, offered trouble-shooting tips around optimizing student engagement and preventing and managing behavioral issues and gave recommendations for how to help students practice their social-emotional skills between weekly lessons. The provision of engaging, user-friendly program materials was intended to minimize teacher lesson planning time: an identified barrier to delivering other classroom-based SEL curricula.

Several strategies were integrated into the curriculum to support students to apply their social-emotional skills in class and at break times. These included a rotational peer partner system, where teachers paired up class members to help and support each other to use the skills they were learning in SAS-WOC in class and at break times. Teachers also used a Class Friendship Chart to encourage students to take note of when their SAS partners used target skills from the program each day. Teachers wrote a prescribed target skill for each week of the program on the chart, and checked in with students at ideally the same time each day (e.g. after lunch break) to see when they or their SAS partner used the target skill (e.g. applying the Conversation Code steps when talking to others). Teachers prompted students to do the target skill at appropriate times each day (e.g. before recess) and added a secret symbol (e.g. a smiley face) to the chart for each example that students provided. If the class reached the symbol target for the week (e.g. 12), they received a class-based reward (e.g. free time).

These skill generalization strategies are examples of how situated learning (prompting students to practice skills in relevant situations), experiential learning (embedding skills practice in everyday social contexts), peer social scaffolding (peers supporting each other to use skills) and behaviorist principles (reinforcement for skill usage) were applied as part of the hybrid instructional framework used in the program. Children were also given weekly "missions" or skills practice tasks that aligned with the Friendship Force Chart behavioral goals. For example, they were asked to use the steps of the D.E.C.O.D.E.R problem-solving formula to solve a problem with adult support, and to fill out the prompts shown in their Cadet Handbook or featured in the Journal section of the SAS computer game to help them to complete this mission. If students had an aversion to doing homework, they often preferred to complete the Mission questions within the computer game, as this allowed them to create pictures to answer the questions using a scene generator device, rather than having to hand-write their responses in their Cadet Handbooks. With adult permission, students were advised that they could also use other means to answer the Mission questions, such as creating voice memos or taking photos or brief video clips with a smart device to collect 'evidence'. These are further examples of how the program incorporated multiple learning approaches, including constructivist (generating new understandings of concepts through skill application), collaborative (learning with others) and experiential (learning from experience). Figure 3 illustrates the eclectic learning theories (Naismith, Lonsdale, Vavoula & Sharples, 2004) that the SAS-WOC Program draws on and the core program components related to each.

SAS-WOC ONLINE TRAINING COURSE

The teacher training course for SAS-WOC was originally developed and delivered as a face-to-face two-day curriculum to Grade 5 teachers and school support staff at Queensland schools. However, the geographical spread of teachers across multiple states in Australia in Year 2 of the trial made this training model cost- and time-prohibitive. Therefore, with the support of funding provided through the Coopera-

Figure 3. Instructional design frameworks integrated into the SAS-WOC Program

tive Research Centre for Living with Autism (Autism CRC) and the Australian Government's Positive Partnerships Program, the PD course was translated into an online format to make it more accessible and affordable. Requiring only internet access and a viewing device (i.e. laptop, tablet, smartphone), school staff were encouraged to complete online training in groups, using either one device or multiple devices at the same time – depending on personal preference. Consistent with collaborative learning theory, this enabled them to discuss how the SAS-WOC curriculum would be optimally delivered to the students within their school. School staff were also encouraged to brainstorm ideas for how they could prompt and support their students to apply the social-emotional skills that they were learning in the program in relevant situations in daily school life (in accordance with the principles of situated learning).

The school staff were supported in real time both during the online training and throughout delivery of the SAS-WOC Program by research personnel. Teachers connected with research staff via email or teleconference each week to discuss the successes and challenges that they had faced in delivering their weekly SAS-WOC lesson and in supporting students to apply their SEL skills. Research staff helped them to trouble-shoot any difficulties. Research staff subsequently helped teachers to plan for the next week's lesson, reminding them of salient tasks. Further details regarding the duration and content of these weekly implementation support sessions are provided in this chapter.

Development of the SAS-WOC Online Teacher Training Course

The development of the online training interface was divided into four main phases:

Identification of Learning Delivery Platform

The SAS-WOC online training environment was created using Moodle: a popular, open-source learning platform allowing the creation of customized learning environments. Encompassing a wide range of built-in features, and a simple user-interface, Moodle is web-based, making the learning platform easily accessible from any location with internet access. With a default-mobile compatible interface and cross-browser compatibility, Moodle is consistent across different web browsers and devices, making it an excellent choice for supporting delivery to diverse users (teachers) across Australia.

Creation of Multimedia Resources for Translation Into the Online Environment

Given the flexibility of Moodle, little limitation was placed on the type of resources that could be incorporated into the learning platform. As previously delivered 'in-person' workshops had received very high teacher satisfaction ratings, the mirroring of the in-person components within the online interface was considered essential to maintaining both fidelity of instruction, and high levels of participant satisfaction (Schroeder, 2016). To support this approach, the following, novel multi-media resources were developed and included in the online interface:

1. **Narrated Videos:** The core content of SAS-WOC was delivered by combining PowerPoint slides, images, case studies and voice overs into MP4 video files. This multi-media approach used best practice demonstration and modelling to provide key elements of successful PD (Schroeder, 2016; Zheng, 2015). The video format allowed participants to pause or stop instruction at any time – offering great flexibility in delivery, considered a significant asset of online training (Collins & Liang, 2015). A screen shot from the introductory video included in the online training course is shown in Figure 4.
2. **Video Simulations:** Video simulations were created to operationalize key concepts and strategies using child and adult actors in mock classroom or school yard situations to optimize teacher engagement (Schroeder, 2016; Zheng, 2015). The inclusion of video simulations was also an attempt to improve teachers' attitudes and beliefs about the potential benefits of facilitating a gaming-based resilience program with their class, by allowing them to see the 'technology in action' (Tweed, 2013). For continuity in delivery, simulations were incorporated into the narrated video presentations. Screenshots from a selection of video simulations included in the online training course are shown in Figure 5.
3. **Activity Handbook:** A downloadable Activity Handbook was created to encourage participants to create personalized training materials as they worked through the program. The use of an 'off-line' resource was also deliberately employed as a means of encouraging colleagues to work collaboratively when undertaking course activities. This emphasizes the important element of constructive dialogue (Prestridge &Tondeur, 2015) and also the finding that teachers prefer to have access to both visual and written resources (Zheng, 2015). The handbook contained a range of interactive activities that teachers were asked to do throughout the online course. These included discussing

how they would apply the curriculum to the students in their class/school, role playing the delivery of core program elements with their colleagues (where possible) and problem-solving anticipated barriers to program delivery (e.g. program tailoring for students with attentional difficulties). These activities were intended to create active and collaborative learning opportunities, as recommended by Desimone (2009).

4. **Online Content Expert Assistance:** While the online training course was developed to function without synchronous content expert assistance, for the purpose of the evaluation trial, a course administrator/content expert was available during all courses for both functionality support and content enquiries.
5. **Online Forum:** An online 'chat room' was created using the functionality within Moodle to allow participants to create discussion topics and ask questions. This forum was monitored by the course administrator (who was also a content expert) and visible to all school staff (i.e. across schools) completing the SAS-WOC Online Training Course. This chat room feature of the course was included to encourage teachers to offer peer support during the online training course and subsequent SAS-WOC Program delivery.
6. **Private Messaging:** For participants who had questions or queries they wished to ask the content expert privately (or confidentially), private message functioning was also enabled within the online course. Messages forwarded via this function were not visible to anyone other than the inquirer and the course administrator.

Figure 4. Screenshot from a video included in the online teacher training course
Reproduced with permission from the author and publisher

Figure 5. Screenshots from video simulations included in the online teacher training course
Reproduced with permission from the author and publisher

Creation of Training Delivery Structure

The final stage of the creation of the online training course was to combine the components of the course into one, cohesive delivery structure. Upon review, the modular format used in the original 'in person' training course was considered an equally suitable format for the online environment. Accordingly, the content was split into six core modules (refer to Table 2), which were then further broken down into individual sub-components, each approximately five to 10 minutes in duration.

Table 2. SAS-WOC Program Online Teacher Training modules

Module	Name
1	An Introduction to SAS-WOC
2	Evidence Base & Current Research
3	Content & Assessment Measures
4	Core Program Elements
5	Program Content & Delivery
6	Planning for Delivery

Testing of Online Interface

Post completion of the delivery structure, functional testing was conducted by novice users against the following criteria:

1. **Mainline Functions:** Functionality was tested by completing all modules, verifying the output and comparing the actual results with the expected results.
2. **Basic Usability:** Was the user able to navigate through the training program without difficulty?
3. **Accessibility:** Was the user able to access the system without difficulty?

The testing process verified that the SAS-WOC Online Teacher Training Course worked against the functional requirements.

Completing the Online Training Course

School staff accessed the online training course via an administrator-generated user name and system-generated password. Participants could choose to work through the program on their own individual device or to share devices. Regardless of the preferred approach to accessing online content, a collegiate approach to learning was encouraged at the school level, i.e. colleagues completed training at the same time and were instructed to pause training at any time to discuss key points and strategies.

Implementation and Evaluation

The online teacher training course was hosted on an internal server. The online course administrator issued participants with user IDs, passwords and instructions for registration the day before training was scheduled to commence, and users were encouraged to check accessibility upon receipt of their registration details.

For the evaluation phase, training was scheduled over two days with guidance placed in the course introduction (Module 1) as to the content goals for each day. A flexible approach to commencement times was adopted, however schools were advised of the times that the course administrator would be available online each day (8am – 4pm) and asked to complete their training within this window each day.

Following completion of the two-day training, the school staff (22 teachers and 7 teacher aides) completed an evaluation form that asked about their satisfaction with the training. They were also asked about aspects of the training that they liked and about those that they did not like or found problematic. Generally, the teachers rated the online training as 'satisfactory' with a mean of 5 on a scale of 1 to 7 (range 4-7). This is a lower rating than the face-to-face training courses offered in Year 1 of the trial (mean rating of 6.5/7) and the research team were keen to discover whether there were any actual problems encountered during the training. In every instance the program had run without technical difficulties, all teachers were able to readily access the training and the system functioned well. Teachers were also able to access the online administrator and ask questions in real time. Many teachers, however, did not take advantage of this.

Several teachers did give very high ratings for the online training (7/7) and these teachers indicated that doing the training online was their preferred medium. The reasons these teachers gave were that they could work at their own pace, and at times that suited them. They could stop and discuss when they needed to and they could move more quickly through sections that they found straightforward. They also mentioned that the training could be done either at school or at home and that they could stop and pick it up again. These teachers found the video clips helpful and enjoyable and they did access the online administrator when they needed to do so. These types of comments are consistent with those made by teachers in other studies when asked about the benefits of online PD (Collins & Liang, 2015; Zheng, 2015).

Issues, Controversies, Problems

The teachers who gave lower ratings (4-5/7) in some cases had completed the training alone rather than with a colleague and reported that they found that more difficult. Some of these teachers also reported that completing training online was not their preferred medium and they would rather have done face-to-face training. The major drawback cited about online training was the inability to interact face-to-face and to ask questions immediately and have them answered or strategies demonstrated further. These teachers did acknowledge that the online administrator could address questions but this was not a preferred option. Several teachers mentioned that they thought the video clips were rather artificial, not like a real classroom and they would like to see video footage of real teachers and students engaged with the program. Despite this, they all indicated that the training had equipped them to adequately deliver the program to their classrooms. This feedback is very similar to that found in a large teacher survey conducted in the US (Collins & Liang, 2015).

Solutions and Recommendations

The authors suggest that despite creation of high quality online PD modules, these modules are frequently not perceived by teachers as directly useful to their specific classroom context. In the case of the SAS-WOC Online Teacher Training Course, it is important going forward to adequately respond to the feedback given and to work directly with teachers to ensure their needs are met. Follow-up data is being collected from school staff participating in the project to further understand the dimensions underlying individual preferences for online versus face-to-face training, and to explore methods to bridge the gap between the two approaches to workforce development.

Weekly Online Implementation Support Calls

Once schools had indicated their intention to trial the SAS-WOC Program, they were given details of the requirements to run the computer game. A time was set up for the school's technical support person to liaise with the research team to talk through the set-up requirements. This was done well in advance of teachers being trained so that the game was functional in the school and teachers could log in and play the game in order to understand the look and feel of the program. Teachers were also given the contact details for an IT support person at the Social Skills Training Institute (the organization that publishes and distributes the computer game) to provide further technical assistance if needed.

Some schools had computer labs that could be used and in some schools, students had their own laptops. In these cases, each student played the game individually and this was the approach that students preferred. In other schools where students had iPads (which the current game is not compatible with), students either played the game in small groups or the teacher showed the game on a single laptop or Smart Board and engaged students in turn. Feedback from teachers suggested that this was the least preferred option from the students' perspective.

Once the teachers began program delivery, they were contacted each week either by Skype, phone or email to briefly discuss how the program was going and to brainstorm and problem solve any issues as they arose. These calls tended to be quite brief and teachers were encouraged to share positive stories as well as to discuss problems. The problems that did arise were for the most part concerned with efforts to engage difficult students. Sometimes these were students on the Autism Spectrum or with other SEL difficulties but often the issues were behavioral and not related to characteristics of autism. Behavioral issues (e.g. lack of focus, frustration at having to wait their turn) appeared to be more common when the computer game was being played as a whole of class activity or within small groups, compared to when students played it individually in a computer lab or on their own computer.

SAS-WOC PROGRAM OUTCOMES

Quantitative data from students, teachers and parents regarding SAS-WOC Program satisfaction and outcomes is still being collected as part of the randomized controlled trial. In this trial, SAS-WOC is being compared to 'curriculum as usual' in Grade 5 classes across three Australian states, with three-month follow-up data collected for all students, and one-year follow-up data collected for a cohort of students on the Autism Spectrum.

Qualitative feedback provided by teachers during the weekly implementation support sessions suggested that many students used the SEL skills that they learned in the SAS-WOC Program to stay calm during challenging situations (e.g. when taking tests or attending school camp) and to make and maintain new friendships with peers.

FUTURE RESEARCH DIRECTIONS

With regards to the SAS-WOC curriculum itself, informal feedback from students and teachers suggests that it could be improved by making the computer game tablet-compatible, and delivering the content over a longer period of time (e.g. two school terms instead of one) to allow more time for students to process and discuss the concepts taught. Students and teachers have also recommended that it would be valuable for students to have an SAS app on their smart devices (where available) to prompt and track their social-emotional skill usage (i.e. to allow more situated learning). The app could also allow students to collect digital evidence of their skill usage and reinforce them for doing so (consistent with a behaviorist learning framework). Teachers indicated that they would find it valuable if there was a back-end reporting function added to the computer game and app that collated student data on computer game play and SEL skill usage.

While the SAS-WOC Program is aligned with Australian National Curriculum general capabilities, it does not currently sit within a designated school subject. Therefore, teachers who participated in the trial had to squeeze the program into an already full timetable, which contributed to feelings of stress and exhaustion at times. To alleviate this burden, the research team is working with teachers to map the SAS-WOC Program content to a specific school subject. Currently the best fit in an Australian context is the Health and Physical Education (HPE) subject that is part of the core curriculum in primary schools. This curriculum mapping exercise is intended to help school leaders and educators to perceive SAS-WOC as a core part of the curriculum for improving students' learning outcomes and work readiness for the future, rather than as an 'optional extra' program offering.

With regards to the SAS-WOC Online Teacher Training Course, ideally, the teachers trained using the online program would be those who state a preference for the online medium, i.e. they would have choice. When teachers did prefer the online medium, they also indicated that they enjoyed the training and felt well equipped to deliver the program in their classrooms, although the accuracy of these testimonials is yet to be empirically determined. In light of the fact that many teachers indicated that they preferred face-to-face to online training, it is crucial to find ways to ensure that the training is as user-friendly, enjoyable and valuable as possible to facilitate optimum outcomes in program engagement and delivery. Before teachers commit to delivering the program, it is also important to ensure that they are intrinsically motivated to do so, have appropriate school leader support and have access to sufficient computers and internet to set them up for success.

Based on feedback from the teachers who completed the SAS-WOC Online Teacher Training Course, it is recommended that where possible, teachers work in pairs or groups rather than individually when completing the course to encourage active participation and discussion. Where this is not possible, teachers should be encouraged to post ideas and questions on the online forum to create a virtual peer support community. Where more than one teacher at a school is trained to deliver the SAS-WOC Program, there may be benefits in encouraging colleagues to co-facilitate some lessons from the program, and/or to sit in on each other's lessons to offer peer encouragement, support and feedback.

Arrangements have been made to film teachers at trial schools delivering the SAS-WOC Program to supplement staged video clips that are included in the current online teacher training course. This will help to address concerns raised by some teachers about the realism of certain classroom and schoolyard clips. Students will also be filmed talking about their experiences with the program and the strategies that they have found helpful in an effort to enhance teacher engagement and enthusiasm about delivering the program from the outset.

Through the incorporation of these ideas and ongoing research in the coming years, it is hoped that the SAS-WOC Program and associated online teacher training course will be a valuable addition to the school curriculum for Grade 5 students. If the program proves to be user-friendly and effective, the intention is for variants to be developed and delivered to other school years, transforming the curriculum into a whole-of-school initiative.

CONCLUSION

In conclusion, online training holds promise as a PD tool to support school staff in integrating technology into their teaching practices. The preliminary results from the current project, together with those of other researchers (e.g., Bates et al., 2016; Edinger, 2017), suggest that the effectiveness of online

PD courses in engaging school staff and transforming teaching methods can be enhanced by providing opportunities for active and collaborative learning. Providing access to footage of real-life classroom demonstrations of content delivery may also help to improve the conversion rate of online teacher PD to positive changes in teaching practices. Offering ongoing professional and technical support when teachers attempt to implement new tools and activities in their classrooms is likely to further enhance their commitment to adopting new tech-programs, and improve the fidelity with which they do so. However, outside of a research trial context, the provision of ongoing tailored implementation support comes at a cost that many schools may struggle to afford. The provision of individualized coaching and support also serves as a potential barrier to the scalability of many online PD programs. These are issues that warrant further consideration.

In the current study and many others cited in this chapter (e.g., Bates et al., 2016; Collins & Liang, 2015), the recommendations for the development and implementation of online PD courses for teachers are largely based on what school staff say they want, and what they believe to be effective. There is little research available that objectively evaluates the impact of these recommendations on the rate or quality with which educators adopt new teaching methods or practices after completing online PD courses, particularly in the long term. It is also likely that teachers who participate in formal research evaluations will deliver new teaching practices at a higher rate and with greater fidelity than those who participate in online PD programs outside of a research context (where 'big brother' is not watching). School leaders need to think creatively about how to incentivize and support teachers to experiment with new teaching methods that have been introduced in online training courses, and to provide a clear rationale for how these practices align with whole-of-school objectives. At present, there appears to be many more questions than answers in understanding the contribution that online PD can make to workforce development in the primary education sector. However, the research to date suggests that hybrid PD models that integrate online training components hold great promise in transforming todays' teachers into tomorrow's technology educators for our future generation.

NOTE

Conflict of Interest: Dr. Renae Beaumont is the author of the Secret Agent Society – Whole of Class Program and receives royalties on all Secret Agent Society Program materials and practitioner training courses sold. Dr. Sharon Hinton declares that she has no conflict of interest. Associate Professor Kate Sofronoff declares that she has no conflict of interest.

ACKNOWLEDGMENT

This research was supported by the Cooperative Research Centre for Living with Autism (Autism CRC), Long Pocket, Brisbane, Queensland, Australia and the Australian Government's Positive Partnerships Program.

REFERENCES

Abeles, P., Verduyn, C., Robinson, A., Smith, P., Yule, W., & Proudfoot, J. (2009). Computerized CBT for adolescent depression ("Stressbusters") and its initial evaluation through an extended case series. *Behavioural and Cognitive Psychotherapy*, *37*(02), 151–165. doi:10.1017/S1352465808005067 PMID:19364416

An, Y., & Reigeluth, C. (2012). Creating technology-enhanced, learner-centered classrooms: K-12 teachers' beliefs, perceptions, barriers, and support needs. *Journal of Digital Learning in Teacher Education*, *28*(2), 54–62. doi:10.1080/21532974.2011.10784681

Bates, M. S., Phalen, L., & Moran, C. (2016). Online professional development: A primer. *Phi Delta Kappan*, *97*(5), 70–73. doi:10.1177/0031721716629662

Beaumont, R., Rotolone, C., & Sofronoff, K. (2015). The Secret Agent Society social skills program for children with high-functioning autism spectrum disorders: A comparison of two brief versions for schools. *Psychology in the Schools*, *52*(4), 290–402. doi:10.1002/pits.21831

Bicer, A., & Carpraro, R. (2017). Logitudinal effects of technology integration and teacher professional development on students' mathematics achievement. *EURASIA Journal of Mathematics Science and Technology Education*, *13*(3), 815–833. doi:10.12973/eurasia.2017.00645a

Chamerland, K., Sanchez, M., Panahi, S., Provencher, V., Gagnon, J., & Drapeau, V. (2017). The impact of an innovative web-based school nutrition intervention to increase fruits and vegetables and milk and alternatives in adolescents: A clustered randomized trial. *The International Journal of Behavioral Nutrition and Physical Activity*, *14*(1), 140. doi:10.118612966-017-0595-7 PMID:29037203

Collins, L. J., & Liang, X. (2015). Examining high quality online teacher professional development: Teachers' voices. *International Journal of Teacher Leadership*, *6*(1), 18–34.

Desimone, L. M. (2009). Improving impact studies of teachers' professional development: Toward better conceptualizations and measures. *Educational Researcher*, *38*(3), 181–199. doi:10.3102/0013189X08331140

Edinger, M. J. (2017). Online teacher professional development for gifted education: Examining the impact of a new pedagogical model. *Gifted Child Quarterly*, *61*(4), 300–312. doi:10.1177/0016986217722616

Einfeld, S., Beaumont, R., Clark, T., Clarke, K., Costley, D., Gray, K. M., ... Howlin, P. (in press). School-based social skills training for young people with Autism Spectrum Disorders. *The Journal of Special Education*.

Fisher, J. B., Schumaker, J. B., Culbertson, J., & Deshler, D. D. (2010). Effects of a computerized professional development program on teacher and student outcomes. *Journal of Teacher Education*, *61*(4), 302–312. doi:10.1177/0022487110369556

Fishman, B. J., Konstantopoulos, S., Kubitskey, B. W., Vath, R., Park, G., Johnson, H., & Edelson, D. C. (2013). Comparing the impact of online and face-to-face professional development in the context of curriculum implementation. *Journal of Teacher Education*, *64*(5), 426–438. doi:10.1177/0022487113494413

Gray, L., Thomas, N., & Lewis, L. (2010). *Teachers' use of educational technology in U.S. public schools: 2009 (NCES, 2010)*. Washington, DC: National Center for Education Statistics, Institute of Education Sciences, U.S. Department of Education.

Khanna, M. S., & Kendall, P. C. (2008). Computer-assisted CBT for child anxiety: The Coping Cat CD-ROM. *Cognitive and Behavioral Practice*, *15*(2), 159–165. doi:10.1016/j.cbpra.2008.02.002

Livet, M., Yannayon, M., Sheppard, K., Kocher, K., Upright, J., & McMillen, J. (2017). Exploring provider use of a digital implementation support system for school mental health: A pilot study. *Administration and Policy in Mental Health and Mental Health Services,* 1-19. DOI:10.100710488-017-0829-7

Masters, J., de Kramer, R. M., O'Dwyer, L. M., Dash, S., & Russell, M. (2012). The effects of online teacher professional development on fourth grade students' knowledge and practices in English language arts. *Journal of Technology and Teacher Education*, *20*(1), 21–46.

Naismith, L., Lonsdale, P., Vavoula, G., & Sharples, M. (2004). *Literature review in mobile technologies and Learning*. NESTA Futurelab Series.

Pittman, T., & Gaines, T. (2015). Technology integration in third, fourth and fifth grade classrooms in a Florida school district. *Educational Technology Research and Development*, *63*(4), 539–554. doi:10.100711423-015-9391-8

Prestridge, S., & Tondeur, J. (2015). Exploring elements that support teachers engagement in online professional development. *Education in Science*, *5*(3), 199–219. doi:10.3390/educsci5030199

Schroeder, H. (2016). Four essential tips for professional development success. *eLearn Magazine*. Retrieved from http://elearnmag.acm.org/featured.cfm?aid=3018842

Trentin, G. (1997). Telematics and on-line teacher training: The POLARIS project. *Journal of Computer Assisted Learning*, *13*(4), 261–270. doi:10.1046/j.1365-2729.1997.00029.x

Tweed, S. R. (2013). Technology implementation: teacher age, experience, self-efficacy, and professional development as related to classroom technology integration. *Electronic Theses and Dissertations*. Paper 1109. Retrieved from http://dc.etsu.edu/etd/1109

Wang, C. Y., Wu, H. K., Lee, S. W. Y., Hwang, F. K., Chang, H.-Y., Wu, Y. T., ... Tsai, C. C. (2014). A review of research on technology-assisted school science laboratories. *Journal of Educational Technology & Society*, *17*(2), 307–320.

Weng, C., & Tang, Y. (2014). The relationship between technology leadership strategies and effectiveness of school administration: An empirical study. *Computers & Education*, *76*, 91–107. doi:10.1016/j.compedu.2014.03.010

Zheng, L. (2015). In-service professional development in an online environment: What are South Australian English as an additional language or dialect teachers' views? *Professional Development in Education*, *41*(3), 527–545. doi:10.1080/19415257.2014.902860

KEY TERMS AND DEFINITIONS

Behavioral Learning Theory: Focuses on observable and measurable aspects of behavior, and how rewards and punishment can be used to shape behavior.

Collaborative Learning Framework: An educational approach to teaching and learning that involves working together in groups to solve problems and complete tasks.

Constructivist Learning Theory: A learning theory which explains how people might acquire knowledge and learn. The theory suggests that humans construct knowledge and meaning from their experiences.

Cooperative Research Centre for Living With Autism (Autism CRC): The Autism CRC was established in 2013 and is the world's first national, cooperative research effort focused on autism. The Autism CRC is based in Brisbane, Queensland, Australia.

Experiential Learning: Learning through reflection on doing. It is a method of education via first-hand experience.

ICT: Information and communication technology.

Professional Development: Learning to earn or maintain professional expertise and/or qualifications.

Secret Agent Society (SAS): The SAS Small Group Program is an evidence-based, multimedia program to help children (aged 8-12 years) improve their emotional resilience and social skills.

Secret Agent Society – Whole of Class Program: Extends SAS from a small group context to include a whole-of-class program delivery model.

Situated Learning: Argues learning is a function of the activity, context and culture in which it occurs (i.e., it is situated).

Social-Emotional Learning (SEL): Considered key to classroom success, SEL teaches children to effectively and ethically manage feelings, resolve conflict, manage friendships, solve problems and make decisions.

Chapter 3
Urban Education Technological Disparities:
The Debilitating Impact on Our Students for Twenty-First Century Employment

Angelia K. Swinton
Texas A&M University, USA

Lauren Ashley Williams
Texas A&M University, USA

ABSTRACT

Technological advancements of the twenty-first century have yielded a greater need for systematic aptitude and preparedness for the current U.S. and global workforce. To meet the needs of the twenty-first century workforce requires teachers to alter the manner in which they teach because student needs call for a new type of learning. The twenty-first century teacher has to assist students develop their technological identities for a new culture of learning shaped by technological advances and globalization. This new culture of learning is less teacher focused and more student centered. As more employers demand graduates who have the requisite twenty-first century skills, higher education must also continue to develop students who are problem solvers, effective communicators, critical thinkers, and effective collaborators. This chapter focuses upon the existing digital divide in historically marginalized communities.

INTRODUCTION

The systemic challenges facing urban schools and communities can impede and/or preclude the realization of future critical thinkers and problem solvers. Employers expect applicants to possess these skills as they are crucial to career alacrity and adaptability. As such, technological advancements of the 21st century have yielded a greater need for systematic aptitude and preparedness for the current U.S. and global workforce. The use of technology in urban settings can provide another layer of quality education that equips students of color with the necessary tools to meet the challenges of global citizenry. Yet,

DOI: 10.4018/978-1-5225-4111-0.ch003

urban centers do not sufficiently provide the resources to students or teachers to adequately meet these demands, therein increasing the divide between low-income students suffering the affliction of poverty and their suburban counterparts. This chapter will discuss capacity building for access to educational technology for students, teachers, and parents, as well as the lack of infrastructure, embedded assumptions, and systemic implications. The information presented will appeal to parents, educators, career counselors, and employers.

BACKGROUND

While advancements in technology have dramatically shifted the ways in which we navigate our day-to-day existence, these advancements remain inconsistently reflected in our Pre-K–12 educational institutions. There are unique factors shaping our country's urban centers that particularly contribute to the ways in which technology plays a role in city schools. The authors find that technology is in fact underutilized in the Pre K–12 student populations. While teachers and students of color suffer disproportionately from equal and quality access to educational technology, teachers of color in these urban school contexts are twice as likely to possess inadequate technology and information literacy training, skills, and knowledge to work with technology in a classroom context that would benefit students of color in these environments (Owen, Song, & Kidd, 2007; Champion, 2017). There is a systemic discrepancy that persists between expectations for the effective integration of technology within the classroom and the teacher training provided. This further exacerbates the under-utilization or misuse of technology, since teachers' capacities are underdeveloped to effectively integrate technology.

21st CENTURY TECHNOLOGY IN URBAN EDUCATION

Issues, Controversies, Problems

Administration and Teacher Capacity

Building teacher capacity takes time and continuous development by experts in a given field. Schools, which cultivate practitioner training and expertise through innovative programming, are often in tandem with rising student outcomes (Cox, 2015). Training in technology is no exception. However, as with any model seeking to develop teacher capacity, there are challenges. Educational technology, as a continuously developing field, creates an additional set of psychosocial and economic barriers preventing the development of teachers' and practitioners' capacities in effective and appropriate use of technology. Advances in the technological revolution have had a significant impact upon expectations placed upon K-12 teachers in the development of 21st-century skills such as creativity, collaboration, and critical thinking among student learners (Oneal, Gibson, & Cotten, 2017). Although arming our students with the technological know-how will prepare them for 21st century employment (Mouza, 2011; O'Sullivan, & Dallas, 2017), barriers exists that hinder the full integration of technology in the classroom. Evidence indicates that teachers' integration of digital tools into instruction remains sporadic and less than optimal (Day, 2017; Ertmer & Ottenbreit-Leftwich, 2010). In fact, while personal use of various technologies such as handheld smart phones or some form of social media remains high for both teachers and their students, (O'Bannon, Waters, Lubke, Cady, & Rearden, 2017; The Nellie Mae Education Foundation, 2011), external and internal obstacles impede complete integration. Oneal et al. (2017) contend that external obstacles include those outside of the teacher's control, such as lack of computer resources or administrative support, whereas, internal barriers, are more teacher related, such as attitudes toward computers or perceived technology skills. Increasingly, as the expectations for teachers multiply to become facilitators of technology integration in the classroom to help in the development of 21st century skills, teachers have to change their own approach to using technology in the classroom that cultivates the development of creativity, collaboration, and critical thinking which are necessary for the workforce and post-secondary education (Oneal et al., 2017). The following four areas affect the development of teachers' capacities in the appropriate and effective use of technology: mindset, logistical feasibility, visibility, and alignment and relevance.

Mindset

Many teachers and administrators, especially those with a number of years in teaching, do not see the value in utilizing technology within the classroom (Blanchard et al., 2016). The mindset of an aversion to educational digital technology use, can be influenced by professional development opportunities, but most likely stems from teacher education programs that play a significant role in developing the pedagogical beliefs of pre-service teachers (Sadaf, Newby, & Ertmer, 2012; Tondeur, van Braak, Ertmer, & Ottenbreit-Leftwich, 2017). For veteran teachers and administrators whose teacher preparation did not include technology as method, this means that their lack of technology integration courses has a major effect on their readiness and willingness to incorporate new teaching methods into their classrooms. For novice teachers being mentored by these more veteran teachers, this means that they see less than sufficient technology usage in "expert" teacher classrooms, therefore affecting their own technology usage as they try to mimic and follow the lead of their mentors (Dexter, Doering, & Ridel, 2006; Irby, 2017). In short, the cycle continues as novices, trained by insufficient technology mentors, grow and become mentors themselves to the next generation of novice teachers.

Professional development opportunities for novice and veteran teachers must offer meaningful and ongoing training in the implementation of technology (Ciampi, 2017). While teachers in higher socioeconomic (SES) classrooms are more likely to have more technology training than teachers in lower SES classrooms (Mouza & Barrett-Greely, 2015 in Ciampi, 2017), a divide arises where lower SES schools have more teachers that are less prepared to integrate technology into instructional practices. Professional development for these teachers, especially, must provide opportunities to assist teachers increase their competence for integrating technology as indicated in Figure 1.

The better prepared teachers are to infuse technology into daily instruction, the greater the likelihood for racially minoritized students to use technology that advance 21st century skills needed for college and today's workforce. Ciampi (2017) offers that effective professional development can assist teachers in low SES schools coordinate technology in ways that support student learning and increase empowerment. Training that does not help teachers to effectively integrate the use of technology in ways that help students to become college and career ready is pointless. With effective professional development in the use of technology, teachers can effectively differentiate lessons that promote students' critical thinking, productivity, and organizational skills (Ciampi, 2017), all capabilities necessary to work in today's technology driven workforce.

Figure 1.
(Adapted from: Davis, 1993)

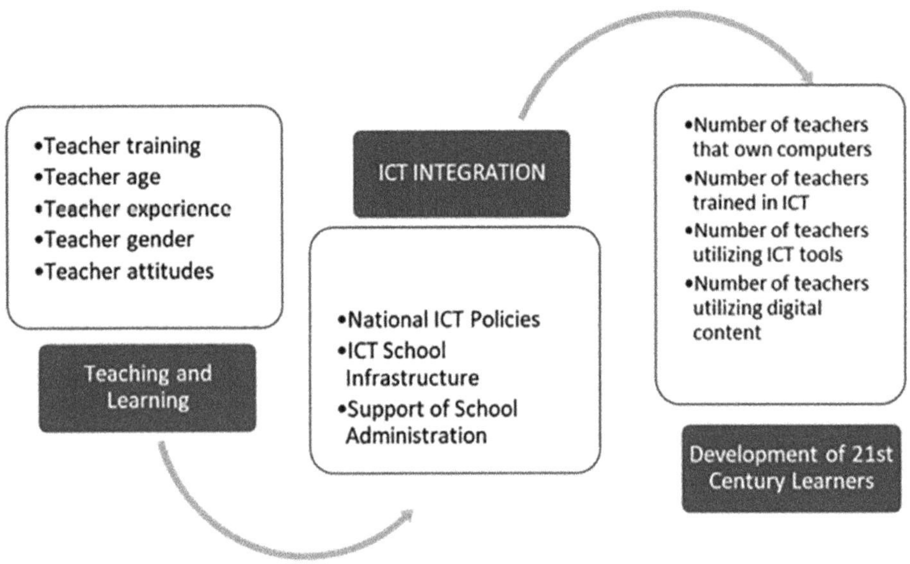

Logistical Feasibility

Building pedagogues' capacities in technology integration relies distinctly upon a school or district's structural and economic supports (NEA, 2017). Schools often face setbacks in terms of physical space and structural capability to support technology integration. Low-income schools face the added challenge of affording the resources themselves; typically, schools which can afford said resources have student populations which are less likely to lack access to technologies at home (Mirazchiyski, 2016).

However, thanks to the Title I, Part A of the Elementary and Secondary Education Act (ESEA), extra funding is provided to schools that serve disadvantaged populations. These funds are provided specifically for improving the learning of students labeled "at-risk" for educational failure. This includes students in high-poverty schools where a large percentage of the student population come from low-income families. Because Title I funding must be focused on serving these particular students, if 40 percent or more of a school's student population is found to come from low-income families, the funding is allowed to be used for school-wide programs that are specifically intended to improve student achievement for the entire student population (U.S. Department of Education, 2015). Typically, these school-wide programs are technology-focused, allowing for more technological resources to be available at a higher student-to-device ratio.

The issue, though, is that higher technology availability does not always determine higher technology competency in terms of teacher-readiness to incorporate effective technology into lessons. If professional developers are not successful in their ability to encourage, or even mandate, effective technology usage by the teachers with the students, no amount of technology available will make a difference if students are not experiencing its capabilities in their learning environment (Comi et al., 2017). In order for technology to become highly effective in the classroom, it is important that it be used as a cognitive tool, encouraging higher levels of thinking and doing, rather than a method of direct instruction (Ashman, & Conway, 2017; Tamim et al., 2011).

Visibility

The task of integrating computing across the curriculum is challenging in spite of research findings that computer usage in the classroom is beneficial for students and for teachers (Coleman, Cotten, Gibson, Stringer, & Howell-Moroney (2016). A survey done by Greaves, Hayes, Wilson, & Gielniak (2010) of nearly 1,000 school principals and technology coordinators indicated that "80 percent of the schools under-utilize technology they have already purchased" (The Nellie Mae Education Foundation, 2011, p. 13). While much of this may be attributed to accessibility, ease and timeliness of logistical tactics to integrate technology, and teacher and administrator efficacy in incorporating the technology, some of this may also be attributed to visibility of the technology within the school district or institution itself. This is where leadership's role is integral: "A school culture that promotes technology use and the adoption of new teaching practices" (The Nellie Mae Education Foundation, 2011, p. 13) is a crucial aspect of organizing support for technology integration within a school. Globally, countries are embracing different forms of ICT to engage and educate students. ICT such as mobile devices specifically are expanding to assist with teaching and learning in urban and rural schools (Sharehu & Emmanuel, 2015). Mobile devices like laptops, iPads/tablets, and smartphones are expanding its use in the field of education due to its portability, less need for physical infrastructure, and the capability to do nearly the same operations as desktop computers (Sharehu & Emmanuel, 2015). With external and internal barriers contributing to some of the challenges in creating technology rich learning environments, teachers' exposure to computing technology, training received for technology use, and support in their use of technology can assist in reducing these impediments (Coleman et al., 2016). The resulting outcome is a teaching force that is better prepared to execute technology integration in seamless alignment with the curriculum to be taught.

Alignment and Relevance

Curriculum should be aligned to clear and predetermined learning outcomes grounded in identified standards. A fundamental issue in terms of technology training for teachers is in terms of the applicability, or relevance, of the content of the training itself to whether teachers find the training particularly useful in delivering their content. A primary issue remains in planning for such technology-based trainings: teachers themselves are at different stages of comfort and familiarity with technology therein affecting their own levels of competence/proficiency which invariably influences both the way in which the trainings themselves could be delivered and the way such trainings will be perceived and received by each individual teacher. Increased use of and familiarity with technology correlates to higher levels of efficacy and therein a substantially more open mindset to engaging in high quality technology professional development that will push attendees thinking (Mentor; 2016; Slaouti & Barton, 2007). Conversely, increased use of and familiarity with technology can adversely influence an attendee's openness if the professional development itself is not engaging and/or does not push the attendees thinking (Hennessy, Haßler, & Hofmann, 2015). An appropriate analogy might be giving an experienced driver a lesson on the functions and use of the indicator: such a lesson would be boring and unnecessary for the driver and therein become a waste of time; whereas a new driver will need that skill before moving onto a more difficult one and therein the lesson becomes less of a waste of time. Mouza (2011) states,

The limited and mostly low-level (e.g., word processing, Internet research) uses of technology in teaching can be largely attributed to the shortage of high-quality professional development programs available

to teachers (Lawless & Pellegrino, 2007). Many technology professional development programs are still short term and focus primarily on technical skills required to use specific software packages (see NCEE, 2007).

Mouza goes on to suggest that the development of effective and high-quality professional development programs that combine content and pedagogical knowledge with technology methods is required to deepen teachers' understandings of technology integration and therein make the trainings relevant. In a recent study, Coleman et al. (2016) suggest that teacher attitudes and participation in an intensive computer based training results in a positive effect on computer integration practices and improves the quality of computer integration in the classroom (p. 275). "Teachers who engage in professional development tend to use computers more often, put more emphasis on the teaching of ICT skills and have a stronger sense of ICT self-efficacy than their skeptical colleagues" (Drossel & Eickelmann, 2017, p. 11).

Moreover, while these categories appear to be applicable to any school district, they are more prevalent in urban environments. The lack of a cultivation of teacher capacities will detrimentally impact the long-term development of urban youths' skills that would adequately prepare them to meet the demands of the 21st century workforce, therein further relegating historically oppressed populations to economically depressed positions that principally deny entrance into areas of substantive influence to affect socially systemic changes.

Parent Capacity

Parental involvement plays an integral role in students' educational career. Building parents' technological capacities is potentially as equally important to that of improving teachers' capacities. To address historical and institutionalized systemic issues of inequality in order to foster a level playing field for our students in order to increase their opportunities to meet the demands of a 21st century workforce, parents' technological lives, including proficiency and general technological awareness, must invariably be developed (Mirazchiqyski, 2016). Impediments, or more aptly, barriers, exist that prevent or make increasingly difficult the involvement of parents in their children's educational technological lives. The authors contend that these obstacles may be deconstructed into four primary areas: poverty; language barriers; required commitments; and perceived needs of their children based on age.

Poverty

The impact of poverty on child development and academic performance has been long studied. While several theories explain the manifestations of poverty, any lack of resources (where resources may be qualified in terms of human contact, home stability, emotional support, food, clothing, knowledge, etc.) impacts child development and academic performance. Jensen (2009) points out that "children raised in poverty rarely choose to behave differently, but they are faced daily with overwhelming challenges that affluent children never have to confront, and their brains have adapted to suboptimal conditions in ways that undermine good school performance." Low-income parents often have decreased exposure to education than their more affluent counterparts (Jensen, 2009). Minimal or sub-standard education for parents in poverty affects child development and in turn often perpetuates the cycle of poverty. Parents facing the challenges of poverty often lack the resources that logistically enable parental support of their children's technological lives (Costley, Bell, & Leggett, 2014). Parents either do not have the educational foundations themselves to access trainings' content or the trainings that are given are structured towards and cater those with access to technology at home and/or minimal competencies with technology for education. Further, low-income parents often struggle to attend any on-site trainings; this is due to a host of factors including inability to find or afford transportation to and from the site, as well as conflicts related to the timing which the training is offered (Watson & Bogotch, 2015). Stigma plays a role in voluntary attendance of parent workshops: parents make decisions pertaining to their involvement in offered school functions such as workshops based upon prior experiences and/or predicted reactions to socioeconomic status. This is particularly relevant in climates wherein explicit attitudes and policies demonize poor and low-income communities and communities of color.

Language Barriers

Raising school and community awareness and fostering authentic learning experiences which develop a person or group's collective social consciousness in terms of cultural competencies is necessary if welcoming folks from non-English speaking backgrounds into the school culture. Parents from non-English speaking backgrounds, immigrant and/or undocumented contexts, and/or of different cultural contexts may come to fear involvement in their children's school community, especially under any administration or government entity which cultivates a demagogic or xenophobic climate (Landsman & Lewis, 2011; Manning, Baruth, & Lee, 2017). Further, school-sponsored parent workshops in technology may not account for parents' inability to access the curriculum due to their inability to access any technologies, presentations, or texts provided in English. Schools wishing to improve parental involvement in their children's technological lives should conduct an equity audit to ascertain approximate numbers of non-English speaking parents and adjust offered trainings and school sponsored events accordingly; schools should do so in consideration of both local and broader social and economic contexts that potentially contribute to parental reticence in audit/survey participation.

Required Commitments

Parents' attendance at school functions expressly designed to increase their aptitude in the use of technology and its impact on their children's daily living and future employment opportunities may be at odds with the parents' own lives and personal obligations. Schools and districts should therefore be mindful of the time commitments and resources needed when designing technology-based parental trainings to maximize the greatest possible gains in achieving learning targets within the shortest amount of times. For instance, during any planning of a parent training involving technology, the school should explicitly plan for the provision of the required technologies for that workshop.

Perceived Needs of Their Children Based on Age

Parents often presuppose their children's capabilities related to social cues and perceptions promoted within their cultural and societal context (Nabi & Krcmar, 2016). Many parents today recognize the continuing and unceasing inculcation of technology into their children's daily lives and associate the preponderance of technology with having cultivated agency over its application and use toward academic and social successes. In contrast and significantly, other parents concertedly resolve to abstain from the incorporation of technology into their children's lives (Rosin, 2013). Part of the school or district's job, then, is to account for this in their planned trainings for parents, as well as their outreach to their parents.

Without vigilance, parental involvement in their children's educational lives will be ultimately affected. The more parents are made knowledgeable about a given area, the better their ability is to reinforce taught concepts and good habits that will increase a child's success that will yield rewards in workplace environments.

Student Capacity

Recent research has shown that young people have adopted technology more rapidly and more universally than adults (Li, Snow, & White, 2015). Given 21st century competencies needed to enter the workforce such as analyzing, evaluating, and synthesizing information, not only from print, but across multimodal sources, there is a need for young people to understand academically appropriate applications of technology in educational settings (Goldman, 2012; Good, & Lavigne, 2017; Lesaux, 2012; Mentor, 2016). These skills are crucial for students to succeed individually and professionally to develop global competence. The goal of technological advancements is to ultimately prepare students in academic settings for successful employment. The four areas that will be explored regarding student capacity include: access, appropriate engagement through various devices, training, and accountability/understanding of the relevant technology to that of the workforce.

Access

Educators find that "students often fall prey to the allure of technology as a tool for reversing the historical influences of poverty, discrimination, inequity, chronic underachievement, and lack of opportunity" because it has the potential narrow the achievement gap, if equally distributed or widen the gap if only accessible to selected groups in the educational system (Edyburn, Higgins, & Boone, 2005; Aydin, Ozfidan, & Carothers, 2017). Studies have shown that it is important for students to have access to technology in a physical manner, as well as in a contextual manner that applies to their cultural and educational needs (Good, & Lavigne, 2017; Pöntinen, Dillon, & Väisänen, 2017; Song & Owens 2011). Studies have also shown "the socio-cultural disparity in technology availability and use in the classroom" (Song & Owens, 2011, p. 24). Students from high-income families have computers and high-speed Internet in their homes and schools in contrast to students of color from low-income families (Owens & Waxman, 1994; American Council on Education, 2003; Sheninger, & Murray, 2017; Toven-Lindsey, 2017).

In building student capacity to maximize the full potential of technology infused learning, teachers' technology content knowledge (TCK) will require professional development that embeds content focus, active learning, coherence, and collective participation (Desimone, 2009 in Ciampi, 2017). With an increase in TCK, teachers can effectively evaluate websites and apps that support their students' content learning. Thus, allowing teachers to choose the appropriate technology tools that correlate to their content goals which will permit more students access to technology that connects learning between home and school.

Infrastructure

In an ideal world, new technologies would be supported and applied uniformly to benefit all children in their academic endeavors to prepare them for the workforce. Many believe that with powerful and cost-effective technologies, minority and poor children will be able to receive education of the same quality as their more fortunate peers (Du, Sansing, & Yu, 2004; Sheninger, & Murray, 2017; Toven-Lindsey, 2017). However, access to technology is not equitable across sociodemographic categories, since it is determined by resources available to the schools, communities, and households. New technologies seem to best accommodate those who already take advantage of available educational opportunities (Barley, 1997; Good, & Lavigne, 2017). It is possible that use of these technologies may widen the educational gap in such a way that "advantage magnifies advantage," as the advantaged benefit most from cutting-edge technologies whereas the neediest benefit the least (Du, Sansing, & Yu, 2004; Carlson, 2017). Skeptics question whether new technologies per se, have the capability to improve educational equity since both access to and use of technologies are socially stratified. This chapter will explore issues of infrastructure such as funding (i.e. grants), physical space to support operating systems and equipment and ways in which to circumvent these issues through mobile learning.

History of the Digital Divide in Urban Schools

For the many students that attend schools where technology and advanced courses which integrate technology are insufficient or under resourced, the long-held belief that merely having access and use of affordable computer technology would level inequalities, enhance learning opportunities, and lead to academic and career success for urban youth (Shank & Cotton, 2013; Mentor, 2016). The initial scholarship of the digital divide centered primarily around access and ownership based upon large gender, racial, and age inequalities in the ownership of computers and access to computers and the Internet (Gunkel, 2003; van Dijk, 2006; Wilkin, Davies, & Eynon, 2017). Although the literature reflects less stratification of access and ownership, racially marginalized students are disproportionately more likely to continue to be stratified when it comes to access and ownership (Cotten, Hale, Moroney, O'Neal, & Borch, 2011; Robinson, 2009, 2011; Wilkin, Davies, & Eynon, 2017).

A secondary examination of the digital divide, or now more commonly known as digital inequality, by current research focuses on the specific ways computers and the Internet are used (Attewell, 2001; Champion, 2017; Good, & Lavigne, 2017), as well as forms of access (Davison & Cotten, 2010; Wilkin, Davies, & Eynon, 2017), computer skills (Hargittai, 2002, 2010; Mentor; 2016), and web-based content creation (Blank, 2013; Mentor; 2015; Schradie, 2011). The most prevalent manner in which young people use computers and the Internet involve the areas of communication, social networking, content creation, and multimedia entertainment (Lenhart, Madden, Smith, & Macgill, 2007; Lenhart et al., 2010; Mentor; 2015; Rideout et al., 2010).

It is important to remember that the digital divide and its impact upon racially minoritized students remains a critical issue in the 21st century (Rogers, 2016). The digital divide has evolved from initially meaning "inequitable access to computers and the Internet" to a more expanded definition that includes inequalities in technology skills and how it is used. A more evolved definition of the gap that persists primarily for historically marginalized people of color, and economically disadvantaged groups are further marginalized due to the control of digital production tools and the appropriate context to use those tools (Rogers, 2016).

Technology and Racially Minoritized Communities

Rogers (2016) argues that technology is a powerful force that can be both transformative and positive for students given that technology instructional methods are inclusive (p. 199). Educators teaching technology have a responsibility to not only introduce a variety of information and computer technology (ICT) experiences to their students, but they must also be cognizant that students who are racially minoritized will have limited exposure to technological experiences. ICT denotes technologies that offer access to information via telecommunications media, such as the internet, wireless networks, computers and computing devices, such as "smart" phones (TechTerms.com, 2012; Mentor, 2014). Therefore, it is critical that any instructional methods utilized to teach ICT must consider the characteristics, cultural perspectives, needs, and attitudes of students (Rogers, 2016) so that access is equitable and open to all students. Without inclusiveness, especially on including computer programming in curricula, students who are racially, ethnically, culturally, linguistically, and socioeconomically different will be further marginalized (Corcino &Mentor, 2018; Mentor, 2018). It is essential for urban educators to contemplate the moral, social, and cultural implications of teaching technology to diverse groups of learners.

At the intersection of race, class, culture, and computer technology, is an urgency for preservice and in-service teachers to find a cultural congruency with their students to assist them in developing necessary technological skills needed for the digital age (CEO Forum, 1997; Education Week, 2007; Frederick and Shockley, 2008; Mentor, 2014). Fully understanding the history and lived experiences of their racially minoritized students is the beginning of becoming culturally congruent for teachers in urban schools wanting to cultivate student technology identities.

In the computer age, there is an urgent need to prepare all students (National Education Association, 2008; Champion, 2017; Russell-Bennett, Drennan, Raciti, Bowring, & Burns, 2017) to be digitally proficient. The need takes on a critical urgency in preparing racially minoritized students to work in an industrialized society where in every facet computers and related technologies is ordinary (Frederick & Shockley, 2008; Mentor 2015, 2016). Technology skills are critical for individuals to improve their social position and capital (Rogers, 2016). The question begs, "How can an educational system where all schools are not equitable and maintain institutionally oppressive barriers adequately prepare racially minoritized students to develop necessary technological skills?" Besides racially minoritized students attending schools that suffer from a disproportional allocation of resources and other institutionalized policies (Darling-Hammond, 1995; Fine, 1991; Kozol, 1991; Conchas, Gottfried, Hinga, & Oseguera, 2017) and inadequate or inferior access to computer and interrelated technologies (Frederick & Shockley, 2012; Mulcahy, 2017).

Often the problem is not a lack of technology, but teachers who are generally ill-equipped or under-prepared to fully integrate technology seamlessly into their instructional practices at the level needed that would help racially minoritized students acquire computer and other related technology skills needed to fully develop a technological identity. Hughes & Ooms (2004) convey that the central factor for teachers' lack of preparation in using technology innovatively to support student learning is attributed to a lack of ongoing focused professional learning opportunities (p. 398). A position that Mentor (2016) echoes through the inclusion of questions on technology use in teacher surveys taken by students in seven week cycles. With urban schools being one of a few points of access to computer-related technology for racially minoritized students (US Department of Commerce, 1999; Conchas, Gottfried, Hinga, & Oseguera, 2017), it is incumbent upon schools to help these students develop identities as technology users, so that they can create and learn with technology (Degennaro & Brown 2009). Improving how teachers integrate technology can potentially lead to positive outcomes such as assisting students in using technology to think critically, connecting students with culturally relevant resources that relate to students' lived experiences, and connecting students' racialized experiences with instruction (See the Appendix). To affect change, teachers require more school administrators to fund, as well as, provide professional development opportunities.

This greater investment in professional development to equip teachers with a broader technology foundation will help to expand teachers' technology capacity. Domine (2007) reiterates, "Widespread lack of professional development poses a serious risk to the nation's investment in educational technology, as many teachers have neither the experience nor the skills in using technology in meaningful ways that support, rather than detract from, their curricula" (p. 210). A view which Champion (2017) and also Carlson (2017) share. Effective technology integration can be accomplished with more time and financial resources devoted to raising the level of faculty expertise in technology utilization (Champion, 2017; Staples, Pugach, & Himes, 2005; Mentor, 2016).

Nationwide, school districts are challenged by the demand to graduate students who are equipped with the skills desirable to meet the needs that today's global markets demand. The challenges confronted by educators in closing the "digital divide" is no longer whether computers are connected to the internet but in preparing students for the demands of a society that is increasingly becoming more technological and widely globalized. The current digital divide that impacts racially minoritized students is further expanded by the ratio of computers to students, as well as newer software programs that require high-speed connections, and Wi Fi-dependent devices such as iPads that are integrating more classrooms (Cheema & Hang, 2013; Eickelmann, Gerick, & Koop, 2016). WiFi dependent devices like tablets and smartphones to name a few, allow schools to adopt policies such as *Bring Your Own Technology* to school which expands equitable access to educational opportunities by allowing students to bring their own mobile technology to school (Johnson et al., 2014; UNESCO, 2012; West & Vosloo, 2013). "Fueled by wireless Internet connectivity, mobile devices (e.g., tablets, smartphones, handhelds) and associated software programs designed to run on mobile devices called apps, are becoming a gateway to mobile learning" (Mouza & Barrett-Greenly, 2015). While a vast array of commercial apps has dramatically expanded the capabilities of mobile devices, Common Sense Media (2011) alludes to some theorizing as to a developing *app gap* related to the availability, access, and use of educational apps by low-income underserved students. More importantly, teachers not receiving the appropriate professional development or support in the use of mobile devices will likely fail to optimize the full use of the technology.

Even with schools in large urban districts spending millions of dollars on technology, a great disparity still exists according to Susan Patrick (2012), and Darling-Hammond, 2015). Patrick notes that some schools have access to one-to-one pilots while other schools have outdated and barely functional infrastructure which prevents students from having access to computers. Consequently, students are failing to build their technology skills which are closing off access to courses and supplemental materials that would help them advance and be successful. With technology spending in schools being vastly different from one district to another across the country, some districts reap the benefits of grants and parental donations, while others tap local, state and federal funding (Black, 2017; Harris, 2017, Lemos, & Charles, 2017; Pandolfo, 2012).

According to the National Center for Education Statistics report (2010), nearly every U.S. school has at least one instructional computer with Internet access which also found a ratio of 3.1 students for every computer connected to the Internet, but for schools with higher numbers of racially minoritized students eligible for free or reduced-price lunch, student ratios were much more dismal. Inadequate access to technology serves to further marginalize racially minoritized students. Subramony (2007, 2011, 2014) and Wamuyu (2017) contend that lack of access and skills in ICT can further disenfranchise already marginalized groups. As Davis, Ambrose, & Orand (2017) assert, sociopolitical forces can shape how technology is utilized in urban schools. For urban students, use of personal devices is restricted, and school-owned technologies are often used more as a classroom management tool than as a tool for learning or creation. Having limited access combined with a lack of exposure to using computer technology could easily force racially marginalized students to the sidelines of the 21st century playing field (Sersion & Stevens, 2012). It will take a concerted effort on the part of policymakers at the federal, state, and local levels to work toward removing the existing barriers that limit the computer mediated learning experiences of students in urban schools (Davis et al., 2017).

SOLUTIONS AND RECOMMENDATIONS

Technology Readiness in the 21st Century and Beyond

Rogers (2016) posits, "developing skills in technology is vital for students in K-12 and beyond" (p. 197). In developing their technology identity, racially minoritized students will need to be equipped with technology skills that translate into marketable employment skills. Twenty-first century knowledge emerged from increasing evidence that the labor force necessary for an increasingly globalized economy mandated a completely different model of education—one that surpasses the 20th century skills of repetition, basic applied knowledge, and limited literacy (Kereluik, Mishra, Fahnone, & Terry, 2013; Mentor, 2016; 2018). Although there are numerous frameworks to justify the need to rethink the kinds of knowledge required for learning in the 21st century, the commonality that emerges is *technological modernization* and *globalization* (Kereluik et al., 2013). While technological modernization is a shift from manual and routine jobs to an intellectual and knowledge economy, globalization encompasses a newly interconnected and diverse global society where technological modernization is quickly changing the employment and education sectors.

For students to be better equipped, there are three central skills needed by students: core content knowledge, a degree of digital and information literacy, the ability to problem solve and think critically, along with being creative and innovative (Kereluik et al., 2013). The skills students need to attain for 21st century readiness require that educators move from a teacher centered focus to a more student-centered focus. With more student focused learning, students are self-directed, construct their own learning personal experiences, and work collaboratively with other students (Abas, 2015).

With schools sometimes serving as the only access point to technology for students (US Department of Commerce, 1999), teachers find themselves in the position to help facilitate students' acquisition of technology skills. This presents the question of "what type of teachers are needed for 21st century learning?" Faulkner & Latham (2016) indicate that teachers need to have the capacity to think outside normative practices, exhibit resilience to deal with complications and preserve self-efficacy and confidence, (p. 143-144) and lastly, be creative and problem solve (Wyn, 2009) to meet the needs of the 21st century learner. To assist students with the acquisition of technology skills, teacher professional development should encompass teachers using a wide variety of technology tools that can provide different learning opportunities for students through project based learning. Buczynski and Mathews (2016) elaborate further by adding that the curriculum needs to incorporate communication, collaboration, critical thinking, creativity, as well as, project based learning (p. 40).

Embedded Assumptions

The digital divide in underserved communities continues to marginalize racially minoritized students across the nation (Black, 2017; Harris, 2017). Fairlie (2007) writes, "Blacks and Latinos are substantially less likely to have a home computer and Internet access," (p. 268). Having access to technology holds the promise of transforming lives economically, socially, and culturally. Far removed from having equitable access to ICT, marginalized groups have been unable to take advantage of the ICT revolution and the opportunities it has provided educationally, professionally, and socially. Morino (1997) writes, "Most in this neighborhood have been completely passed by. Unable to climb over the wall to a better future, they have been further shunted aside by those with the means and knowledge to forge ahead," (p. 5). And a sad phenomenon that continues according to Black (2017), Harris (2017) and Mentor (2018).

Technology disparities continue to exist between lower socioeconomic communities and more affluent suburban enclaves because stakeholders living, learning, and working in tony communities have extensive access to ICT infrastructure through schools, businesses, and government agencies. Kvasny (2006) asserts, "Digital inequality reflects not only disparities in the structure of access to and use of ICT; it also reflects the ways in which longstanding social inequities shape beliefs and expectations regarding ICT and its impacts on life chances" (p. 160). Although the US government has initiated several policy responses to alleviate the digital divide to promote broader engagement with ICT, community technology interventions are somewhat weakened if the benefits of ICT used, fail to accrue to those who are disadvantaged (Kvasny, 2006).

With most initiatives looking to assist those from lower socioeconomic levels, or 'people on the wrong side of the divide', and provide them with free or low-cost Internet access and computing skills, Kvasny (2006) surmises that these same 'people' are historically disadvantaged in many social spheres such as transportation, health, employment, and education, (p. 163).

FUTURE RESEARCH

Kvasny (2006) and Mentor (2016; 2018) theorizes that with future ICT initiatives designers of future initiatives should define the digital divide more broadly, and recognize the divide is one of longstanding inequities in access to basic life chances such as education, safety, housing and healthcare, (p. 177). The digital divide is a byproduct of social, economic, and cultural factors that have propagated a technology chasm of the "haves" and the "have nots". Technology integration in urban schools is a matter of access and equity (Kress, 2011). Although technology integration in U.S. schools has made tremendous progress, there is still work to do in the area of teacher training to help teachers utilize a wide range of technology tools as well as seamlessly integrate technology through cross-curriculum, project-based learning (Buczynski, & Mathews, 2016; Carlson, 2017; Mentor, 2016; Sersion & Stevens, 2012;).

Even with a great percentage of the nation's schools having Internet access and a lower student-to-computer ratio, technology has not been integrated into the lives of all U.S. students to the same extent and in the same manner (Black, 2017; Harris, 2017; Kvasny & Keil, 2006; Mentor; 2016). Equitable access to technology for urban learners continues to be limited or non-existent and merely reflects social stratification. The perpetuation of inequitable access is furthered by insufficient facilities and financial resources. Although intended to reduce digital inequality, ICT helps to sustain a hegemonic system of exclusion for historically marginalized groups from social, economic, technical, and cultural advancement.

Efforts by urban schools to achieve equitable access are further impeded by additional pressing issues such as overcrowding, teacher and administrator turnover, insufficient Annual Yearly Progress (AYP), contending with high drop-out rates, complying with mandated curriculum reforms, and school safety, (Black, 2017; Harris; Kress, 2011). Domine (2010) adds, "Failure to resolve issues of internet access denies urban students and their families access to information, knowledge, opportunity, and the world at large," (p. 148).

For urban students and educators, pedagogically, lack of access to the internet denies urban teachers and their students a global connection to a worldwide community and ultimately denies them an identity that extends beyond the self, something educators argue is essential to achieving social and political democracy, (Domine, 2010; Harris; 2017). More commonplace in today's urban schools are technologically advanced equipment used for surveillance, but students and educators are relegated to using limited or antiquated technology and equipment. In an age of zero-tolerance and a culture of surveillance, urban schools are more likely to have advanced technology that monitors and scrutinizes student and educator actions. Weiss (2007) argues, "Surveillance policies are subjecting urban youth to a stunning lack of privacy, conditioning them to being watched at all times and from every angle, and offering students with disciplinary and academic problems mostly punitive options" (p. 60).

The integration of technology is rarely seamless for most urban schools which primarily serve racially minoritized students as little technology is available for teaching and learning compared to the technology that is available for surveillance and security. The technology that may be available is generally isolated in one or two computer labs made available for use to large student bodies. Kress (2011) writes of one urban New York City high school, "Despite this enormous investment in surveillance technologies, technology for teaching and learning was largely unavailable," (p. 18).

To achieve digital equity, that is, "ensuring that every student has equitable access to advanced technologies, communication and information resources, and the learning experiences they provide," (Clark, 2003, p. 1), more initiatives need to focus upon applying technology to achieve meaningful outcomes in economic opportunities, education, and employment of underserved populations (Morino, 2001; Mentor 2018). The emergence of schools that focus upon Science, Technology, Engineering, and Math is one initiative that can help in preparing students for today's workforce to address the quickly changing technological advances and world globalization. Take for example, Lake View High School in Chicago, IL. The school offers four tracks — STEM (Science, Technology, Engineering, and Mathematics), humanities, music/visual arts, and world language — though its overall STEM focus integrates computer science, design process, and project-based learning into all areas of the curriculum. Through this approach, students in every track develop technological literacy, critical thinking, and collaborative skills (Noodle Staff, 2015; Cubbage, 2017).

It has ongoing partnerships with Northwestern University, DePaul University, Microsoft, and Lumity (a local nonprofit providing digital support to Chicago philanthropies) to facilitate the integrated use of technology, college and career mentoring, and technology-oriented extracurricular and summer activities (Noodle Staff, 2015; Cubbage, 2017). Other schools that set the trend in technology integration and preparation for the workforce include Pathways in Technology Early College High School in Brooklyn, NY; Quest to Learn, New York, NY; and Young Women's Leadership School of East Harlem in New York, NY (Noodle Staff, 2015). Reducing the digital divide is a necessity for students of color to gain access to technology and effectively use it to cultivate technological competency to enter a workforce that is part of an "information rich society" (Clark, 2005; Wamuyu, 2017). STEM high schools may be one way to address how to prepare racially minoritized students for a technology based workforce. Technology education must do more than provide service economy employment but instead foster the development of professionals who can design, develop, and evaluate a myriad of information technologies (Tettegah & Mayo, 2005; Frank & Castek, 2017). Connors-Kellgren, Parker, Blustein, & Barnett (2016) maintain, schools play a critical role in developing STEM talent and creating learning environments that encourage students onto pathways toward STEM careers, but has proven to be difficult in disenfranchised communities. Therefore, it is important for educators to continue to increase the self-efficacy of students from racially minoritized communities and offer them positive experiences with STEM and other technologically related fields as a part of workforce development efforts (Connors-Kellgren et al., 2016).

For racially minoritized students, there will be little chance of developing their technological competency due to the limited use of computers for anything other than remediation purposes. Zielezinski (2016) argues, "When given access, digital tools are being used for "remediation" in low-income environments, which isn't working" (p. 31). Additional research conducted by Zeilezinski suggests,

Using the computers and other technology related devices provided to urban schools will require teachers to utilize technology in a manner that support problem solving, inferencing, analyzing and synthesizing information; give students opportunities to be content creators rather than content consumers; use technology that promotes interactivity and discovery; allow students to share their expertise with an authentic audience; employ blended learning where there is the right blend of teacher and technology (pp. 31-33).

While access to computers at school is important for racially minoritized students, having home access is just as critical for these students *and* their parents. Home access is critical to the full integration of technology in the lives of residents in underserved communities (Black, 2017; Harris; Bishop, Tidline, Shoemaker, & Salela, 1999). A lack of home access to technology is characteristic of homes in underserved communities. Clark (2005) describes an underserved community as one in which, "persons who lack adequate access to resources, in this case technology, because of their geographic location, income level, education level, race, ethnicity, or physical capability" (p. 431). Tettegah & Mayo (2005) write that urban schools and communities must deliver effective technology education to nurture greater diversity in the field of information and educational technology to broaden the potential sources of employment for students and to meet the rapidly growing demands of tomorrow's information technology workforce (p. 364). Positions that Davis, Ambrose, and Orand, (2017), Mentor, (2018) as well as Nabi, and Krcmar (2016) contend.

CONCLUSION

A global society in the 21st century requires a repertoire of skills including information-based digital literacy. Using technology in classrooms helps to prepare all students for the digital future and the learned skills are essential to be successful in our current society. Jobs that may not have had a digital component in the past, may have one now. Education is no longer about the memorization of facts and vocabulary words; it is about solving complex problems and collaborating with others in the workforce. Abas (2015) contends, "A world-class education is necessary to produce the right set of graduates for a world that is constantly changing. 21st century students are also no longer motivated by just any learning and teaching approach," (p. 14). Without equal access to the workforce of an increasing digital economy, there is great risk in increasing the opportunity gap. Unequal access to technology and the skills needed to use it means risking isolation of marginalized groups from the larger society's economic, social, educational, and political life.

Moving education to the future calls for educators to possess efficient technology skills that will allow them to teach in an effective manner for a new learning culture. The demand for students who are effective communicators, critical thinkers, collaborative workers, creative and innovative means teachers need to be equipped to address the needs of students who are adept at using mobile technology like smartphones and social media. Learners want teachers who can bring together traditional instruction that is innovative and highly engaging (Abas, 2015). Building 21st century skills means moving from teacher centered instruction to a more student- centered focus. Student centered learning allows students to be more self-directed, constructivist of their own learning based on personal or lived experiences, and collaborative team members.

Despite an increasing number of schools closing the digital divide, students who have been historically marginalized, i.e. students of color, continue to be far removed from the opportunities afforded to White students who have ready access to countless computers and other related technological devices. Urban schools have managed to acquire computers and related technological devices but have failed to raise student engagement and academic achievement using the technology. Zielezinski (2016) writes that a number of U.S. schools have received an influx of new technology (p. 30). For a large segment of students, there is little to no engagement with technology outside of remediation purposes. Noting previous educational technology research, Zielezinski (2016) posits, "overall, students who are Black, Hispanic, or low income are more likely to use computers for drill-and-practice, whereas students who are White or high income are more likely to use computers for simulations or authentic applications" (p. 31).

Computers for remediation is primarily the case in urban schools where the students are primarily racially minoritized and generally have a socioeconomic status at or below the poverty level. Using computers and other related technology for remediation in no way helps the achievement of racially minoritized students. Using technology solely for remediation and drill-and-practice activities can and often does negatively affect student achievement, not to mention engagement, motivation, and self-esteem (Zielezinski, 2016). Computers and other technology have been available in urban schools since the earliest of computers. Although available in urban schools since the 1980s, computer technology which holds the promise to change the conditions and life choices of urban school populations has yet to be fully tapped (Tettegah & Mayo, 2005).

To be competitively prepared for employment in a workforce that demands highly developed technological skills, racially minoritized students must be educated in a way that helps to establish a technological identity, build problem solving capacity, critical thinking, effective communication skills, and the ability to work collaboratively with others ((Tettegah & Mayo, 2005). A paradigm shift among teachers and teacher educators has to occur to move instructional methods from teacher centered to more student centered. If not, racially minoritized students will be unable to participate in a technology-based workforce that values technologically literate workers. Access to technology and using it effectively are becoming requisite competencies in our information-rich society (Goslee, 1998; Servon & Nelson, 2001). The gap between those who have access, those who can effectively use technology, and those who do not may lead to larger gaps in educational and employment opportunities (Blakely, Hadi, & Johnson, 1995; Servon & Nelson, 2001). Without shrinking the digital divide in underserved communities, a situation where those who have and do not have access will continue to expand and further marginalize communities that have been historically marginalized.

REFERENCES

Abas, Z. W. (2015). 21st century education: strategies to ignite and engage students. Special Edition on Teaching and Learning in South East Asia, 13(2), 5-16.

Ashman, A. F., & Conway, R. N. (2017). Using cognitive methods in the classroom. New York, NY: Routledge.

Aydin, H., Ozfidan, B., & Carothers, D. (2017). Meeting the Challenges of Curriculum and Instruction in School Settings in the United States. *Journal of Social Studies Education Research*, *8*(3), 76–92.

Barley, S. R. (1997). Competence without credentials: The promise and potential problems of computer-based distance education. Paper presented at the workshop on Competence Without Credential, Washington, DC.

Black, D. W. (2017). *Preferencing Educational Choice: The Constitutional Limits*. Academic Press.

Blakely, E. J., Hadi, S., & Johnson, P. (1995). *Information city and the ghetto: The LA experience*. Los Angeles, CA: Lusk Center Research Institute.

Blank, G. (2013). Who creates content? stratification and content creation on the internet. *Information Communication and Society*, *16*(4), 590–612. doi:10.1080/1369118X.2013.777758

Buczynski, S., & Mathews, K. (2016). An urban school district's 21st century teaching vision: Integration and readiness to incorporate technology. *Annual International Conference On Education & E-Learning*, 34. 10.5176/2251-1814_EeL16.12

Carlson, D. (2017). Teachers and crisis: Urban school reform and teachers' work culture (Vol. 1). New York, NY: Routledge.

CEO Forum on Education and Technology. (2001). *Key building blocks for student achievement in the 21" century: Assessment, alignment, accountability, access and analysis. Year 4 report.* Retrieved May 24, 2017 from http://www.ceoforum.org

Champion, R. (2017). *Digitally Sound? Teachers' Use of Digital Literacies in Predominantly African American Classrooms in a Low SES Urban School Setting*. Academic Press.

Cheema, J. R., & Hang, B. (2013). Quantity and quality of computer use and academic achievement: Evidence from a large-scale international test program. *International Journal of Education and Development Using Information and Communication Technology*, *9*(2), 95–106.

Ciampa, K. (2017). Building bridges between technology and content literacy in special education: Lessons learned from special educators' use of integrated technology and perceived benefits for students. *Literacy Research and Instruction*, *56*(2), 85–113. doi:10.1080/19388071.2017.1280863

Clark, A. (2003). *Natural born cyborgs: why minds and technologies are made to merge*. Oxford University Press.

Clark, K. (2005). Serving underserved communities with instructional technologies: Giving them what they need, not what you want. *Urban Education*, *40*(4), 430–445. doi:10.1177/0042085905276388

Coleman, L., Cotten, S., Gibson, P., Stringer, K., & Howell-Moroney, M. (2016). Integrating computing across the curriculum. *Journal of Educational Computing Research*, *54*(2), 275–294. doi:10.1177/0735633115616645

Comi, S. L., Argentin, G., Gui, M., Origo, F., & Pagani, L. (2017). Is it the way they use it? Teachers, ICT, and student achievement. *Economics of Education Review*, *56*, 24–39. doi:10.1016/j.econedurev.2016.11.007

Common Sense Media. (2011). *Zero to eight: Children's media use in America*. Retrieved September 2017 from http://www.commonsensemedia.org/sites/default/files/research/zerotoeightfinal2011.pdf

Conchas, G. Q., Gottfried, M., Hinga, B. M., & Oseguera, L. (Eds.). (2017). *Educational Policy Goes to School: Case Studies on the Limitations and Possibilities of Educational Innovation*. Routledge.

Connors-Kellgren, A., Parker, C. E., Blustein, D. L., & Barnett, M. (2016). Innovations and challenges in project-based STEM education: Lessons from ITEST. *Journal of Science Education and Technology*, *25*(6), 825–832. doi:10.1007/10956-016-9658-9

Corcino, C., & Mentor, D. (2018). *Computer programming for under resourced communities - from training, to internship, to employment.* Presented at the UNESCO Mobile Learning Week's Skills for a connected world 2018, UNESCO Headquarters, Paris, France.

Costley, K.C., Bell, D., & Legett, T. (2014). *More feet hitting the road: Ten ways to get impoverished childrens' test scores up*. Academic Press.

Cotten, S. R., Hale, T. M., Moroney, M. H., O'Neal, L., & Borch, C. (2011). Using affordable technology to decrease digital inequality: Results from birmingham's one laptop per child XO laptop project. *Information Communication and Society*, *14*(4), 424–444. doi:10.1080/1369118X.2011.559266

Cox, J. (2015, December). *Benefits of technology in the classroom*. Retrieved from http://www.teachhub.com/benefits-technology-classroom

Cuban, L. (2001). *Oversold and underused: Computers in the classroom*. Cambridge, MA: Harvard University Press.

Cubbage, A. (2017, January 20). Northwestern university forms new partnership with lake view high school. *Northwestern Now*. Retrieved from https://news.northwestern.edu/stories/ 2017/01/northwestern-university-new-partnership-lake-view-high-school-cps/

Darling-Hammond, L. (1995). Inequality and access to knowledge. In J. A. Banks & C. A. Banks (Eds.), *The handbook of multicultural education* (pp. 465–483). New York: MacMillan.

Darling-Hammond, L. (2015). *The flat world and education: How America's commitment to equity will determine our future*. Teachers College Press.

Davis, F. D. (1993). User acceptance of information technology: System characteristics, user perceptions and behavioral impacts. *International Journal of Man-Machine Studies*, *38*(3), 475–487. doi:10.1006/imms.1993.1022

Davis, K. K., Ambrose, A., & Orand, M. (2017). Identity and agency in school and afterschool settings: Investigating digital media's supporting role. *Digital Culture & Education*, *9*, 31–47.

Davison, E., & Cotten, S. R. (2010). Connection disparities: The importance of broadband connections in understanding today's digital divide. In E. Ferro (Ed.), *Handbook of research on overcoming the digital divides: constructing an equitable and competitive information society* (Vol. 1). IGI Global. doi:10.4018/978-1-60566-699-0.ch019

Day, V. (2017). *Understanding the Relationship Between K-12 Teachers' Perceptions of Their Levels of Teaching Innovation and Their Experiences with Technology-Driven Professional Development*. Drexel University.

Degennaro, D., & Brown, T. L. (2009). Youth voices: Connections between history, enacted culture and identity in a digital divide initiative. *Cultural Studies of Science Education*, *4*(1), 13–19. doi:10.100711422-008-9108-y

Dexter, S., Doering, A. H., & Riedel, E. S. (2006). Content area specific technology integration: A model for educating teachers. *Journal of Technology and Teacher Education*, (2): 325.

Domine, V. (2010). Chapter ten: How important is technology in urban education? *Counterpoints*, *215*, 145–158.

Drossel, K., & Eickelmann, B. (2017). Teachers' participation in professional development concerning the implementation of new technologies in class: A latent class analysis of teachers and the relationship with the use of computers, ICT self-efficacy and emphasis on teaching ICT skills. *Large-Scale Assessments in Education*, *5*(1), 1–13. doi:10.118640536-017-0053-7

Du, J., Sansing, W., & Yu, C. (2004). The impact of technology use on low-income and minority students' academic achievements: Educational longitudinal study of 2002. *Association For Educational Communications and Technology*, 274-283.

Education Week. (2007, March 30). *Technology counts 2007: A digital decade*. Retrieved May 20, 2017, from http://www.edweek.org/ew/toc/2007/03/29/

Edyburn, D., Higgins, K., & Boone, R. (2005). *Handbook of special education technology research and practice*. Whitefish Bay, WI: Knowledge by Design, Inc.

Eickelmann, B., Gerick, J., & Koop, C. (2016). ICT use in mathematics lessons and the mathematics achievement of secondary school students by international comparison: Which role do school level factors play? *Education and Information Technologies*, 1–25.

Ertmer, P. A., & Ottenbreit-Leftwich, A. (2013). Removing obstacles to the pedagogical changes required by Jonassen's vision of authentic technology-enabled learning. *Computers & Education*, *64*, 175–182. doi:10.1016/j.compedu.2012.10.008

Fairlie, R. (2007). Explaining differences in access to home computers and the Internet: A comparison of Latino groups to other ethnic and racial groups. *Electronic Commerce Research*, *7*(3/4), 265–291. doi:10.100710660-007-9006-5

Fairlie, R. W. (2012). Academic achievement, technology and race: Experimental evidence. *Economics of Education Review*, *31*(5), 663–679. doi:10.1016/j.econedurev.2012.04.003

Fine, M. (1991). *Framing dropouts: notes on the politics of an urban public high school*. New York: State University of New York Press.

Frank, T. J., & Castek, J. (2017). From digital literacies to digital problem solving: Expanding technology-rich learning opportunities for adults. *Journal of Research and Practice for Adult Literacy, Secondary, and Basic Education*, *6*(2), 66–70.

Frederick, R., & Shockley, K. (2008). Culturally responsive uses of computer technology: A portrait of three teachers working in urban schools. *Electronic Journal for Instructional Technology*, *7*(3), 3–21.

Good, T. L., & Lavigne, A. L. (2017). *Looking in classrooms*. Routledge.

Goslee, S. (1998). *Losing ground bit by bit: low-income communities in the information age*. Washington, DC: Benton Foundation.

Gunkel, D. J. (2003). Second thoughts: Toward a critique of the digital divide. *New Media & Society*, *5*(4), 499–522. doi:10.1177/146144480354003

Hargittai, E. (2002). Second-level digital divide: Differences in people's online skills. *First Monday*, *7*(4), 1–14. doi:10.5210/fm.v7i4.942

Hargittai, E. (2010). Digital na(t)ives? variation in internet skills and uses among members of the "net generation". *Sociological Inquiry*, *80*(1), 92–113. doi:10.1111/j.1475-682X.2009.00317.x

Harris, M. (2017). *Kids These Days: Human Capital and the Making of Millennials*. Little, Brown.

Hennessy, S., Haßler, B., & Hofmann, R. (2015). Challenges and opportunities for teacher professional development in interactive use of technology in African schools. *Technology, Pedagogy and Education*, *24*(5), 1–28. doi:10.1080/1475939X.2015.1092466

Hughs, J. E., & Ooms, A. (2004). Content-focused technology inquiry groups: Preparing urban teachers to integrate technology to transform student learning. *Journal of Research on Technology in Education*, *36*(4), 397–411. doi:10.1080/15391523.2004.10782422

Irby, D. R. (2017). *Middle School Student and Teacher Perceptions About the Effectiveness of the Technology Integration in the Classroom*. Academic Press.

Johnson, L., Adams Becker, S., Cummins, M., Estrada, V., & Freeman, A. (2014). *NMC Horizon report: 2014 K- 12*. Austin, TX: The New Media Consortium.

Kereluik, K., Mishra, P., Fahnoe, C., & Terry, L. (2013). What knowledge is of most worth: Teacher knowledge for 21st century learning. *Journal Of Digital Learning In Teacher Education*, *29*(4), 127–140. doi:10.1080/21532974.2013.10784716

Kerr, S. (2011). High school online: Pedagogy, preferences, and practices of three online teachers. *Journal of Educational Technology Systems*, *39*(3), 221–244. doi:10.2190/ET.39.3.b

Kolesinski, M., Nelson-Weaver, E., & Diamond, D. (2013). Digital solidarity in education: promoting equity, diversity, and academic excellence through innovative instructional programs. In T. Bastiaens & G. Marks (Eds.), *proceedings of e-learn: world conference on e-learning in corporate, government, healthcare, and higher education 2013* (pp. 2228–2233). Association for the Advancement of Computing in Education.

Kozol, J. (1991). *Savage inequalities: children in America's schools*. New York: Harper-Collins.

Kress, T. (2011). Going high tech under high surveillance: Technology integration, zero tolerance, and implications for access and equity. *Radical Teacher*, *90*(1), 15–24. doi:10.5406/radicalteacher.90.0015

Kvasny, L. (2006). Cultural (re) production of digital inequality in a US community technology initiative. *Information Communication and Society*, *9*(2), 160–181. doi:10.1080/13691180600630740

Kvasny, L., & Mark, K. (2006). The challenges of redressing the digital divide: A tale of two US cities. *Information Systems Journal*, *16*(1), 23–53. doi:10.1111/j.1365-2575.2006.00207.x

Landsman, J., & Lewis, C. W. (2011). *White teachers, diverse classrooms: Creating inclusive schools, building on students' diversity, and providing true educational equity*. Sterling, VA: Stylus Pub.

Lawless, K., & Pellegrino, J. W. (2007). Technology into teaching and learning: Knowns, unknowns, and ways to pursue better questions and answers. *Review of Educational Research*, *77*(4), 575–614. doi:10.3102/0034654307309921

Leffler, M. E. (2015). Digitally divided in Jackson: Are students getting the digital literacy skills they need to succeed? (Doctoral dissertation). Jackson State University.

Lemos, M. H., & Charles, G. U. E. (2017). *Patriotic Philanthropy?* Financing the State with Gifts to Government.

Lenhart, A., Madden, M., Smith, A., & Macgill, A. (2007). *Teens and social media. Pew Internet & American Life Project*. Pew Research Center.

Lenhart, A., Purcell, K., Smith, A., & Zickuhn, K. (2010). Social media & mobile internet use among teens and young adults. *Pew Internet & American Life Project*. Pew Research Center. Retrieved 5/20/17 from http://www.pewinternet.org/w/media//Files/Reports/2010/PIP_Social_Media_and_Young_Adults_Report_Final_with_toplines.pdf

Li, J., Snow, C., & White, C. (2015). Teen culture, technology and literacy instruction: Urban adolescent students' perspectives. *Canadian Journal of Learning and Technology*, *41*(3), 1–36. doi:10.21432/T2004H

Manning, M. L., Baruth, L. G., & Lee, G. L. (2017). *Multicultural education of children and adolescents*. Taylor & Francis.

Mentor, D. (2014a). *M-Class: Mobile learning for teaching and learning*. Presented at the UNESCO Mobile Learning Week 2014, UNESCO Headquarters, Paris, France.

Mentor, D. (2014b). *Creating e- and m-learning Foundations to Engage Young Adults in a Workforce Development Program*. Presentation delivered at the ICELW 2014: The International Conference on E-Learning in the Workplace at Columbia University, New York, NY.

Mentor, D. (2015). From App Attack to Goal-Oriented Tablet Use. In H. An, S. Alon, & D. Fuentes (Eds.), *Tablets in K-12 Education: Integrated Experiences and Implications* (pp. 1–21). Hershey, PA: IGI Global. doi:10.4018/978-1-4666-6300-8.ch001

Mentor, D. (2016a). EMxC3 = e&mLearning Cultivating Connected Communities: Sustainable Workforce Talent Development. In D. Mentor (Ed.), *Handbook of Research on Mobile Learning in Contemporary Classrooms* (pp. 240–259). Hershey, PA: IGI Global. doi:10.4018/978-1-5225-0251-7.ch012

Mentor, D. (2016b). Empowering Urban Students through Regular and Collated Student Evaluation of Teachers. Presented at the *International Conference on Urban Education (ICUE 2016) Biennial Meeting San Juan*, Puerto Rico.

Mentor, D. (2018a). *Creating Teaching and Learning Communities for MOOCS*. Presented at the UNESCO Mobile Learning Week's Skills for a connected world 2018, UNESCO Headquarters, Paris, France.

Mentor, D. (2018b). *Including Digital Proficiency for Teaching and Learning benchmarks in a workforce development program*. Presented at the UNESCO Mobile Learning Week's Skills for a connected world 2018, UNESCO Headquarters, Paris, France.

Mirazchiqyski, P. (2016). The digital divide: The role of socioeconomic status across countries. *Solsko Polje*, *27*(3/4), 23–52.

Morino, M. (1997). *The impact of technology on youth in the 21st century*. Morino Institute.

Morino, M. (2001). *From access to outcomes: raising the aspirations for technology initiatives in low-income communities*. Morino Institute.

Mouza, C. (2011). Promoting urban teachers' understanding of technology, content, and pedagogy in the context of case development. *Journal of Research on Technology in Education*, *44*(1), 1–29. doi:10.1080/15391523.2011.10782577

Mouza, C., & Barrett-Greenly, T. (2015). Bridging the app gap: An examination of a professional development initiative on mobile learning in urban schools. *Computers & Education*, *88*, 1–14. doi:10.1016/j.compedu.2015.04.009

Mulcahy, K. L. J. (2017). *The effects of resources on the performance of competitive high school marching bands* (Doctoral dissertation). Boston University.

Nabi, R. L., & Krcmar, M. (2016). It takes two: The effect of child characteristics on U.S. parents' motivations for allowing electronic media use. *Journal of Children and Media, 10*(3), 285–303. doi:10.1080/17482798.2016.1162185

NEA Education Policy and Practice Department. (2017). *Technology in school: The ongoing challenge of access, adequacy, and equity*. Retrieved from http://www.nea.org/assets/docs/PB19_Technology08.pdf

Noodle Staff. (2015). *41 most innovative K–12 schools in America*. Retrieved from.

O'Bannon, B. W., Waters, S., Lubke, J., Cady, J., & Rearden, K. (2017). Teachers and Students Poised to Use Mobile Phones in the Classroom. *Computers in the Schools, 34*(3), 125–141. doi:10.1080/07380569.2017.1347454

O'Hara, K. E. (2014). From filtering to Freire: Critical use of the internet in urban classrooms. *Journal of Arts and Humanities, 3*(6), 9.

O'Neal, L., Gibson, P., & Cotten, S. (2017). Elementary school teachers' beliefs about the role of technology in 21st-century teaching and learning. *Computers In The Schools*, 1–15. doi:10.1080/07380569.2017.1347443

O'Sullivan, M. K., & Dallas, K. B. (2017). A collaborative approach to implementing 21st century skills in a high school senior research class. *Education Libraries, 33*(1), 3–9. doi:10.26443/el.v33i1.284

Owen, E., Song, H., & Kidd, T. (2007). Re-examining the sociocultural factors affecting technology use in mathematics classroom practices. *International Journal of Web-Based Learning and Teaching Technologies, 4*(2), 37–45.

Pandolfo, N. (2012). As some schools plunge into technology, poor schools are left behind. *The Hechinger Report, 24.*

Pöntinen, S., Dillon, P., & Väisänen, P. (2017). Student teachers' discourse about digital technologies and transitions between formal and informal learning contexts. *Education and Information Technologies, 22*(1), 317–335. doi:10.100710639-015-9450-0

Rideout, V. J., Foehr, U. G., & Roberts, D. F. (2010). *Generation m2: Media in the lives of 8- to 18-year-olds.* Kaiser Family Foundation. Retrieved 5/20/17 from http://www.kff.org/entmedia/upload/8010.pdf

Robinson, L. (2009). A taste for the necessary: A bourdieuian approach to digital inequality. *Information Communication and Society, 12*(4), 488–507. doi:10.1080/13691180902857678

Robinson, L. (2011). Information-channel preferences and information-opportunity structures. *Information Communication and Society, 14*(4), 472–494. doi:10.1080/1369118X.2011.562224

Rogers, S. (2016). Bridging the 21st century digital divide. *TechTrends, 60*(3), 197–199. doi:10.100711528-016-0057-0

Rosin, H. (2013). The touch-screen generation: Young children-- even toddlers-- are spending more and more time with digital technology. Thousands of apps appealing to kids just out of diapers are now released every year. Should parents recoil or rejoice? An examination of what tablets and smartphones are doing to young children's brains. *Atlantic (Boston, Mass.)*, (3), 56.

Russell-Bennett, R., Drennan, J., Raciti, M., Bowring, N., & Burns, K. (2017). *Using service design to understand digital preferences for improving applications for tertiary education in under-represented markets*. Academic Press.

Schradie, J. (2011). The digital production gap: The digital divide and web 2.0 collide. *Poetics*, *39*(2), 145–168. doi:10.1016/j.poetic.2011.02.003

Servon, L. J., & Nelson, M. K. (2001). Community technology centers and the urban technology gap. *International Journal of Urban and Regional Research*, *25*(2), 419–426. doi:10.1111/1468-2427.00320

Shank, D. B., & Cotten, S. R. (2014). Does technology empower urban youth? The relationship of technology use to self-efficacy. *Computers & Education*, *70*, 184–193. doi:10.1016/j.compedu.2013.08.018

Sharehu, A. L., & Emmanuel, A. E. (2015). Readiness of Teachers and Pupils for Use of Mobile Devices as Support for Effective Pedagogy in Nigeria: Could Location be a Major Determinant? *Online Journal of Distance Learning Administration*, *18*(3), 1–11.

Sheninger, E. C., & Murray, T. C. (2017). *Learning Transformed: 8 Keys to Designing Tomorrow's Schools, Today*. ASCD.

Slaouti, D., & Barton, A. (2007). Opportunities for practice and development: Newly qualified teachers and the use of information and communications technologies in teaching foreign languages in English secondary school contexts. *Journal of In-service Education*, *33*(4), 405–424. doi:10.1080/13674580701687807

Song, S., & Owens, E. (2011). Rethinking technology disparities and instructional practices within urban schools: Recommendations for school leadership and teacher training. *Journal of Technology Integration in The Classroom*, *3*(2), 23–36.

Staples, A., Pugach, M. C., & Himes, D. J. (2005). Rethinking the technology integration challenge: Cases from three urban elementary schools. *Journal of Research on Technology in Education*, *37*(3), 285–311. doi:10.1080/15391523.2005.10782438

Subramony, D. (2011). Socio-cultural issues in educational technology integration. *Colleagues, 6*(1), Article 10. Retrieved from http://scholarworks.gvsu.edu/colleagues/vol6/iss1/10

Subramony, D. P. (2007). Understanding the complex dimensions of the digital divide: Lessons learned in the Alaskan arctic. *The Journal of Negro Education*, *76*(1), 57–67.

Subramony, D. P. (2014). *Reframing the digital divide within a 'Flat World' Context: The McJulien Lecture*. Paper presented at the Association for Educational Communications and Technology (AECT) International Convention 2014, Jacksonville, FL.

Tamim, R. M., Bernard, R. M., Borokhovski, E., Abrami, P. C., & Schmid, R. F. (2011). What forty years of research says about the impact of technology on learning a second-order meta-analysis and validation study. *Review of Educational Research*, *81*(1), 4–28. doi:10.3102/0034654310393361

Tettegah, S. Y., & Mayo, C. (2005). Urban education and technology in the digital Age. *Urban Education*, *40*(4), 363–367. doi:10.1177/0042085905276374

The Nellie Mae Education Foundation. (2011). *Integrating technology with student-centered learning.* Quincy, MA: Author.

Tondeur, J., van Braak, J., Ertmer, P. A., & Ottenbreit-Leftwich, A. (2017). Understanding the relationship between teachers' pedagogical beliefs and technology use in education: A systematic review of qualitative evidence. *Educational Technology Research and Development, 65*(3), 555–575. doi:10.100711423-016-9481-2

Toven-Lindsey, B. A. (2017). *Digital Literacy and Career Capital: How College Experiences are Preparing Students for the Transition to Work* (Doctoral dissertation). UCLA.

United Nations *Educational, Scientific and Cultural Organization (UNESCO). (2012). Turning* on mobile learning: Global themes. UNESCO working paper series on mobile learning. Retrieved September 2017, from http://unesdoc.unesco.org/ images/0021/002164/216451E.pdf

United Nations Educational, Scientific and Cultural Organization (UNESCO). (2017). Retrieved from http://www.unesco.org/new/en/unesco/themes/icts/

U.S. Department of Commerce. (1999). *Falling through the net: Defining the digital divide.* Washington, DC: National Telecommunications and Information Administration.

U.S. Department of Education, Office of Elementary and Secondary Education, Office of State Support. (2015). *Improving basic programs operated by local educational agencies (Title I, Part A).* Author.

van Dijk, J. A. G. M. (2006). Digital divide research, achievements and shortcomings. *Poetics, 34*(4-5), 221–235. doi:10.1016/j.poetic.2006.05.004

Wamuyu, P. K. (2017). Bridging the digital divide among low income urban communities. Leveraging use of community technology centers. *Telematics and Informatics, 34*(8), 1709–1720. doi:10.1016/j.tele.2017.08.004

Watson, T. N., & Bogotch, I. (2015). Reframing parent involvement: What should urban school leaders do differently? *Leadership and Policy in Schools, 14*(3), 257–278. doi:10.1080/15700763.2015.1024327

Weiss, J. (2007). "Eyes on me regardless": Youth responses to high school surveillance. *Educational Foundations, 21*(1-2), 47–69.

West, W., & Vosloo, S. (2013). *UNESCO policy guidelines for mobile learning.* Paris: UNESCO. Retrieved from http://unesdoc.unesco.org/images/0021/002196/219641e.pdf

Wilkin, S., Davies, H., & Eynon, R. (2017). Addressing digital inequalities amongst young people: Conflicting discourses and complex outcomes. *Oxford Review of Education, 43*(3), 332–347. doi:10.1080/03054985.2017.1305058

Wyn, J. (2009). *Touching the future: Building skills for life and work.* Melbourne, Australia: Australian Council for Educational Research Press.

Zielezinski, M. B. (2016). Using technology to support underserved students. *Education Digest, 82*(3), 30–34.

KEY TERMS AND DEFINITIONS

Digital Divide: The way in which social status (e.g., race, ethnicity, class) hinders student access to technology and its educational capabilities and potential.

Racially Minoritized: Groups that are different in race, religion, nation of origin, culture, sexuality, and gender, and due to social constructs have less power or representation compared to other members or groups in society.

Technology: Tools used in and out of the classroom to assist in student knowledge gain, content understanding, and skill practice.

Technology Integration: School and teacher implementation of technology into content skills and knowledge through multiple facets of classroom learning (i.e., lesson delivery, content practice, student formative/summative assessment, classroom management).

Underserved Community: A group of people within society that have been historically oppressed based on specific, often inherent, attributes and characteristics (gender, age, ethnicity, race, sexuality, etc.) who experience greater challenges than other demographic groups.

Urban Education: Urban education is an edifying experience during which information and knowledge is shared in a systematic way, in a setting that is rich in culture, languages, perspective, and overall diversity.

APPENDIX

High School Online: Pedagogy, Preferences, and Practices of Three Online Teachers

Kerr (2011) undertakes a multiple case study approach to explore how three online, high school teachers utilized technological tools to create meaningful learning activities for their students. Findings reveal that teachers used a wide variety of tools and approaches to online learning. Tools are categorized as content, communication, and management tools.

Chapter 4
Plugged In:
Youth Computer-Mediated Civic Action and Interaction Through the Prism of Modern Protest Movements

Dino Sossi
Columbia University, USA

ABSTRACT

In the aftermath of the Arab Spring, Occupy Wall Street, and Black Lives Matter movements, protest has become the default response to social problems. As students and youth become more involved in political upheaval, they turn to the technology that surrounds them. This chapter focuses on computer-mediated youth civic action and interaction. It examines past trends in youth activism and how social media skills acquired through activism could help these same youth activists transition to the workforce.

INTRODUCTION

In the aftermath of the Arab Spring, Occupy Wall Street, and Black Lives Matter movements, youth public protest seems to have become the default response to a litany of social problems. As students and youth become more involved in political upheaval due to this unrest (Day, 2015; Epstein, 2015; Lotan et al., 2011) as well as precarious economic prospects (Milkman, Luce, & Lewis, 2013), they tend to turn to the technology that constantly surrounds them, namely smartphones and desktop computers. Similarly, communications technology (e.g., email, internet forums), videotelephony technology (e.g., Facetime), communications platforms (e.g., Google Hangouts) and free or inexpensive social media apps (e.g., Facebook, Periscope, Twitter, WeChat, WhatsApp) leverage seemingly perpetual internet connectivity. This allows youth to constantly communicate in real-time, remain mobile, and give them the ability to react to current events. These developments have invariably changed the frequency, tone, and ability of youth to participate in the body politic, including broad social movements like the recently occurring ones mentioned above. Technology gives youth who are interested in these movements the ability to express themselves and their deep personal concerns in ways that were unimaginable even a decade ago.

DOI: 10.4018/978-1-5225-4111-0.ch004

From these developments, this chapter focuses on computer-mediated youth civic action and interaction. It examines past trends in youth activism and the current deployment of technology during the present moment. Finally, it discusses how social media skills acquired through activism could help these same young people gain skills to help them successfully transition to the workforce.

BACKGROUND

The 1960s was a defining time for youth protest in the United States. Following the peaceful prosperity following the Second World War as the country recovered from its prolonged participation in warfare, the 1960s became filled with wrenching civil unrest. This is in sharp contrast to the more genteel atmosphere characterizing the nation following the previous world war. The Vietnam War's protracted length, the corresponding increasing number of casualties over its duration, and its reliance on the draft drew deep demarcations among the American populace based on race, class, gender, and socioeconomics. As an example, this includes those who chose to enlist as well as draftees were disproportionally poor and people of color who were less able to employ various strategies to avoid service (e.g., successfully apply for deferment). This combination of difficult issues became a major instigator for the student protests and riots that occurred throughout the 1960s (Barton, 1968; Bradley, 2003; Hariman & Lucaites, 2001; Lewis & Hensley, 1998; Moore, 1999), their reverberations still echoing during the present moment as seen in the Charlottesville protests (DeVega, 2017; Owen, 2017; Sinclair, 2017).

One of the most popular and pervasive categories of technology of that time, telecommunications, was one of the key modes of organizing protests during this tumultuous period (Gladwell, 2010). With just a name, a phone number, and loose pocket change, youth activists could communicate to organize and engage in civic action courtesy of widely available pay phones (Gladwell, 2010). In short, the revolution will be dialed.

In addition, the growing dissemination of broadcast television brought immediate real-time visuals that could not be easily avoided by the general population. For example, think of the stunning images of soldiers' coffins being flown back from Vietnam. Explosive student revolts at Columbia University (Barton, 1968; Bradley, 2003; Moore, 1999) or at Kent State more tragically also roiled the nation (Hariman & Lucaites, 2001; Lewis & Hensley, 1998; Moore, 1999). If the medium was the message (McLuhan, 1964), the affordances of television and radio as well as their omnipresent nature could no longer simply be ignored. Mass media's immediacy and visually/aurally stunning nature brought the realities of youth dissent, and even death, into the warm and comfortable environs of American living rooms. In the aftermath of the primacy of traditional mass media in rapidly delivering news from around the world, today's youth have arguably accelerated the pace of dissemination. In short, youth bearing smartphones that are constantly connected to the internet deploy social media platforms and associated tools to amplify the reach of news.

With the recent advent of smartphones and social media, youth can easily communicate with these devices to plan, aggregate, and adjust their protest activities in real-time, even as they are on the go as these civic actions are in the midst of erupting. This was probably most clearly seen in the unfolding of the Arab Spring (Abouzeid, 2011; Cottle, 2011; Lotan et al, 2011; Lynch, 2011; McNally, 2010, 2011; Mentor, 2015; Mohyeldin, 2011; Ryan, 2011; Sadiki, 2010), Occupy Wall Street (Besser, 2012; Byrne, 2012; Calhoun, 2013; Conover, Ferrara, Menczer, & Flammini, 2013; Earle, 2012; Epstein, 2015; Gillham, Edwards, & Noakes, 2013; Lazar, 2011; Milkman, Luce, & Lewis, 2013; Occupy Wall Street,

2011; Penney & Dadas, 2013; Schechter, 2011; Schneider, 2011; Skinner, 2011), and Black Lives Matter (Bailey & Leonard, 2015; Black Lives Matter, 2017; Day, 2015; Eligon, 2015; Olteanu, Weber, & Gatica-Perez, 2016; Ross 2015) as well as subsequent forms of protest. Individually, these movements demonstrate the power of technology to foster youth participation, activism, and dissent during these protests. Collectively, they speak to much broader trends of youth disquiet with respect to explosive social issues, both in the United States and abroad, as demonstrated by similar forms of movements occurring globally.

Youth also face a challenging economic moment in developing work prospects given present employment market challenges. For example, the global youth unemployment rate was expected to reach 13.1% in 2016, which is extremely close to its historic peak of 13.2% back in 2013 (International Labor Organization, 2016). In raw numbers, youth unemployment impacts over 73.3 million young people globally, a decrease from its peak of 76.6 million in 2009 (International Labor Organization, 2015). These statistics have reached crisis levels in some countries. For example, the rate of youth unemployment in some developed countries exceeds 50% (United Nations World Youth Report, 2016). Perhaps the social media skills that students deploy so successfully to meet their activist needs can help them develop valuable communication skills that could transfer beyond advocacy to help them successfully transition into the workplace.

MAIN FOCUS OF THE CHAPTER

Issues, Controversies, and Problems

From this introduction, it is clear that there is a broad array of subjects that need to be examined regarding computer-mediated youth civic action and interaction. The main issues this chapter investigates include:

- What were some of the main modes of youth civic action during earlier protests (e.g., the 1960s)?
- What role does social network theory play in explaining forms of youth civic action and interaction?
- What have been the major developments in technology, such as social media channels, especially their impact on the most recent wave of social movements involving youth activists?
- What lessons have we learned, or have yet to learn, when it comes to using digital devices for civic action?
- How were technologies such as smartphones and social media deployed during recent forms of civil participation and unrest (e.g., the Arab Spring, Occupy Wall Street, Black Lives Matter, etc.)?
- What connections are there between using digital devices and developing skills that can later be used in youth transition into the workforce? This would include the development of communications skills, organization skills, and logistics, etc. (Daskal, 2016; Deiser & Newton, 2013; DeMers, 2015; Fullam, 2017; Lombard, 2016; Lua, 2017)

Due to their relatively recent occurrence and high profile, the complex nature of the Arab Spring, Occupy Wall Street, and Black Lives Matter will be explored in detail.

The History of Using Media in Social Protests

As mentioned earlier, the 1960s was a pivotal period for American youth protest. A great portion of the 1960s was filled with civil unrest. One of the main drivers was the Vietnam War. Another was the start of the Civil Rights Movement that also had an incredibly significant impact on the country that lasts until today. This is seen most clearly through the Black Lives Matter movement (Bailey & Leonard, 2015; Black Lives Matter, 2017; Day, 2015; Eligon, 2015; Olteanu, Weber, & Gatica-Perez, 2016; Ross 2015). Given current events, the fight for civil rights also shows no sides of letting up as per the actions of counter-protestors to the recent Charlottesville white supremacist protests (DeVega, 2017; Owen, 2017; Sinclair, 2017).

Arguments Against Social Media Use in Activism

Writer Malcolm Gladwell has investigated the role of media in civil rights fights during the 1960s (Gladwell, 2010). He makes the provocative case that these earlier movements were largely driven by a characteristic of social networks identified as *strong ties*. Gladwell's use of this concept is an application of the earlier classic work of sociologist Mark Granovetter (1973).

In his seminal research, Granovetter makes the distinction between two types of social connections: *strong ties* and *weak ties* (1973). Strong ties are characterized by more immediate, personal, face-to-face interactions among individuals and groups of people. In contrast, weak ties are more distant relationships among people who are comparably less deeply associated.

To clarify these concepts within the context of today's multitude of networking opportunities, strong ties could be exemplified by our direct human relationships situated in the analog world. These are the types of deep, long-term, and intimate relations that can be formed among physically proximate friends, family, and close acquaintances, etc. In contrast, weak ties are demonstrated by the often less personally impactful relationships we have in digitally mediated spaces. For example, think about your many Facebook *friends*, unsolicited Twitter *followers*, YouTube channel *fans*, etc.

Gladwell makes the case that weak ties are important and have profound strength. People who are our most remote acquaintances can still become useful sources of new data or novel information (Gladwell, 2010). The internet generally, and social media specifically, both allow us to leverage remote interpersonal connections easily, virtually instantaneously at the click of a button (Gladwell, 2010).

However, in the high-stakes world of youth activism where lives are threatened, Gladwell believes that weak ties are less useful than their strong counterparts. He states that even if there were more advanced forms of social media tools in place during the Civil Rights Movement (e.g., similar to those available today), they would be less important in helping youth participate in these incidents of civil unrest in places like the U.S. South. During the 1960s, deep levels of racism in the South threatened unwelcome activists with physical harm or death. As Gladwell states, "[t]he instruments of social media are well suited to making the existing social order more efficient" (2010) and internet-based activism only "makes it easier for activists to express themselves, and harder for that expression to have any impact" (2010). Although social media technologies have distinct affordances, Gladwell firmly believes and succinctly states that, "the revolution will not be tweeted" (2010). However, in a paradoxical development, the recent past few administrations arguably may have been inadvertently inspiring it. Specifically, the administrations' social media postings have apparently inspired such developments as the national women's marches and other forms of public dissent. The push and pull between governance and the governed will continue to shift and develop.

Harvard Law professor Cass Sunstein supports Gladwell's less optimistic view of social media, at least with respect to their impact on our democracy (Sunstein, 2017). In his book *#Republic: Divided Democracy in the Age of Social Media* (2017), Sunstein speaks about the problem of siloing social media users into their own groups. As a result, these media channels are devoid of the thoughts of others outside of their preferred acquaintances through social media, contributing to their own echo chambers (Sunstein, 2017). People become more politically fragmented, polarized, and can even be pushed toward political extremism (Sunstein, 2017). In addition, there is evidence that bots (a computer program that engages in automated tasks) interfered in the 2016 U.S. presidential election by spreading fake news, among other forms of interference (O'Connor & Schneider, 2017; Watson, 2017; Williams, 2017). Bot-directed communications on social media can further manipulate and distort our public discursive spaces, especially if their influences are undetected or left unchecked.

In spite of Gladwell and Sunstein's negative views toward social media's impact on civic participation and action, other scholars disagree and see more beneficial aspects.

Arguments in Favor of Social Media Use in Activism

Zeynep Tufecki believes that Gladwell misdiagnoses the issue of social media activism by making two points. First, there is a mismatch between the problems that activists want to solve (e.g., global issues) and how we naturally socialize (e.g., local interactions) (Tufekci, 2010). Due to their ability to transcend time and space, social media potentially play a profound role in dealing with worldwide concerns, but potentially engaged with on a more manageable local level. Second, Granovetter's ideas of strong and weak ties are not binary opposites, but can be interpreted as being complementary (Tufekci, 2010). If activists can commandeer large pools of weak ties, they could possibly be leveraged into more robust networks of stronger ties (Tufecki, 2010). From this, the affordances of one form of ties feed into another, weak ties can eventually lead to the formation of strong ties.

Mathew Ingram supports Tufecki's interpretation of the power of social networks (2011). He also believes that social media's weak ties can become connected to our strong tie relationships (Ingram, 2011). Also, on somewhat of an ironic note, adopting Gladwell's own language from his book *The Tipping Point* (2000) and using it to comment on its author, Ingram believes that the tools of the internet or social media and the activity that they beget can lead to a *tipping point* that might not have occurred without the presence of these same tools (Ingram, 2011). One of the key factors in amplifying a message to meet this critical point is the internet and its constituent social media channels.

Other intellectuals also mirror Tufecki's criticism of Gladwell's take on strong and weak ties with regard to youth civic activism. For example, Leo Mirani believes that the significance of social media is their affordances of quickly sharing a great deal of information that involves alternative viewpoints that, without the internet and social media, would not reach a mass audience (2010). Again, social media plays an important role in amplification.

Additionally, Ben Brandzel believes that Gladwell also misses the broader point about social media and activism (Brandzel, 2010). Although social media may not be perfectly equivalent to the power of real friendships, they help motivate strong tie relationships that have previously been formed offline. Again, social media can amplify or diffuse key activist information. Brandzel argues that social media have begun changing the nature of relationships between individuals and their weak tie social networks.

Social media leverages weak tie networks to make them more interactive and accessible than ever before. As a result, weak ties can be used so that people can broadcast their endorsement for activism by *liking* a particular movement (e.g., on the official Facebook page for a social movement).

Another positive aspect of online activism is that it can apply across a diverse range of different subject areas. For example, social media and hip-hop culture have been used to end a war against youth (Porfilio, Roychoudhury, & Gardner, 2013). Also, Latina/o students have leveraged online activism for human rights and social justice against anti-immigrant sentiment (Vélez, Perez Huber, Benavides Lopez, de la Luz, & Solórzano, 2008). As such, social media tools can shift and change to meet the needs of a seemingly infinite variety of constantly changing youth interests.

In the next few sections, recent youth-inspired social movements will be analyzed. Specifically, the Arab Spring, Occupy Wall Street, and the Black Lives Matter movements are examined regarding factors like their provenance, key actors, organization, resolution, their implications regarding the economic security of youth, and the impact of technology at all of these levels. Unfortunately, due to a lack of academic research on more recent social upheavals like the re-emergence of neo-Nazism and Antifa as a form of resistance as demonstrated by incidents at Charlottesville, these later movements will not be discussed.

What Is the Arab Spring?

How can we explain the recent wave of progressive social mobilizations from Tahir Square to Occupy Wall Street? Surely there are many factors, starting of course with the multiple structural crises of new liberalism (inequality, wage stagnation/job loss, reductions in social spending, privatization). Moreover, small but core constituencies of young, college educated, typically underemployed/unemployed youth who with computer access and savvy, have acted as catalysts igniting but not really "leading" protests and mobilizations. The youth who first mobilized articulated wide spread grievances and soon attracted many other groups, classes and age cohorts. But this much is clear to most people who are experiencing the consequences of the adverse consequences of neoliberalism over the past decades, the 2007 meltdown, and subsequent "global slump." (McNally, 2010, 2011)

Early in 2011 was a tumultuous time in the world's history, especially regarding youth protest and activism. The global epicenter for this tumult was concentrated in two regions, namely the Middle East and North Africa. To simplify the nomenclature for such large regions, broad swaths of countries, diverse peoples, and idiosyncratic political uprisings, these movements were collectively referred to as the *Arab Spring* by the mass media, politicians, scholars, and members of civil society.

The Arab Spring as a concept arguably started even earlier than spring 2011. Marc Lynch wrote a prescient story in January 2011 titled "Obama's 'Arab Spring.'" It chronicled the rise of conflicts beginning in a number of nations including Algeria, Egypt, Jordan, Kuwait, and Tunisia (Lynch, 2011). Rather prophetically, Lynch posited the following provocative question especially given the focus of this chapter.

Are we seeing the beginnings of the Obama administration equivalent of the 2005 "Arab Spring," when the protests in Beirut captured popular attention and driven in part by newly powerful satellite television images inspired popular mobilization across the region that some hoped might finally break through the stagnation of Arab autocracy? Will social media play the role of [A]l Jazeera this time? Will the outcome be any different? (Lynch, 2011, para. 1)

As a result, it appears that youth with technical skills helped to drive momentum in the Arab Spring due to a combination of political and economic factors. This included income inequality, stagnant wages, job losses, and a general state of underemployment or unemployment.

Who Started the Arab Spring?

It takes time to begin analyzing large scale and complex political movements as broad-based and swiftly moving as the Arab Spring. However, the array of countries that would become involved in the Arab Spring became readily clear during the beginning of 2011. The movement started through massive protests across North African countries like Tunisia and Egypt. Following these examples, several other states became involved in rising up against anti-democratic, authoritarian, and repressive rule. These countries include Algeria, Bahrain, Iran, Lebanon, Libya, Morocco, Oman, Syria, and Yemen. On a more tentative basis, even Saudi Arabia was included to some extent in this uprising. The defining moment of the Arab Spring was the death of street vendor Mohamed Bouazizi on December 17, 2010.

Bouazizi sold vegetables from a cart in Sidi Bouzid, Tunisia. He was a popular man who was well known in his area. Bouazizi often gave away free fruits and vegetables to locals despite having to support his own family that included eight children (Ryan, 2011). On December 17, a policewoman took away his unlicensed food cart and goods. Although Bouazizi tried to pay a 10-dinar fine (approximately equal to $7 USD and a relatively high amount of daily earnings in Tunisia), the officer slapped him, spat in his face, and even insulted his deceased father (Abouzeid, 2011). Subsequent to this humiliating incident, local officials refused to see Bouazizi regarding his complaint about this mistreatment. Feeling a strong sense of dejection and humiliation, he poured fuel on his body and set himself ablaze. The resulting outcry of anger was so great that Tunisian President Zine el Abidine Ben Ali visited Bouazizi in an attempt to make amends for the situation and quiet the unrest. Bouazizi died on January 4, 2011. Ben Ali's 23-year-long reign as the dictator who ruled Tunisia with an iron fist ended 10 days later on January 14, 2011.

What Were the Demands of the Arab Spring Movement?

The instruments of governmental rule can easily devolve into absolute control. Oppression often varies by country. The forms of resistance that build to fight this oppression correspondingly become localized in their tactics, including differing forms of protest. Given this point, there was a broad call across the Arab Spring countries for greater protection regarding human rights, governmental anti-corruption reform, as well as improved living standards and better conditions for these populations (Cottle, 2011). Other scholars who study the developments in the region note the role of the inflated price of food, high unemployment, a lack of free speech, and a minimum degree of other basic political freedoms (Mohyeldin, 2011; Sadiki, 2010). From this, it is clear that both political and economic concerns helped drive the Arab Spring.

Youth Involvement in the Arab Spring

As per Lynch's prophetic story regarding the possible role of new technology impacting the Arab Spring (2011), social media has in fact played a distinct role in this movement that has been subsequently corroborated by scholars.

For example, Lotan et al. (2011) examined information flows during the revolutions that took place in Egypt and Tunisia, countries deeply involved in the Arab Spring movement. They investigated the role of Twitter tweets among different groups of stakeholders during these periods of unrest. Participants included journalists, bloggers, mainstream media outlets, miscellaneous organizations, influential regional and global actors, and most importantly, on the ground activists who worked to make these revolutions a reality.

Lotan et al. (2011) found that those who are interested in phenomena like the Arab Spring break down into three categories. The first is those people who are directly connected to the incident (e.g., those who are on the ground participants). The second category is mainstream media outlets that want to know current developments to share with their media channels and hold their audiences' attention. Finally, the third includes those general interest media consumers who want to know about these types of protest events as they happen. As a result, youth leveraged the power of social media among multiple types of roles and categories of participation. As such, they have different forms of influence on the ground, organizationally, and even as commentators. This allows youth multiple entry points to express solidarity in activism. This also attests to young people's quick acquisition of highly useful technical communications skills to help drive their respective movements that could be leveraged in other fields as well, even including future employment as addressed later in this chapter.

A Counter Case to the Twitter Revolution

As above, it appears that Twitter and analogous social media services have the potential to play some constructive or oppositional role in youth activism. However, some scholars do not fully believe this idea.

For example, Evgeny Morozov does not believe that the internet has an exclusively positive impact on democracy and liberty. Regarding the Middle East, he writes about the state of affairs in Iran from late 2009 to early 2010 (Morozov, 2010). He highlights the banning of Gmail, Google's popular email service, by Iranian authorities. This tool of communication, and presumably of liberty, apparently does not necessarily have any impact on freedom depending on its actual reach. Its effect is contingent on the whims of authoritarian governments like Iran that can stop its national use (Morozov, 2010). Other research shows the impact of censoring political content regarding human rights, freedom of expression, minority rights, and religious movements across the globe (OpenNet, 2010). In short, even global technologies have limits in their impacts given strong local actors like authoritarian governments that are desperate to maintain control.

Speaking in more detail about the case of Iran, due to the atomizing impacts of the internet and social media, Morozov theorizes that Iran's Green Movement may have become split into so many competing debate chambers that it was not able to organize on the eve of the 31st anniversary of the Islamic revolution (Morozov, 2010). Instead of *acting*, the Green Movement may have been hamstrung through endless *talking* and *debating* (Morozov, 2010). In a strange development, an excess of *communication* provided by the internet and social media may have led to an absence of *action*.

Morozov is even more doubtful about the power of technology, the internet, and social media for one simple reason: oppressive regimes also have access to these same tools (Morozov, 2009). A rebel's tool of liberty easily becomes an authoritarian regime's mechanism of dominance. Morozov points to the role of Facebook or Twitter as open source information for Iran's secret services (2009). In this case, a potential tool of freedom instead becomes a weapon of state surveillance. The digital panopticon can become real in the analog world and have serious consequences for those who challenge the state, including activist youth who may not know the surveillance implications of these emerging forms of technology.

A Protest Movement Evolves Into Another? What Is Occupy Wall Street?

The Occupy Wall Street movement builds on the momentum of the Arab Spring in many ways (Besser, 2012). In fact, Occupy overtly claims these roots, but focuses more deeply on economic issues than its predecessor.

#ows [the Occupy Wall Street Twitter hashtag] is fighting back against the corrosive power of major banks and multinational corporations over the democratic process, and the role of Wall Street in creating an economic collapse that has caused the greatest recession in generations. The movement is inspired by popular uprisings in Egypt and Tunisia, and aims to fight back against the richest 1% of people that are writing the rules of an unfair global economy that is foreclosing on our future. (Occupy Wall Street, 2011)

Although the respective underlying sociopolitical and economic factors driving these movements differs, the Arab Spring and Occupy Wall Street share a clear link in being efforts at resistance against larger societal forces. Also, depending on the country or region, they also share concerns regarding economic inequality. Due to historically high amounts of global youth unemployment (International Labor Organization, 2016), young activists have been deeply involved in Occupy.

Figure 1. Poster from Occupy Wall Street at Zucotti Park

Who Started Occupy Wall Street?

Occupy Wall Street was originally initiated by two members of Adbusters, a Canadian anti-consumerist publication. Adbusters' Senior Editor Micah White and Co-founder Kalle Lasn thought of the underlying idea for this movement as a peaceful occupation of lower Manhattan that began on September 17, 2011 (Byrne, 2012; Penney & Dadas, 2013). The organization leveraged its 90,000 person list of followers and shared the Twitter hashtag #occupywallstreet (Schneider, 2011). Its dominant slogan, "[w]e are the 99%," indicates the movement's resistance to being economically dominated by an alarmingly small percentage of the population (Besser, 2012; Calhoun, 2013, pp. 8–9). Activist-scholar David Graeber and others who had previously participated in the anti-globalist movement helped with the initial stages of planning Occupy (Earle, 2012, p. 3). It was estimated that between 800 and 2,000 protesters marched through downtown Manhattan and convened at the final site during Occupy's initial efforts (Earle, 2012, p. 3). Although the physical takeover of Zucotti Park started on September 17, 2011 (Skinner, 2011), White claimed that Adbusters had been advocating for a citizen movement against corporations for 20 years (Lazar, 2011). They had previously proposed a peaceful occupation of Wall Street well before this particular point in time during the aftermath of the global 2008 recession.

What Were the Demands of Occupy Wall Street?

The Occupy movement had a limited set of goals for its initial gathering in Manhattan. The attending participants were trying to peacefully protest against corporations' perceived excessive influence on democracy, corporate greed, an increase in wealth disparity, and a lack of legal consequences for those who were allegedly responsible for the 2008 global monetary insolvency crisis (Gillham, Edwards, & Noakes, 2013, p. 3; Lazar, 2011). Another articulated rationale for Occupy included the perception of an inadequate federal response to problems that were caused by the 2008 financial crisis (Gillham, Edwards, & Noakes, 2013, p. 3).

Due to having less of a sure footing in the global economy, these negative economic factors impacted youth disproportionately, including historic levels of youth unemployment (International Labor Organization, 2016). As a result, there was a relatively focused set of financially motivated demands by the Occupy protestors, especially among youth participants.

Youth Involvement in Occupy Wall Street

Technology helped drive Occupy Wall Street, including the movement having its own app (Schechter, 2011, p. 66). The presence of technology among participants resulted in multiple outcomes. For example, different technology helped drive the movement's base of supporters, communicate its messages, and provide a countervailing force against the mainstream media (Schechter, 2011, p. 66). A time-series of Twitter, Facebook, and onsite protests was used to assess the causality between social media streams and what subsequently developed onsite at protests across the globe (Bastos, Mercea, & Charpentier, 2015). From this study, it was clear that there was a distinct relationship between social media communications and global protests in amplifying a message from one city across the world. Being able to instigate global protests via social media is a very powerful skill that can be applied to other fields across the planet.

The plaza where the movement physically occurred became a generator-powered media center for sending Twitter tweets, blog posts, YouTube videos, etc. for dissemination (Schneider, 2011). Some research included a variety of protests including the Indignados, Brazilian vinegar, and the Occupy protests. The researchers found that communications that were contentious in nature on either Facebook or Twitter forecasted onsite protests (Bastos, Mercea, & Charpentier, 2015). There appears to be a temporal lag between online communications and youth civic protests. This seems to imply some correlation between these two factors and that youth social media use can amplify efforts at protest. As such, protesting youth were able to successfully communicate with large and diverse global audiences through social media.

The Occupy movement produced a tremendous amount of social media content. It was estimated that more than half-a-million individual photos had been posted to Flickr with the tag #Occupy (Besser, 2012, p. 2). Approximately 169,000 individual postings to YouTube were tagged with #Occupy (Besser, 2012, p. 2). From this, these youth activists learned valuable media capture, post-production, writing, editing, and other miscellaneous communication skills. As a result, youth participants were able to communicate about the Occupy movement among themselves to update each other and with a broader global audience without depending on traditional mass media outlets like television, radio, or newspapers. These youth could, at least partially through these social media channels, craft their own stories about Occupy to the broader public, shaping their own message.

The composition of people who occupied the Zucotti Park site trended toward the young. Those who stayed on site included many underemployed or unemployed young adults (Milkman, Luce, & Lewis, 2013). These young people were also highly educated, but saddled with significant indebtedness (Milkman, Luce, & Lewis, 2013). Their shared precarious economic situation fed their opposition to the current trend toward the economic concentration of wealth (Stiglitz, 2011), which is also thematically linked to the overall Occupy movement that focuses on economics.

It is widely recognized that the New York City Zucotti Park site was organized through a combination of participant autonomy, social media, and participatory General Assembly forums to organize and conduct business (Epstein, 2015). As such, the daily operation of Occupy Wall Street would not have been as successful if it was not for the use of social media. Although the Zucotti site is not geographically large, it was dense enough that social media and other forms of digital communication allowed the participants to communicate with one another more effectively. Organizing diverse groups of stakeholders using social media tools is useful in marshaling scarce resources quickly and effectively.

Also, social media had other ramifications on Occupy. For example, youth activists used these new forms of media to confront the conservative media establishment (Epstein, 2015, p. 104) that had an alternative take on the movement. In addition, youth used social media to advance the idea of "broad institutional reform even if there was strategic disagreement among student ranks regarding the advisability of negotiating with the center-right government" (Epstein, 2015, p. 105). Social media allowed them to communicate with one another as well as broader audiences. Further, on a wider level, social media provides society with an alternative means of being able to communicate with each other (Epstein, 2015, p. 106). As a result, this new form of media allows youth to constantly renegotiate the possible ways they define and engage in collective engagement (Epstein, 2015, p. 106). Finally, the utter power of social media helped to expose the class privilege that the financial industry possessed (Epstein, 2015, p. 107). The affordances of virtually instantaneous global communication created a worldwide movement that legitimized a popular critique of neoliberalism to a level that even the opponents of these global protest efforts had to acknowledge the vibrancy of this opposition (Epstein, 2015, pp. 107–108). By surpassing traditional communication channels, youth were able to make a convincing case for the merits of their protest using new and evolving forms of social media that could ideally be applied much more widely.

A Counter Case to Occupy Wall Street

From above, it is clear that Occupy Wall Street and similar movements can take advantage of the passion of youth activists and leverage their facility with social media communication. You may go as far as to say that these movements would not have occurred were it not for youth participation. However, there may be limits in how much these young people could have helped foster their movements' successes.

For example, a recent study showed that the Occupy Wall Street movement benefited greatly from microblogging sites like Twitter (Conover, Ferrara, Menczer, & Flammini, 2013). In fact, Twitter was able to foster a great deal of interaction between a highly interconnected set of users who already had interests in common before using the site. However, the drop off in interest of research participants appeared to greatly decrease over the length of the study (Conover, Ferrara, Menczer, & Flammini, 2013). From this, although social media did influence Occupy Wall Street, there were some potential limits from the impact of youth and social media technology. Learning how to sustain a successful social media campaign for longer periods could be a challenge in continually motivating youth participation in social movements.

What Is the Black Lives Matter Movement?

As a young person, I was always afraid of the police, and I have had experiences of being mistreated and yelled at and talked down to. I've always been afraid when I see them coming, and so when Black Lives Matter emerged, I joined in immediately. – Russell Simmons (Siddiqui, 2016)

Alicia Garza was devastated by the acquittal of George Zimmerman after his role in the tragic death of teenager Trayvon Martin (Luibrand, 2015). Zimmerman killed Martin. The latter was a 17-year-old African American teenage boy who was unarmed (Guynn, 2015). Garza was also deeply disappointed by people's social media responses to the verdict. She was frustrated that many posts blamed Black people for their own conditions (Guynn, 2015). To fight against her feelings of upset, Garza wrote "Black people. I love you. I love us. Our lives matter" as a closing remark to a Facebook post titled "A Love Note to Black People" (Day, 2015; Ross 2015). Her friend Patrisse Cullors, a Los Angeles-based community organizer, read and shared this post. On subsequent reposts, Cullors added the Twitter hashtag #blacklivesmatter.

As horrific as the Martin shooting was, the problems across the African American community arguably became worse due to additional tragic incidents that followed his death. On August 9, 2014, approximately a year after Zimmerman's acquittal, Officer Darren Wilson fired 12 rounds at 18-year-old Michael Brown. This Ferguson, Missouri teen was unarmed like Martin (Day, 2015). Garza, Cullors, and Opal Tometi, another activist and additional Co-founder of Black Lives Matter, organized a *freedom ride* to Ferguson under the #blacklivesmatter banner (Eligon, 2015). Over 500 people signed up for this initiative across 18 American cities (Day, 2015). Additionally, the death of 43-year-old Eric Garner, who died after being placed in a chokehold in Staten Island during an arrest, also contributed to a sense of a community under siege and feeling an unbelievable amount of stress (Day, 2015; Denby, 2014).

Use of the #blacklivesmatter hashtag peaked repeatedly during 2015. This was largely in response to a number of racially charged incidents. These included the use of police force against Black teens during a Texas pool party (June 5, 2015), the tragedy of nine people slain at a Black church in Charleston, South Carolina (June 17, 2015), the discovery of Sandra Bland found dead in a Texas jail cell (July 13,

2015), and the killing of Samuel Dubose by Cincinnati police (Eligon, 2015). Unfortunately, these deaths highlighted a brutally violent 2015, especially for Black Americans.

Who Was Involved in the Black Lives Matter Movement?

Garza and Cullors were two of the people most responsible for Black Lives Matter due to their initial work. However, the movement subsequently became diffuse due to its unique organizational structure, fostered by youth activists' use of social media like its Twitter hashtag. Similar to Occupy, Black Lives Matter actively tries to avoid hierarchy (Cobb, 2016).

Jelani Cobb further explains the organization of Black Lives Matter.

We associate whole blocks of history with W. E. B. Du Bois or Booker T. Washington or Malcolm X or Dr. King. Entire constellations of history are distilled down to two individuals, almost always male. There's been a certain kind of fatigue, and Black Lives Matter represents that. They find a more horizontal ethic of leadership to be more important, and they find inspiration from Ella Baker, who was one of the most central figures of the [C]ivil [R]ights [M]ovement . . . - partly due to her belief in kind of grassroots and local leadership in her aversion to the spotlight. Her story wasn't as well known. And so they have kind of resurrected her idea that it's better to have 10,000 candles than a single spotlight. (McEvers, 2016)

As a result, Black Lives Matter has become a broad social movement with young activists organizing internationally under its banner.

What Were the Black Lives Matter Movement's Demands?

Black Lives Matter has a narrow focus on protecting the lives of Black people as per the following quotation.

Black Lives Matter is an ideological and political intervention in a world where Black lives are systematically and intentionally targeted for demise. It is an affirmation of Black folks' contributions to this society, our humanity, and our resilience in the face of deadly oppression. (Black Lives Matter, 2017)

From this, it is clear that long, deeply felt frustration with the mistreatment of Black people, motivated in the wake of the Trayvon Martin murder, was a strong instigating factor for Black Lives Matter, which has blossomed from a national concern to an international movement.

Youth Involvement in Black Lives Matter

Black Lives Matter started through the hashtag #blacklivesmatter by Garza's Facebook post, but explicitly shaped by Cullors and her work on Twitter (Garza, 2014). Studies have been conducted to understand the dispersal of this hashtag. One recent piece of research found that more African Americans engage with the #blacklivesmatter hashtag than the broader population (Olteanu, Weber, & Gatica-Perez, 2016). African Americans were also more active with the hashtag than other demographic groups. Given the focus of Black Lives Matter, it seems reasonable that although many individuals and groups thankfully show support, African Americans would be most interested in this movement that directly impacts its own people.

Some of their more societally useful characteristics of Twitter and social media are their allowances of multivocality and dialogicality (Bakhtin, 1981 in Bonilla & Rosa, 2015). Multiple voices and forms of dialogue help increase participation. This is particularly important for young members of the African American community who deeply embrace social media. For example, 96% of African American internet users aged 18- to 29-years-of-age use some form of social network (Day, 2015). On a comparable basis, 40% of this same demographic group use Twitter, while only 12% of young white people engage in the same type of online activities (Day, 2015). Social media can promote the articulation of many voices, increase dialogue, inspire individual identity exploration, and improve socialization among members of the same racial group.

One appeal of social media activism is its fluidity as a form of expression. Its participatory, open-ended, and digital nature allows users to express themselves in multiple ways. Social media's "endless framing and shaping of the immediate present, has been so crucial to the growth of Black Lives Matter. Expressions of love can coexist with expressions of outrage in the space" (Bailey & Leonard, 2015, p. 68). Youth can thereby express themselves in seemingly inconsistent ways as mentioned above. This is vitally important for an embattled community like African Americans that has been violently victimized. The social ties fostered by these new forms of online media allow members to express themselves with one another and heal when faced with a tragedy. This type of use of social media is a positive development that hopefully will continue.

Opposition to the Black Lives Matter Movement: All Lives Matter and Blue Lives Matter

Although there was strong social media and political support for Black Lives Matter, there was also opposition. Support or opposition to Black Lives Matter varies greatly depending on race and ethnicity (Marist College, 2015). Almost two-thirds of African Americans agreed with the motivational reasons informing Black Lives Matter. These supportive respondents believe it is a non-violent civil rights campaign, is more than just a simple slogan, and its work highlights issues of discrimination.

In contrast, only 42% of white respondents are uncertain of Black Lives Matter or have no opinion of it. Carla Shedd believes that the contrasting hashtag #alllivesmatter disrespects the dehumanization and vulnerability of Black people (May, 2016). Even President Obama weighed in on the controversy. He made the following statement.

I think the reason that the organizers used the phrase "Black [L]ives [M]atter" was not because they were suggesting nobody else's lives matter. What they were suggesting was, there is a specific problem that is happening in the African American community that's not happening in other communities. And that is a legitimate issue that we've got to address. (Townes, 2015)

Finally, in response to the shooting of two police officer in Ferguson, Missouri, the hashtag #bluelivesmatter trended on Twitter. The hashtag arose a few hours after this incident and spiked sharply on Twitter. Followers included supporters of gun rights, the police, and conservative causes (Wendling, 2015). This latter movement appears to be an attempt to graft on to the good will engendered by Black Lives Matter, potentially cheapening its predecessor movement and diluting its work. Collectively, these competing forces have led to a heightened focus on police training to minimize shooting (Lantigua-Williams, 2016; Nagin, 2016).

SOLUTIONS AND RECOMMENDATIONS

Connections Between Using Digital Devices and Developing Skills for Transitioning Youth Into the Workforce

Leveraging social media to foster youth civic action has been deeply valuable in fostering social progress like the Arab Spring, Occupy Wall Street, and Black Lives Matter. Additionally, using digital devices to participate in these types of movements can also help youth develop skills for their inevitable transition into the workforce. This is important given the challenging times that young people today face economically, especially their ability to work. As discussed earlier, the global youth unemployment rate was expected to reach 13.1% in 2016, which is extremely close to its historic peak of 13.2% in 2013 (International Labor Organization, 2016). From the extremely effective ways that students have been able to use their social media skills for activist goals, they could transfer these valuable communication skills to successfully transition into the workplace such as digital media production skills that will now be discussed.

Media Production Skills: Video, Photography, and Writing

Youth activists can learn to become media producers by interacting with social media (Deiser & Newton, 2013). Learning about the production of graphics and other visuals through social media activism improves the volume and impact of youth outreach. For example, articles that include imagery receive 94% more views than those without this type of media (DeMers, 2015). Further, tweets that include images garner 18% more clicks than posts without them (DeMers, 2015). Finally, Facebook posts that include photographs receive 120% more engagement than other types of posts (DeMers, 2015).

The dynamic nature of social media, its ubiquity, and audio-visual capture through mobile devices allow youth activists to engage in live events. For example, live engagement through Facebook Live, Periscope, and Snapchat has grown quickly (Lombard, 2016). As a result, real-time events like youth activist driven protests can be easily shared. Participants can also learn to harness this immediacy to quickly capture the intensity of an event and share it with captive audiences.

However, despite the power and importance of live engagement in reinforcing the on-the-ground reality of youth activism, high quality video acquisition as well as photography shooting and editing are also important skills (Lombard, 2016). This includes the ability to distinguish the quality of third-party produced media (Lombard, 2016). For example, managing large dynamic social movements potentially includes an array of media that may not necessarily positively contribute to the movement. As a result, curating this media to meet overall objectives is a valuable editorial skill. Learning the language of visual media production is also important for helping youth activists articulate and achieve their communicative vision to share their message in the broadest way possible (Lombard, 2016).

Although visuals are important, social media can also help youth activists improve their writing fluency (DeMers, 2015). For youth to effectively communicate their social media message, they need to have an effective ability to write (DeMers, 2015). The quickly changing nature of a dynamic protest necessitates the delivery of quick, clear, and powerful prose to accurately describe a movement.

Media Dissemination Skills: Sharing Information in an Impactful Way

Based on the media they produce, youth activists can also learn how to use social media platforms to distribute compelling digital content and make the most effective use of constantly changing modes of dissemination (Deiser & Newton, 2013). Youth activists can use social media to transcend the limits of traditional one-way mass media communication and to know more about the stakeholders they engage with through social media's multiple channels of communication (Daskal, 2016). Youth activists can gather honest feedback from social media participants that can lead to open dialogue about their protest tactics and motivations. This will allow young activists to learn skills in responding to various concerns about their work (Daskal, 2016). By communicating directly with their followers and media consumers who are curious about their social movement, youth activists can develop a genuine voice that resonates with others (Daskal, 2016).

Youth activists can also learn to manage communications overflows when information invariably increases over the trajectory of a movement and becomes difficult to handle (Deiser & Newton, 2013). For youth who engage at the leadership level of social media driven protest movements, they can advise and orchestrate the strategic utilization of social media (Deiser & Newton, 2013). By participating in activist groups, young people can also create an enabling organizational infrastructure (Deiser & Newton, 2013). These activists can learn ways to empower not only those people who are on the ground, but also to critically engage casual media consumers. They also further help to cultivate or offer a sense of belonging as well as social connectedness on a micro and macro level (Mentor, 2018).

Rhetorical Skills: Learning the Powers of Persuasion

Having a committed social media presence (Daskal, 2016) can help youth activists appeal to more stubborn observers who may not initially favor the endorsed social movement. These activists can use their channels to develop a long lasting community by learning deeply about the true interests of followers and observers (Daskal, 2016). Engaging with social media can correspondingly help develop the activist identity of youth. For example, a recent case study examined a New Jersey youth who was able to develop himself as a young activist through different types of social interaction including social media (Fullam, 2017).

Youth activists can even learn the basics of applied behavioral psychology through their social media interactions. For example, by examining data and analytics, young people can begin to understand which types of social media posts engage with different groups of media observers at various levels of effectiveness. Behavioral psychology also helps activists learn about people's attraction to posts, the motivations for sharing certain posts, etc. By doing this, youth can identify successful social media trends (Lua, 2017) and leverage the power of broader communicative movements.

By developing an appreciation for social media analytics, youth activists can also begin to learn search engine optimization (SEO) (DeMers, 2015). With this skill, young people can figure out how social media content, blogs, videos, and photographs impact search rankings (DeMers, 2015). As a result, activists can learn how to better shape their social media messaging for various purposes, which is a valuable skill for a variety of businesses.

Shifting Skills: The Evolving Need to Evolve

Due to its ephemeral nature, social media necessitates constant attention to how this field is evolving (Lombard, 2016). Youth have a tendency to adopt the latest and most talked about digital tools. Keeping current is important for youth activists as they attempt to fully understand online tools, social networks, and digital platforms as they constantly arise and disappear (Lombard, 2016). Continually learning to learn will help youth keep their skill set attuned to the needs of the marketplace and help ensure the best chance at employability.

FUTURE RESEARCH DIRECTIONS

My Brother's Keeper: A Federal Response to Assist Young Men of Color

There are practical initiatives that have been launched to help youth earn gainful entry points into the workforce. For example, President Obama signed a 2014 Presidential Memorandum to establish the My Brother's Keeper (MBK) Task Force (Balfanz, Bridgeland, Hornig Fox, DePaoli, Ingram, & Maushard, 2014; The White House, 2016). This was an initiative to address the painfully persistent opportunity gaps faced by boys and young men of color. In addition to multiple educational and restorative justice goals (e.g., improving reading levels, increasing graduation rates, reducing violence, and providing second chances to justice-involved youth, etc.), it also included training and workforce goals (The White House, 2016). For example, this includes the federal Department of Labor's (DOL) $20 million grant competition for 10 communities to provide innovative approaches for summer and year-round jobs (The White House, 2016). In addition, the DOL awarded almost $22.5 million to seven communities suffering from high unemployment rates. This initiative prepared youth for work in information technology, among other fields. As a result, it is clear that the federal government is aware of not only the systematic patterns of violence that people of color face, but also the disparity in work opportunities and science, technology, engineering, and mathematics (STEM) fields as highlighted by several of the social movements addressed in this chapter. Given the change in administration, this and other initiatives like Year Up (2016) need to be more deeply studied to monitor their efficacy in helping young people successfully transition into the workforce. This is especially important with regards to whether they can lead to wider structural changes in society that can help alleviate problematic factors that have inhibited upward mobility for everyone but the affluent (Harris, 2015).

Unfilled Opportunities: Youth Employment in the Developing World

Unfortunately, there has been a relatively jobless recovery from the 2008 global recession and financial crisis. For example, the youth unemployment rate was 12.6% in 2013, close to the rate that occurred during 2009 (International Labor Organization, 2013, p. 8). To emphasize the gravity of this issue, it has been estimated that approximately one billion additional youth will take their place in the job market (Accenture, 2016). To manage this increase in labor, around 600 million new jobs, the equivalent of five million per month, will need to be created (Accenture, 2016).

Despite these acute labor force demands, there are opportunities for youth to find employment. For example, in emerging countries, there is a marked lack of supply of employees with mid-level skills (Davis, Hansmeyer, Minic, Prakash, & Rangan, 2013). Along a similar vein, today's youth are more entrepreneurially inclined than previous generations, which could help them become more proactive in dealing with their relatively limited employment prospects.

It has been estimated that millennials across the globe are 1.6 times more likely than their counterparts from Generation X and Baby Boomers to be entrepreneurial (Moraru, 2015). As such, if these countries are able to adequately train their youth and promote their entrepreneurial instincts, there are job opportunities for them.

CONCLUSION

Youth activists have taken a leading role in protest movements around the world that has been amplified by the digital technology they constantly interact with and find so simple to use. Youth think of new ideas that contrast with their elders who shaped legacy institutions that benefitted their older interests. These contrasting generational cohorts will perpetually clash. The overtures by youth for revolution are clearly in the air as demonstrated by the broad-based protest movements discussed throughout this chapter. A jobless recovery post-2008 recession fuels dissent that disproportionately impacts the young. In addition to the poor underlying social situations that this protest emphasizes, it also demonstrates how we need to help today's youth make a successful transition into the labor force and channel their discontent for positive ends. The underlying competition for scarce jobs also requires a wider array of skills such as facility with social media in all of its forms. Youth voices will be heard. It is just the modes of communication, their frequency, and their corresponding intensity that are in dispute. The revolution may not be televised. It also may neither be tweeted, nor blogged about. But youth will play some part in civic engagement and dissent as our global society ideally moves forward.

REFERENCES

Abouzeid, R. (2011, January 21). Bouazizi: The man who set himself and Tunisia on fire. *Time*. Retrieved from http://content.time.com/time/magazine/article/0,9171,2044723,00.html

Accenture. (2016). *Contributing solutions to youth unemployment: Perspectives from a skills to succeed insight to action virtual event*. New York, NY: Author.

Adbusters. (2011). #OccupyWallStreet. *Adbusters blog*. Retrieved from https://www.adbusters.org/occupywallstreet/

Anonymous. (2011). #Anonymous #OccupyWallStreet call to action. *Anonymous blog*. Retrieved from http://anonops.blogspot.com/2011/09/anonymous-occupywallstreet-call-to.htm

Bakhtin, M. (1981). *The dialogic imagination: Four essays* (C. Emerson & M. Holquist, Trans.). Austin, TX: University of Texas Press.

Balfanz, R., Bridgeland, J. M., Hornig Fox, J., DePaoli, J. L., Ingram, E. S., & Maushard, M. (2014). *Building a grad nation: Progress and challenge in ending the high school dropout epidemic*. Baltimore, MD: Civic Enterprises, Everyone Graduates Center at the School of Education at Johns Hopkins University, America's Promise Alliance, Alliance for Excellent Education, AT&T, Target.

Barton, A. H. (1968). The Columbia crisis: Campus, Vietnam, and the ghetto. *Public Opinion Quarterly*, *32*(3), 333–351. doi:10.1086/267619

Bastos, M. T., Mercea, D., & Charpentier, A. (2015). Tents, tweets, and events: The interplay between ongoing protests and social media. *Journal of Communication*, *65*(2), 320–350. doi:10.1111/jcom.12145

Besser, H. (2012). Archiving media from the "Occupy" movement: Methods for archives trying to manage large amounts of user-generated audiovisual media. *Imatge I Recerca, 12es Jornades Antoni Varés 2012, Ponencies, Experiencies I Communicacions: Proceedings of conference in Girona*, pp. 106–110.

Bonilla, Y., & Rosa, J. (2015). #Ferguson: Digital protest, hashtag ethnography, and the racial politics of social media in the United States. *American Ethnologist*, *42*(1), 4–16. doi:10.1111/amet.12112

Bradley, S. (2003). "Gym Crow Must Go!" Black student activism at Columbia University, 1967–1968. *Journal of African American History*, *88*(2), 163–181. doi:10.2307/3559064

Brandzel, B. (2010, November 15). What Malcolm Gladwell missed about online organizing and creating big change. *The Nation*. Retrieved from https://www.thenation.com/article/what-malcolm-gladwell-missed-about-online-organizing-and-creating-big-change/

Byrne, J. (2012, April 30). 9 moments in the Occupy movement history. *The Huffington Post*. Retrieved from http://www.huffingtonpost.com/janet-byrne/occupy-movement_b_1464717.html?slideshow=true#gallery/223338/0

Calhoun, C. (2013). Occupy Wall Street in perspective. *The British Journal of Sociology*, *64*(1), 26–38. doi:10.1111/1468-4446.12002 PMID:23488697

Cobb, J. (2016, March 14). The matter of black lives. *The New Yorker*. Retrieved from http://www.newyorker.com/magazine/2016/03/14/where-is-black-lives-matter-headed

Conover, M. D., Ferrara, E., Menczer, F., & Flammini, A. (2013). The digital evolution of Occupy Wall Street. *PLoS One*, *8*(5), e64679. doi:10.1371/journal.pone.0064679 PMID:23734215

Cottle, S. (2011). Media and the Arab uprisings of 2011: Research notes. *Journalism*, *12*(5), 647–659. doi:10.1177/1464884911410017

Daskal, L. (2016, January 28). 6 social media skills every leader must learn. *Inc*. Retrieved from https://www.inc.com/lolly-daskal/6-social-media-skills-you-need-to-lead.html

Davis, N., Hansmeyer, E., Minic, B., Prakash, S., & Rangan, S. (2013). TEN youth: Unlocking enterprise growth by focusing on the fortune at the bottom of the talent pyramid. *Innovations, 8*(1/2), 141–157.

Day, E. (2015, July 19). #BlackLivesMatter: The birth of a new civil rights movement. *The Guardian*. Retrieved from https://www.theguardian.com/world/2015/jul/19/blacklivesmatter-birth-civil-rights-movement

Deiser, R., & Newton, S. (2013, February). Six social media skills every leader needs. *McKinsey Quarterly*. Retrieved from http://www.mckinsey.com/industries/high-tech/our-insights/six-social-media-skills-every-leader-needs

Demby, G. (2014, December 31). The birth of a new civil rights movement. *Politico*. Retrieved from http://www.politico.com/magazine/story/2014/12/ferguson-new-civil-rights-movement-113906

DeMers, J. (2015, June 8). 5 skills your social media manager must have. *Forbes*. Retrieved from https://www.forbes.com/sites/jaysondemers/2015/06/08/5-skills-your-social-media-manager-must-have/#3577b703ac56

DeVega, C. (2017). From "All Lives Matter" to the terror in Charlottesville: How the media's phony fairness got us here. *Salon*. Retrieved from http://www.salon.com/2017/08/14/from-all-lives-matter-to-the-terror-in-charlottesville-how-the-medias-phony-fairness-got-us-here/

Earle, E. (2012). *A brief history of Occupy Wall Street*. New York, NY: Rosa Luxemburg Stiftung.

Eligon, J. (2015, November 18). One slogan, many methods: Black Lives Matter enters politics. *The New York Times*. Retrieved from https://www.nytimes.com/2015/11/19/us/one-slogan-many-methods-black-lives-matter-enters-politics.html?_r=0

Epstein, I. (2015). No turning back, neoliberalism exposed: Youth protest in Chile, Spain and the U.S. In I. Epstein (Ed.), *The whole world is texting: Youth protest in the information age* (pp. 81–114). Rotterdam: Sense Publishers. doi:10.1007/978-94-6300-055-0_5

Fullam, J. (2017). Becoming a youth activist in the internet age: A case study on social media activism and identity development. *International Journal of Qualitative Studies in Education: QSE*, *30*(4), 406–422. doi:10.1080/09518398.2016.1250176

Garza, A. (2014, October 7). A herstory of the #BlackLivesMatter movement. *The Feminist Wire*. Retrieved from http://www.thefeministwire.com/2014/10/blacklivesmatter-2/

Gladwell, M. (2000). *The tipping point: How little things can make a big difference*. Boston, MA: Little, Brown and Company.

Gladwell, M. (2010). Small change: Why the revolution will not be tweeted. *The New Yorker*. Retrieved from http://www.newyorker.com/reporting/2010/10/04/101004fa_fact_gladwell?currentPage=all

Golkar, S. (2015). Student activism, social media, and authoritarian rule in Iran. In I. Epstein (Ed.), *The whole world is texting: Youth protest in the information age* (pp. 61–79). Rotterdam: Sense Publishers. doi:10.1007/978-94-6300-055-0_4

Granovetter, M. S. (1973). The strength of weak ties. *American Journal of Sociology*, *78*(6), 1360–1380. doi:10.1086/225469

Guynn, J. (2015, March 4). Meet the woman who coined #BlackLivesMatter. *USA Today*. Retrieved from https://www.usatoday.com/story/tech/2015/03/04/alicia-garza-black-lives-matter/24341593

Hariman, R., & Lucaites, J. L. (2001). Dissent and emotional management in a liberal-democratic society: The Kent State iconic photograph. *Rhetoric Society Quarterly*, *31*(3), 5–31. doi:10.1080/02773940109391204

Harris, F. C. (2015). *The challenges of My Brother's Keeper*. New York, NY: The Brookings Institute.

Ingram, M. (2011, February 3). Gladwell still missing the point about social media and activism. *GigaOm*. Retrieved from https://gigaom.com/2011/02/03/gladwell-still-missing-the-point-about-social-media-and-activism/

International Labor Organization. (2015). *Global employment trends for youth 2015: Scaling Up investments in decent jobs for youth*. Geneva: Author.

International Labor Organization. (2016). *World employment social outlook: Trends for youth 2016*. Geneva: Author.

International Labour Organization. (2013). *Global employment trends for youth 2013: A generation at risk*. Geneva: Author. Retrieved from http://www.ilo.org/wcmsp5/groups/public/---dgreports/--dcomm/documents/publication/wcms_212423.pdf,1.10.14

Khazraee, E., & Losey, J. (2016). Evolving repertoires: Digital media use in contentious politics. *Communication and the Public*, *1*(1), 39–55. doi:10.1177/2057047315625076

Kohut, A. (2011, October 27). 'Haves' and 'have-nots.' *The New York Times*. Retrieved from http://www.nytimes.com/roomfordebate/2011/10/18/the-psychology-of-/occupy-wallstreet/occupy-wall-streets-taps-into-longstanding-concerns?scp=1&sq=kohut&st=cse

Lantigua-Williams, J. (2016, July 13). How much can better training do to improve policing? *The Atlantic*. Retrieved from https://www.theatlantic.com/politics/archive/2016/07/police-training/490556/

Lazar, S. (2011, October 10). Occupy Wall Street: Interview with Micah White from Adbusters. *The Huffington Post*. Retrieved from http://www.huffingtonpost.com/shira-lazar/micah-white-adbusters-_b_996931.html

Lewis, J. M., & Hensley, T. R. (1998). The May 4 shootings at Kent State University: The search for historical accuracy. *The Ohio Council for the Social Studies Review, 34*(1), 9–21. Retrieved from http://www.stetson.edu/law/conferences/highered/archive/media/higher-ed-archives-2009/ii-lewis-may-4-shooting-at-kent-state-pdf.pdf

Lombard, J. (2016, August 10). Social media skills you need to hone now. *Media Bistro*. Retrieved from https://www.mediabistro.com/climb-the-ladder/social-media-skills/

Lotan, G., Graeff, E., Ananny, M., Gaffney, D., & Pearce, I., & boyd, d. (2011). The Arab Spring: The role of ICTs. The revolutions were tweeted: Information flows during the 2011 Tunisian and Egyptian revolutions. *International Journal of Communication*, *5*(2001), 1375–1405. Retrieved from http://ijoc.org/index.php/ijoc/article/view/1246/643

Lua, A. (2017, February 23). 10 important skills your social media manager needs in 2017. *Entrepreneur*. Retrieved from https://www.entrepreneur.com/article/289363

Luibrand, S. (2015, August 7). Black Lives Matter: How the events in Ferguson sparked a movement in America. *CBS News*. Retrieved from http://www.cbsnews.com/news/how-the-black-lives-matter-movement-changed-america-one-year-later/

Lynch, M. (2011, January 6). Obama's 'Arab Spring'? *Foreign Policy*. Retrieved from http://foreignpolicy.com/2011/01/06/obamas-arab-spring/

Marist College Institute for Public Opinion. (2015, September). Summary of National Survey findings. *PBS NewsHour/Marist Poll*. Retrieved from https://www.dropbox.com/s/j64c02ub48aipja/PBS-NEWSHOUR-MARIST-POLL-SEP2015.pdf

May, A. (2016, July 13). #AllLivesMatter hashtag is racist, critics say. *USA Today*. Retrieved from https://www.usatoday.com/story/news/nation-now/2016/07/13/why-saying-all-lives-matter-opposite-black-lives-matter/87025190/

McEvers, K. (2016a, March 7). 'The New Yorker' charts the rise of Black Lives Matter – with Jelani Cobb. *NPR* [Radio podcast]. Retrieved from http://www.npr.org/sections/codeswitch/2016/03/2016/03/07/469545405/the-new-yorker-charts-the-rise-of-black-lives-matter?live=1

McEvers, K. (2016b, March 9). Jelani Cobb on his epic 'New Yorker' piece on Black Lives Matter. *NPR* [Radio podcast]. Retrieved from http://www.npr.org/sections/codeswitch/2016/03/09/469627740/jelani-cobb-on-his-epic-new-yorker-piece-on-black-lives-matter

McLuhan, M. (1964). *Understanding media: The extensions of man*. New York, NY: McGraw Hill.

McNally, D. (2010). *Global slump: The economics and politics of crisis and resistance*. Oakland, CA: PM Press.

Mentor, D. (2015). Mobile phone use enhances social connectedness. In Z. Yan (Ed.), *Encyclopedia of mobile phone behavior* (pp. 673–693). Hershey, PA: IGI Global; doi:10.4018/978-1-4666-8239-9.ch056

Mentor, D. (2018). Micro to macro social connectedness through mobile phone engagement. In D. B. A. Mehdi Khosrow-Pour (Ed.), *Encyclopedia of information science and technology* (4th ed.; pp. 6184–6194). Hershey, PA: IGI Global. doi:10.4018/978-1-5225-2255-3.ch537

Milkman, R., Luce, S., & Lewis, P. (2013). *Changing the subject: A bottom-up account of Occupy Wall Street in New York City*. New York, NY: Murphy Institute of the City University of New York. Retrieved from https://media.sps.cuny.edu/filestore/1/5/7/1_a05051d2117901d/1571_92f562221b8041e.pdf

Mirani, L. (2010, October 2). Sorry, Malcolm Gladwell, the revolution may well be tweeted. *The Guardian*. Retrieved from https://www.theguardian.com/commentisfree/cifamerica/2010/oct/02/malcolm-gladwell-social-networking-kashmir

Mohyeldin, A. (2011, January 20). Suicide sparked Tunisia revolution. *Al Jazeera English, Opinion*. Retrieved from http://english.aljazeera.net/indepth/opinion/2010/12/20101231161958792947.html

Moore, K. (1999). Political protest and institutional change: The anti-Vietnam war movement and American science. In M. Giugni, D. McAdam, & C. Tilly (Eds.), *How social movements matter* (pp. 97–118). Minneapolis, MN: University of Minnesota Press.

Moraru, I. (2015, October 1). *Insight byte*. New York, NY: Accenture.

Morozov, E. (2009). Iran: Downside to the "Twitter revolution.". *Dissent*, *56*(4), 10–14. doi:10.1353/dss.0.0092

Morozov, E. (2010, February 20). The digital dictatorship. *The Wall Street Journal*. Retrieved from https://www.wsj.com/articles/SB10001424052748703983004575073911147404540

Muhammad, K. G. (2014, December 1). The revolution will be live-tweeted: Why #BlackLivesMatter is the new model for civil rights. *The Guardian*. Retrieved from https://www.theguardian.com/commentisfree/2014/dec/01/black-lives-matter-civil-rights-movement-ferguson

Nagin, D. (2016, October 16). Police, violence, and data: The BlackLivesMatter movement. *Heinz College News, Carnegie Mellon University*. Retrieved from http://www.heinz.cmu.edu/news/news-detail/index.aspx?nid=3860

O'Connor, G., & Schneider, A. (2017, April 3). *How Russian Twitter bots pumped out fake news during the 2016 election* [Radio podcast]. Retrieved from http://www.npr.org/sections/alltechconsidered/2017/04/03/522503844/how-russian-twitter-bots-pumped-out-fake-news-during-the-2016-election

Occupy Wall Street. (2011a). About Occupy Wall Street. *Occupy Wall Street website*. Retrieved from http://occupywallst.org/about/

Occupy Wall Street. (2011b). You can't evict an idea whose time has come. *Occupy Wall Street website*. Retrieved from http://occupywallst.org/article/you-cant-evict-idea-whose-time-has-come/

Olteanu, A., Weber, I., & Gatica-Perez, D. (2016, March). *Characterizing the demographics behind the #BlackLivesMatter movement.* AAAI Spring Symposia on Observational Studies Through Social Media and Other Human-Generated Content, Stanford, CA. Retrieved from http://arxiv.org/abs/1512.05671

OpenNet. (2010). Global internet filtering map. *OpenNet*. Retrieved from http://map.opennet.net/filtering- pol.html

Owen, T. (2017, August 12). Car slams into Black Lives Matter protesters. *Vice News*. Retrieved from https://news.vice.com/story/armed-white-supremacists-march-in-charlottesville-state-of-emergency-declared

Penney, J., & Dadas, C. (2013). (Re)tweeting in the service of protest: Digital composition and circulation in the Occupy Wall Street movement. *New Media & Society,* 1–17. doi:10.1177/1461444813479593

Porfilio, B. J., Roychoudhury, D., & Gardner, L. (2013). Ending the "war against youth": Social media and hip-hop culture as sites of resistance, transformation and (re)conceptualization. *The Journal for Critical Education Policy Studies*, *11*(4), 85–105. Retrieved from https://www.researchgate.net/publication/306107562

Ross, J. (2015, August 19). How Black Lives Matter moved from a hashtag to a real political force. *The Washington Post*. Retrieved from https://www.washingtonpost.com/news/the-fix/wp/2015/08/19/how-black-lives-matter-moved-from-a-hashtag-to-a-real-political-force/?utm_term=.26d7d0519fdc

Ryan, Y. (2011, January 20). The tragic life of a street vendor. *Al Jazeera*. Retrieved from http://www.aljazeera.com/indepth/features/2011/01/201111684242518839.html

Sadiki, L. (2010, December 27). Tunisia: The battle of Sidi Bouzid. *Al Jazeera English, Opinion*. Retrieved from http://english.aljazeera.net/indepth/opinion/2010/12/20101227142811755739.html

Schechter, D. (2011). Occupy: Dissecting Occupy Wall Street. New York, NY: NewsDissector.org & Coldtype.net.

Schneider, N. (2011, October 12). From Occupy Wall Street to occupy everywhere. *The Nation*. Retrieved from https://www.thenation.com/article/occupy-wall-street-occupy-everywhere/

Siddiqui, S. (2016, August 9). Russell Simmons talks Black Lives Matter. *Politics for humans podcast. Guardian U.S.* [Radio podcast]. Retrieved from https://www.theguardian.com/politics/audio/2016/aug/09/russell-simmons-black-lives-matter-clinton-trump-politics-for-humans

Sinclair, H. (2017, August 15). Black Lives Matter calls for U.S. to ban Confederate symbols after Charlottesville violence. *Newsweek*. Retrieved from http://www.newsweek.com/black-lives-matter-ban-confederate-symbols-charlottesville-violence-651106

Stiglitz, J. E. (2011, May). Of the 1%, by the 1%, for the 1%. *Vanity Fair*. Retrieved from http://www.vanityfair.com/news/2011/05/top-one-percent-201105

Sunstein, C. R. (2017). *Republic: Divided democracy in the age of social media*. Princeton, NJ: Princeton University Press. doi:10.1515/9781400884711

The White House. (2016). *My Brother's Keeper 2016 progress report. Two years of expanding opportunity and creating pathways to success*. Washington, DC: Author. Retrieved from wh.gov/mbk

Townes, C. (2015, October 22). Obama explains the problem with "All Lives Matter." *ThinkProgress*. Retrieved from https://thinkprogress.org/obama-explains-the-problem-with-all-lives-matter-780912d54888

Tufekci, Z. (2010, September 27). What Gladwell gets wrong: The real problem is scale mismatch (Plus, weak and strong ties are complementary and supportive). *Technosociology*. Retrieved from http://technosociology.org/?p=178

United Nations World Youth Report. (2016). *The world youth report on youth civic engagement*. New York, NY: United Nations Department of Economic and Social Affairs.

Vélez, V., Perez Huber, L., Benavides Lopez, C., de la Luz, A., & Solórzano, D. G. (2008). Battling for human rights and social justice: A Latina/o critical race media analysis of Latina/o student youth activism in the wake of 2006 anti-immigrant sentiment. *Social Justice (San Francisco, Calif.)*, *35*(1), 7–27. Retrieved from http://www.jstor.org/stable/29768477

Watson, K. (2017, March 30). Russian bots still interfering in U.S. politics after election, says expert witness. *CBS News*. Retrieved from https://www.cbsnews.com/news/russian-bots-still-interfering-in-u-s-politics-after-election-expert/

Wendling, M. (2015, March 12). "Blue Lives Matter" trends after officers shot. *BBC Trending*. Retrieved from http://www.bbc.com/news/blogs-trending-31853299

Williams, L. C. (2017, May 15). Where are they now? The Russian bots that disrupted the 2016 election. *ThinkProgress* [Radio podcast]. Retrieved from https://thinkprogress.org/russian-bots-where-are-they-now-e2674c19017b/

Year, Up. (2016). *2016 annual report: Betting big on year up*. Boston, MA: Author.

Zuckerman, E. (2011, January 15). The first Twitter revolution? *Foreign Policy*. Retrieved from http://foreignpolicy.com/2011/01/15/the-first-twitter-revolution-2/

KEY TERMS AND DEFINITIONS

Arab Spring: A broad movement of North African and Middle Eastern countries against anti-democratic, authoritarian, and repressive rule.

Black Lives Matter: A social movement against police violence, largely targeting members of the African American community.

Occupy Wall Street: A movement against the power of financial and multinational institutions on the global economy, economic inequality, and democratic institutions.

Protest: A gathering in opposition to an undesirable state of social, political, or economic affairs.

Social Movements: Broad-based participation of citizens in support of, or against, sociopolitical issue(s).

Workforce Participation: The ability of an individual or a group to engage in paid labor.

Youth Activism: The participation of young people in sociopolitical movements.

Youth Unemployment: The phenomenon of young people not having full labor participation in the workforce.

Chapter 5
TLC for MOOCS:
Teaching and Learning Communities for Computer Programming

Dominic Mentor
Teachers College, Columbia University, USA

ABSTRACT

This chapter reports on efforts to create a sustainable model to increase engagement, success, and retention in a MOOC for learning computer programming, for a U.S.-based national vocational program. In 2014, the training organization was one a few national and regional organizations who were awarded scholarships by a telecommunication's company to participate in a MOOC whose curriculum was informed and designed by multinational corporations to try and address the dearth of young computer programming talent. The vocational training program aimed to convert MOOC registrations into active and supported participation, with a view to increase completion. Theoretical frameworks were employed to ramp up knowledge of an unknown subject area and skill. Social connectedness methods were used to create teaching and learning communities (TLC) of support. Key results allowed the organization's trainees to outperform all other participating organizations. Resulting in the organization being awarded 500 more scholarships for computer programming that could be used over a three-year period.

INTRODUCTION

Massive Open Online Courses (MOOCs) offer digitized courses for the masses. MOOCs are still highly popular as indicated by the number of people who register for them. However, less than 5% to 7% of people who register, complete the MOOC course (Dillahunt, Wang, & Teasley,2014; Evans, & Baker, 2016; Rai, & Chunrao, 2016). Using Gartner's hype cycle of emerging technologies (LeHong, Fenn, & Leeb-du Toit, 2013), the verdict seems to still be out on whether it has hit its peak, or levelled out at the trough of disillusionment. One technology training organization was invited amongst 20 others to participate in a sponsored participation of MOOCs on computer scripting and programing. The invitation stemmed from various technology companies and the Obama administration to address the dearth in computer programming talent. Each of the participating organizations worked on their own as they

DOI: 10.4018/978-1-5225-4111-0.ch005

approached the opportunity and challenges of having their young talent participate in a forum that sees massive attrition.

There are many challenges with participating in MOOCs. Separately from the attrition and maintaining interest, other challenges within this specific MOOC endeavor ranged from an intensely high and rapid learning curve, untrained trainers, lack of, or rather less than desirable insight of learning and progress transparency. There was a need to address the challenges as well as the other general MOOC stumbling blocks, which ranged from significant drops in engagement, inability to retain the initial number of participants who registered, as well as a lack of drive to complete the course. These difficulties were further compounded by the physical distance between trainees and their on-the ground-support staff (Atlanta), and the project managers and tutors (New York) guiding and supporting the effort.

To address and to overcome these challenges; the rapid and intense learning curve, untrained or lack of Subject Matter Expertise (SME), and lack of learning transparency, some aspects of social connectedness were employed. Other theories that informed the practical pedagogy were cognitive constructivism in the form of building on prior knowledge and utilizing zones of proximal development methods with collaborative learning engagements. Additionally, a schedule was set up for in-person contact time as well as contact time with the virtual tutors and project managers. The virtual meetings were between the virtual tutors and the on-the-ground instructor, and separately with the students and trainer. The methodologies used, and practices implemented were informed by computer mediated collaborative learning (Beetham, & Sharpe, 2013; Majchrzak, Rice, King, Malhotra, & Ba; 2014; Walther, Hoter, Ganayem, & Shonfeld, 2015), otherwise also known as computer supported collaborative learning theories (Andriessen, Baker, & Suthers, 2013; Fischer, Kollar, Stegmann, & Wecker, 2013; Goodyear, Jones, & Thompson, 2014).

BACKGROUND

This chapter reports on methodologies used to cultivate social connectedness for a learning community participating in a Massively Open Online Course (MOOC). The participating urban young adults were part of a workforce training program that were trying to expand its curriculum offerings to target the labor market need of computer programming. The trainees had little, to no computer programming knowledge or skill. Data from a preliminary survey indicated that only a few of the 21 students had some HTML exposure, but have not necessarily fully built out a website with HTML and CSS, not any JavaScript or Java exposure. This first phase of 21 students were part of an initial pilot to determine the viability of a concept product and process. If successful, a larger scale second phase of hundreds of students would be considered to participate in similar MOOC models to learn programming skills.

The modus operandi of the national training organization, was to offer a six-month, intensive training program, followed by a six-month internship. The Learning and Development (L&D) organization provides urban young adults, ages 18-24, with a combination of technical and professional skills, college credits, an educational stipend, and an internship. At the time of the TLC MOOC effort, the organization operated in 14 cities, with 8 of those cities having their own instructors. The L&D organization was invited to participate in the MOOC with the backdrop of trying to provide a workforce with skills to hit current and future labor market programming needs.

Further background information is that the national training organization had been in existence for about 12 years, and only piloted blended e-learning models in 2012, with a 100% soft roll-out in 2013 to eight of its eight training sites (Mentor, 2016). Blended learning is the thoughtful integration of classroom face-to-face learning experiences with online learning experiences; it integrates the strengths of synchronous (face-to-face) and asynchronous (text-based Internet) learning activities (Garrison & Kanuka, 2004). After introducing a blended e-&-m-Learning approach as part of its academic vision and strategy (Mentor, 2016; 2017), the national training organization could execute a broad and nimble curriculum alignment and deployment. As well as starting to explore and expand its e-Learning initiatives. Through simulations, the organization was also able to incorporate social collaboration, m-Learning, and interoperable learning tools that cultivated and accommodated differentiated learning styles. Ultimately growing an ecology of collaboration, shared material and planting seeds for further growth and innovation needed by the training organization. Further setting itself up to introduce and utilize distance learning for workforce training.

In 2014, the entry level national training organization (NTO) piloted participation in a MOOC prompted by scholarship offerings from major telecommunications, and technology companies, as well as long-standing MOOC service provider. The telecommunications and technology company worked with the MOOC service provider to design the curricula content that they identified as needs from labor market data.

The technology companies involved who were funding the effort hailed the endeavor as preparing the workforce of the future (Smith, 2013). These companies stated that they were "moving to a software-defined, mobile-first environment" (ibid.) and the need for employable talent in this regard was considered "more critical than ever" (ibid). Hence the need to develop a pipeline of employees with the skills to succeed, be retained and to aid the technology companies in their desire to have skilled talent to address the current dearth in programmers as well as the anticipated labor market needs. This view was part of an even broader scope and need, and that was to "prepare students, as well as those already in the workforce — no matter where they work — for success in 21st century careers" (ibid).

The labor market need served as the push for the technology companies to partner with one another to seek out a MOOC service, to write, review or contract the construction of computer programming curricula. The MOOC service was hosted by MOOC service provider who, in conjunction with the technology companies, helped launched what they termed a Nanodegree credential for the basic skills needed for the Software Economy (Smith, 2013). It must be noted that at the launch of, and at the time of writing this chapter, the Nanodegree was not a recognized credential by higher education (Guzdial, 2014; Lazaroiu, Popescu, & Nica, 2016; Popenici, 2015), but was envisioned to become a key part of the telecommunication and technology companies' training and talent acquisition model.

Theoretical Framework

The national workforce development organization was invited amongst other regional and national nonprofit after school and training organizations to participate in a funded MOOC on computer science to grow entry level computer scripting and programming talent. The theoretical concepts of self-directed learning (Candy, 2004), cognitive constructivism, social connectedness and a growth mindset approach was used to head off potential high attrition of MOOCs as well as to foster higher engagement, learning success and completion. Social connectedness speaks to the sense of affiliation that the students had through on-the-ground and virtual remote support, whereas the growth mindset challenged trainees about concepts of natural talent and the ability to learn (Hsu, Baldwin, & Ching, 2017; Loksa, Ko, Jernigan, Oleson, Mendez, & Burnett, 2016; Mercer, & Ryan, 2010).

To overcome the high and rapid learning curve for these students who had never participated in any computer programming training, had minimal online learning, or participated in a MOOC for that matter, a learning ramp was created to cultivate a supportive zone of proximal development. The learning ramp consisted of pre-work that was specifically created from pre-information supplied by the technology and MOOC companies. The pre-work consisted of virtual and on-the-ground support structures. Why was this attractive? The baseline was that the national academic training department knew this could be a scalable solution to develop highly marketable skills in a way that reduces our cost and investment in curriculum development with a clear connection to internships and jobs. The ramp-up was attractive because it included informal learning and professional development of the on-the-ground facilitators. The ramp-up period could help interested students ascertain whether they have an interest, affinity or proclivity for learning and working with computer scripting and programming languages. The ramp-up period was set up to act as an empowerment training force, to establish with students and teachers that they could reach out to the provided virtual support at any time, without fear of mockery or signals of their reaching out was a bootless errand.

The MOOC service provider indicated they would provide two virtual Coaches that will work with students during scheduled hours or by appointment. Coaches could be contacted by email or web conference, and they may possibly provide support by hosting class forums or webinars for larger groups. It was envisioned that the coaches would help with the following aspects:

- Help students set the pace at which they want to work (e.g., set a date by which they want to finish the required courses completely) and help students stick to that pace by setting goals and benchmarks with them.
- Answer students' questions about the content or subject matter.
- Provide feedback on students' final projects as they build out their portfolio.

MAIN FOCUS OF THE CHAPTER: ISSUES, CONTROVERSIES, PROBLEMS

There were many challenges that ranged from an intensely high and rapid learning curve, untrained trainers, lack of learning versus progress transparency within the MOOC system, as well as general foreseen MOOC challenges that needed to be combatted like significant drops in engagement, high attrition, and lack of completion. To address and to overcome these challenges; ranging from the rapid and intense learning curve, untrained trainers, and lack of learning transparency, aspects of social connectedness were employed through virtual digital means, for both virtual trainers, students and on-the- ground training staff. The methodologies and practices implemented aided the trainees who had never done any computer scripting or programming, as well as the on-the-ground trainer, who had limited experience in this regard, but had a huge passion, and dedication to seeing the trainees succeed.

The proposed plan was to mitigate risks and over exposure of slim resources was to host the MOOC as an elective, where a local branch could opt-in with full national support. The latter would offer valuable lessons, and maximize value without creating a burden for national or local academic teams.

There were other broader challenges from an organizational point of view. The organization had a focus on hardware and desktop support and hired trainers in this regard. On a larger scale, outside of the training department, there were no subject matter expertise, nor knowledge of what computer programming or training entails. Some misunderstandings that needed to be overcome was that Information and

Communication Technologies (ICT) was a massively wide field with specialized subject matter expertise. There was the assumption that just because a trainer taught a course on hardware, it meant that they could teach computer coding. This lack of basic and deep knowledge had some narrow interpretations in that the learning and training of computer science elements could be put down to either critical thinking, music ability, or mathematical ability. The understandable inability to talk with a proper comprehension of computer programming, other than repeating or reciting some terminology that was heard, bled into the inability to articulate and grow these new business opportunities at the rapid pace at which it was needed. This then fed in to the inability to foresee or understand the needs or challenges shared by subject matter experts regarding required training within this specific context. Initially, due to this lack of knowledge within affiliated departments, it was also determined by other departmental stakeholders, that there was not enough information here to determine if and how this curriculum may form the basis of a new career track or multiple tracks for the organization to offer the trainees, or for current and potential corporate partners. There was a need to build the understanding of the entire organization, and each department, on how much we needed to know and how much we could deliver in the short, intensive training period, as well as a need to know more about the skillsets required and career progression associated with possible full-time jobs in this regard. The latter information could aid the firming up of the number of corporate partner needs, over which time period and for how long, as well as the number of trainees needed per location. Furthermore, deeper knowledge and less of a superficial understanding of computer programming would also have helped identify the expansive scope of intern and internship commitments. From the many challenges and perceived limitations, other channels of the organization were also considered. Both of which consisted of alumni of the training program.

Notwithstanding the misperception that the consideration of the alumni of the program to make use of this opportunity was further predicated on the misunderstanding or rather limited perception that all ITC skills, be it hardware in this case, was easily and highly transferable to the training and learning of computer coding languages. But, the alumni phase of the program was deemed by the National Academic Department to not be ready for this MOOC opportunity for a number of reasons;

1. There was no proper infrastructure or processes in place yet to support the training and learning needs, and
2. It was deemed that there was no subject matter expertise in these channels to execute on this opportunity that were initiated by technology companies.
3. Moreover, the participant, program and organizational desired outcomes needed to be met.

Those organizational outcomes were paid internships, job placements from the main sponsoring ICT company, as well as other corporate partners. However, given the monitoring of labor market data and subject matter expertise by the national academic learning and development department, as well as an alignment with their vision and strategy of blended, mobile and eLearning, a rationale could be developed to proceed after an evaluation of the MOOC service provider and the course content.

MOOC Service Provider and Course Assessment

Comparing various MOOC vendors according to a criteria taxonomy helped decide between multiple requests from MOOC vendors to use their platform. Criteria used as part of the decision-making process ranged from user interface, user experience (UI/UX), pedagogical and technological criteria (Yousef,

Chatti, Schroeder, & Wosnitza, 2014) as well as evaluation from a Service Oriented Software (SOA) approach (Meinel, Totschnig, & Willems, (2013). In context, The MOOC service provider's curriculum was also evaluated to assess whether it met National Academic Training Departments' instructional design standards, and whether it matched our short and long-term goals in terms of online and blended learning.

Moreover, our assessment involved looking at whether the MOOC service provider and the course curriculum measured up to the National Training Organization's standards and needs. The hope was that we would be able to learn from the content and structural placements of the industry informed MOOC courses and to evaluate our own courses against industry asks or guide our current in-development courses accordingly. We chose to evaluate the MOOC service provider's framework as a MOOC for blended and flipped learning by looking at the following aspects:

- Quality of course content
- Delivery and structure of online course materials
- Support from remote coaches provided by the MOOC service provider
- Formative and Summative Assessments

There were some who considered the MOOC provider as a market-leader, and established further value from that notion to agree to, and promote the use of resources to participate in the MOOC as a pilot project. Yet, further evaluation was needed of the en masse hypermedia for learning technology. A MOOC, while similar to other technology enhanced systems, the massive numbers add a different dimension and demands other considerations to ensure the learning is effective and goals are met. Hence, an evaluation of a MOOC requires additional sets of criteria (Yousef, Chatti, Schroeder, & Wosnitza, 2014).

Evaluation of Students' Work

During participation with the MOOC, students would be evaluated based on project and portfolio work that they complete. The MOOC courses hosted project completion benchmarks. Some of the projects in the portfolios would include developing blogging and wiki applications, building websites, interactive resumes, single page applications and games. The projects in the student portfolio would be informed by the learning objectives of the course.

Students will develop particular skills in Front End Development ranging from:

- Designing websites using the latest web standards.
- Testing user design and user experience to transform text into compelling web applications.
- Developing dynamic web applications using JavaScript.
- Using version control software to work effectively with other developers.

As well as in Back End Development with foci on:

- Developing technical projects that will connect the systems that power modern websites.
- Learning the necessary languages to communicate between clients and servers.
- Storing, manipulating, retrieving, and presenting data from a database.
- Developing APIs with effective patterns and techniques.

It was estimated by the industry and MOOC course developers, that each portfolio project would require around 40 hours of work to complete. As a new training opportunity with our urban young adult, we needed to monitor the MOOC assessment system to see how it measures up in terms of fidelity and pedagogy, and add our own checks and balances of support, assessment and feedback to students. The National Academic Team and the local site, along with the TAs, will gather information about the course engagement and the progress of students. The data will be used to evaluate whether the MOOC course fits the short and long-term goals of setting our trainees up for success. The short-term goals included providing students with skills on how to learn online and find other suitable online training in the computer programming field, so that they can acquire the necessary skills to find jobs in related fields.

The on-the-ground instructor could check in with students on a Monday to see what issues they have been dealing with or the concepts that are difficult for them to understand, these were then addressed during a Wednesday brown bag lunch meeting. From what the virtual staff also received or supported, the on-the-ground trainer could then address the concepts or specific problems in greater detail during the in-person Wednesday session. In addition, she could identify students who have been understanding material at a greater pace as peer tutor leads. She could ask such students to demonstrate what they have learned to the rest of the class or to those that are struggling with grasping the material.

Two virtual TAs were provided from the academic fellows' program as support from the National Academic team. The two graduate student TAs, Kristin and Jeanny, served as an invaluable resource for the students and worked with Luci the local site instructor, as well as the potential coaches that would be appointed by the MOOC service provider. The virtual TA's were available to further clarify course content to the students. The TAs was available via email and Skype during scheduled hours, or by appointment. Responsibilities of the supporting fellows included:

- Observe the progress of students during the course by looking over the submitted assignments and projects.
- Assist in clarifying to the students any of the course content available on the MOOC.
- Collaborate with MOOC Coaches and local site instructor to better understand course content.
- Help students with any technical issues that they have in accessing course content.
- Assess the MOOC's curriculum, as a blended and flipped learning environment, to see whether it meets the short term and long-term goals of the National Training Organization.
- Evaluate ways in which MOOC service provider's curriculum can be implemented in future courses on Schoology.

In addition, social connectedness was employed as part of the theoretical framework to offer trainees a sense of belonging and affiliation. Other theories that informed the practical pedagogy were self-directed learning (Candy, 2004) and cognitive constructivism in the form of building on prior knowledge and utilizing zones of proximal development methods with collaborative learning engagements. Often students come up with more effective solutions. The identification and use of students who worked at a faster pace within the self-directed learning framework and who understood the course work, were appointed as leaders within a zone of proximal approach. The aforementioned were combined with social anthropological elements to further establish a community of practice (Carter, & Adkins, 2017; Lave & Wenger, 1998; Wenger, 2010). The same theoretical framework was used to establish a teaching and learning community amongst the instructional staff across the national landscape through webinars. The framework was used to create instructional quality review surveys () and to identify instructional talent and to disseminate that talent.

To establish what cognitive resources and experiential knowledge we have within our trainer pool, the national academic department had been building a skills matrix based on the qualifications, knowledge and expertise of their current instructional staff. It did so as part of its academic vision and strategy, future anticipated growth, and supplied a visual index of trainer talents. In this way, we were able to identify and leverage our instructors' strengths - their experiential knowledge, passion and dedication, as well as their transformational power. From those evaluative frameworks, the national academic department and its local counterparts deemed it possible to pursue a short-term pilot program. Keeping in mind survey results to use as part of its practical application of theoretical frameworks of cultivating social connectedness combined with identifying students' zone of proximal development.

Rationale for Proceeding

An alignment check of organizational goals and mission was done by the national academic team and shared with the other departments, the CEO, President and other senior leaders. Even though the MOOC program measured up to the training organizations goals and business objectives, it was reflective of previously monitored and shared labor market data and needs. However, while the MOOC standards and quality were high and informed, by the technology companies, it did not offer a ramp up to the learning. For example, it did have a scaffolded construction, but it did not necessarily truly match the prior knowledge or scaffolding of the trainees within this particular organization. Nor did it necessarily reflect the scaffolded approach from a delivery, support, or assessments perspectives. There were no formative assessments built into the program and summative assessment of work produced was somewhat non-existent. Completion of tasks and projects was the only measuring tool within the MOOC's platform and was recorded by a simple upload. Personal development was reflected by a progress bar of completion on projects, but not by the quality or correctness of the work. However, there were still many transferrable skills present in the MOOC program for students, and aside from extended internship sales opportunities and integrated ICT knowledge construction, the MOOC content offered valuable skills that could be transferred to job and life skills within the 21st Century economy. Our rationale needed the inclusion of surveying our departmental and organizational strengths, weaknesses, objectives and targets in this regard.

Figure 1. Students' responses to prior knowledge self-assessment

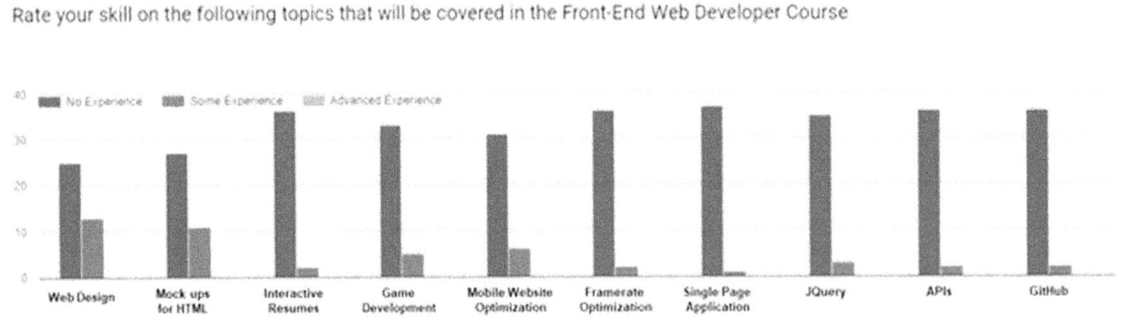

Some believed the primary value of participating in the MOOC was the potential of a trainee earning a nano-sized portion of a degree. Nevertheless, the courses were not accredited in any way, but college were promised if pursued by the student or organization itself. Some of the value in participating in the MOOC was having access to the ICT course content and program itself, as that was commissioned and informed by national telecommunication and technology ICT entities. The MOOC participation created the opportunity to plant further seeds for our academic vision and strategy; first, to weigh our own course structures against industry asks, and second to affirm and develop our potential to close the gap of computer programming labor needs. The participation in the MOOC brought potential upsides for the work and mission of the organization to foster further partnerships and business opportunities based on the subject matter covered. In this instance, the organization had an opportunity in making a further impact in society by training disadvantaged and underrepresented populations, including women, for the computer programming industry. The opportunity had high potential for internship opportunities, and converting those internships into employment for the entry level trainees with the national ICT corporation and many other employers.

A MOOC also presented the idea that the talent organization could deliver the corporate or ICT business partners' required skills. For national telecommunication and technology companies, the training organization could do so to meet these companies' entry level demands across the country in a standardized quality manner. Companies were often looking for the national training organization to prepare students with an identical skill set for similar internship roles across different cities. The claim was that the most effective way to deliver this consistency is with students who have studied in the same class together. Instead of working on constant alignment across sites, a MOOC also offered opportunities to deliver courses across training sites and deliver the same content with high-quality across dozens of cities. This approach could enable the talent development organization to scale up its partnerships with a few of its other national partners, and presented other high value targets to pursue in line with its business objectives.

There were four other values taken into consideration as part of the rationale for proceeding. The MOOC opportunity presented:

1. Possible quality "elective" content for our young trainees, even if it was not currently tied directly to our contractual internship sales or possibilities of converting those internships into full-time hiring opportunities within our current employer partners.
2. Access to this content could serve as well in the future when we eventually we have clients asking us to deliver entry level talent with these required skills.
3. The content in the MOOC program could serve as a foundation for several new computer programming developments as a pipeline of talent for an industry that needs diverse talent.
4. Opportunities to measure, expand, and hasten the testing of our e-learning, blended instruction, and flipped classroom training models.

The initial piloting of the program involved 20 new trainees from one training site. We latched onto the Front-End Web Development portion of the micro course program. In the Fall of 2014, the local academic team of a site and the National Academic team partnered to pilot the participation within one segment of the MOOC. To prepare the students for this new method of engagement, we put together an introductory course that would help acquaint students with the new material, as well as the type of learning material and assignments they would get. The preparatory course was formulated from prior

Figure 2. Specific self-ratings on content covered during the pre-launch ramp up course

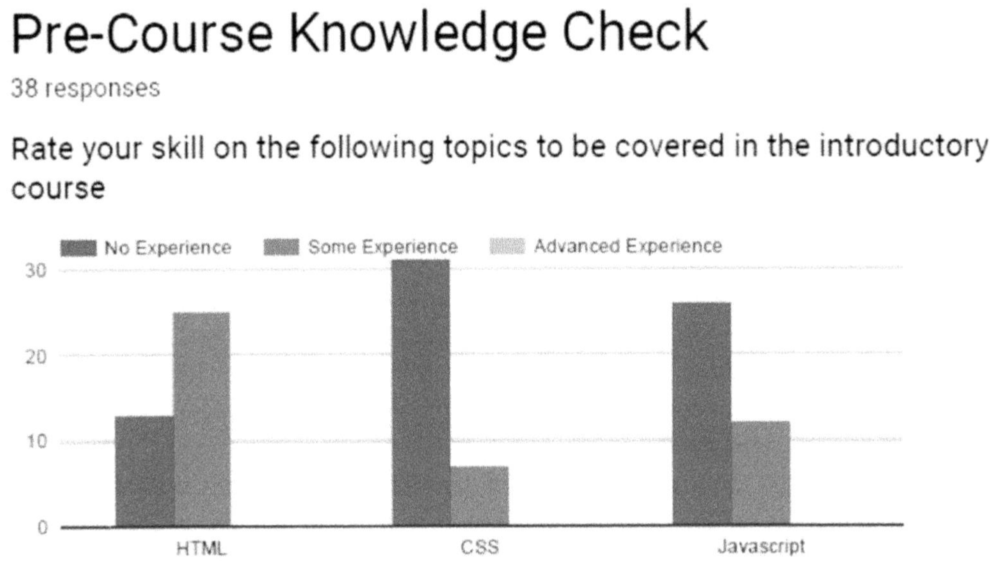

knowledge surveys to help cultivate a social connectedness between the virtual tutors, students and the on-the-ground instructors.

Preparatory Launch Ramp

The national academic learning and development department of the program decided to construct a preparatory course. The course was to be considered a ramp-up platform for trainees, the virtual teaching assistants as well as for the in-person instructor. The preparatory course allowed us to ask what do students, virtual assistants, the instructor and the local site need during the ramp up stage? We correspondingly asked what would serve as lessons, and, what to cater for during the MOOC participation itself?

Participation in the MOOC courses were asynchronous with limited virtual support from the MOOC service provider. Trainees could begin and proceed on the course work and sections whenever they wished and they could finish projects at their own pace. However, for the in-person support and to help prepare the trainees, and our own virtual teaching assistants, as well as the instructor for engaging with the MOOC content, we built in virtual and in-person scheduled times. During which they will meet with the students to answer any questions that the students have about the course content on MOOC service provider. The instructor scheduled once a week, hour long slots for students to gather in-person, to check in on any challenges and work progress. After a video conference call kick-off meeting with the virtual assistants, an LMS as engagement platform for the preparatory course and its communication spaces were highlighted in addition to email options.

We anticipated that there may be some students who will progress further ahead than the rest, could either lean on the virtual teaching assistants for extra support and/or be used as guides and peer tutors to others at their site participating in their MOOC. To accommodate all students and find a middle ground for the instructor to teach, she could suggest a pace at which she will instruct on components of the course, or support the students with their individual progress. For example, she could say that she will answer questions about HTML and CSS during the first two weeks and then move onto answering questions about JavaScript on the third week. This could allow the trainees to progress in a more structured manner even though students might be at different points in the course. Students could then still be encouraged to go ahead with the course at a faster pace if they have the time and desire to do so, while getting feedback on the work that they completed.

We envisioned that to foster a more balanced learning environment between the MOOC and face-to-face class time, the virtual teaching assistant could offer virtual "check ins." Accompanied by the weekly check-ins by the on-the-ground instructor. We asked ourselves, what type of preparation will the in-person instructor (Luci) and the virtual tutors need to instruct or support in this elective? In terms of preparation, both Luci and the two TAs will need access to the material to become familiar with the MOOC course content to act as subject matter supporters. Furthermore, if the coaches at the MOOC service provider has a long delay in responding or are unable to help the students with any specific problem or issue, Luci and the TAs should be able to guide the students in a way that will help them arrive at a solution. Luci and the TAs do not necessarily have to solve the coding problem for the students, instead they should act as a support by providing them with tools or options to explore in order to succeed. Also, the TAs should be able to reach out to the MOOC service provider's coaches and ask the students' questions differently or pick up on global student class issues that could be addressed more holistically by the instructor, the virtual TAs or the MOOC coaches.

From previous experience, the national academic department (NAT) deemed that it would be easier if Luci sets a pace for the overall class, even though students will be going at their own pace. This was based on the fact that the content will be covered mostly in the MOOC provider's online classes. Leveraging Luci's previous lunch room coding club set up, her 'in-person sessions would be structured like workshop sessions where she is able to support the students interactively and listening as a whole group so each one can teach one. Whenever possible, Luci would provide support to students during class, but if she needed support from the virtual TAs, she could call on them before hand or during the in-person session. If students have more intricate issues with coding, she can collaborate with the virtual TAs and possibly the MOOC provider's coaches, if they are available or were to respond. Moreover, student leads could be appointed as the instructor sees fit to assist with class facilitation.

Separately from the MOOC content and service provider evaluation, we needed to establish criteria on how to assess the success of the preparatory course. We needed to use that as a further check to see if our trainees would be able to succeed within the MOOC. Hence, we established that the preparatory course will be assessed on two levels. The students will be assessed on the content of the projects, their progress, and the successful completion of their short course portfolios.

Operationalized

We established what the relationship between the virtual TAs, on-the ground instructor and MOOC service provider partners would be and what the responsibilities of each would entail. The virtual TA's would lead the preparatory short course with the on-the-ground instructor in support. While the on-the-ground instructor would lead the MOOC effort thereafter with the virtual TAs supporting the trainees with their online course work.

The preparatory course was launched by the national academic team with support from Luci on the ground. A video conference call was arranged between the trainees, the virtual tutors and Luci where the short course was introduced via an LMS that had social media characteristics. The short preparatory course was designed to introduce and mimic the MOOC material. Technology needs and lessons learned during the introductory course were addressed before the trainees commenced their participation within the MOOC.

The MOOC participation project was launched through the remote MOOC service provider and our own on the ground instructor, and the two TAs. There was a total of five team members who supported the students, initially including the two course designers, but eventually only one other appointed remote supporter from the MOOC provider. Their coaches operated from Silicon Valley and our two virtual TAs operated from the national headquarters in New York. Other than supporting the trainees, the virtual TAs also supported Luci to help students work with the course content towards their portfolio and project goals. We correctly anticipated that the MOOC provided coach will primarily serve to provide general support with questions and assignments, but that there would be a significant lag time variance in their responses. For one, only a single person was appointed to facilitate and manage hundreds of organizations and thousands of people participating in the MOOC. Hence, we asked our virtual TAs to primarily monitor curriculum implementation, manage technical functionalities, and provide support with course content when needed.

The weekly contact was scheduled in staggered fashion which provided us with timely touch points. The staggered schedule resonated with one of the training organization's mantras which was high support, high expectations.

We wanted to book-end the engagement either side of the F2F time to keep our finger on the pulse of virtual and in-person support of the instructor and the students. The book-ended schedule aided the monitoring of the self-directed progress of the students. The quality of their projects could be observed virtually as the uploaded their work, as well as when they presented their work and progress, informally to their fellow students.

RESULTS

Through the preparatory work and the remote virtual provisions, we were able to increase engagement, retention and completion. The constant, or rather regular contact with the in-person instructor and virtual tutors, aided the trainees who had never done any computer scripting or programming. So much so, that the cohort of 20 outperformed other trainees from other regional and national participating organizations.

Table 1. Schedule of touchpoints with students

Monday	Wednesday	Friday
Introductory Video Conference with Virtual Tutors	Check-in Call with Local Instructor after Working Lunch with trainees	End of Week program evaluation
Goal Setting	Blended Classroom Engagement	Goal Check & Reflection
Progress check on objectives	Weekly to bi-weekly student progress presentations	End of module project presentations

Building on a first pilot with 20 Atlanta students, the MOOC service provider and the telecommunications company sought to partner with the NAT to award them with 500 scholarships to participate further in their micro-degree program. Subsequently being awarded an increase from 20, to 500 scholarships to participate in further computer programming MOOCs at $2000 per student. The scholarship sponsors also encouraged every other participating training organizations and those that were to participate in the MOOC, to adopt these methodologies and practices.

The feedback model proved successful and set plans in motion for scaling to other city sites of the non-profit where there were similarly teaching staff without teacher training. Several partnership needs that emerged in the Atlanta pilot were raised, and shared with the MOOC service provider. From requests for additional virtual support, faster response times to students' queries, as well as feedback on students' project work. The competing demands from students in the MOOC initiative; their non-profit six-month intensive core training program course assignments, part-time work, and/or college course work, impacted participation, completion and quality of work. However, stronger bonds between participating students and staff were observed. Resulting in a technological orientation, cultivating social connectedness as well as support for both on-the-ground training staff as well as the students.

The participation within the MOOC curriculum started creating a community of practice amongst other local instructional staff and students, which spread across sites around the country. A common ground of those who were studying the latest technical curriculum together, started with the adoption of the organization's LMS, and deepened for those who participated in the MOOC initiative. After a highly successful pilot, the MOOC approach was expanded and implemented to other cities in 2015. The MOOC enabled NAT to prepare students more efficiently and effectively, quickly equipping them with the skills that corporate partners demand, while facilitating and incorporating faster feedback turn-around times. The feedback model proved successful and set plans in motion for scaling to other non-teaching program staff without teacher training. The model was shared with and piloted at one of the community college partners, with consideration for expansion to other community college partners.

With the extension of the MOOC approach, the NAT could explore the following:

- The scholarships could be allocated to any of the young adults in our network - including students and alumni across all markets,
- The telecommunications company and MOOC service provider envision awarding the scholarships over multiple years (current thinking is 3 years), and
- Would seek to determine the appropriate pacing with the non-profit training program's schedule.

There was consideration to expand the program to other available nanodegree programs which included Front-End Web Development, Back-End Web Development, Data Analyst, iOS, and Intro to Programming. The telecommunications company and MOOC service provider shared that they anticipated adding four or more new programs in 2015 that were likely to include: Android, Entrepreneurship, Cloud/DevOps, Data Engineer in a junior role capacity compared to a full Data Analyst role requirements.

The MOOC approach also grew to other city sites. See Table 2 for a proposed list that was constructed as a strategic approach to scale the program based on the instructor skills matrix.

Table 2. Strategic approach proposed based on instructor skills matrix

	City Sites	Number of Students and Instructors
1.	San Francisco & Silicon Valley	40 students split between two instructors and two sites
2.	Chicago	20 students working with one instructor and one tutor
3.	Atlanta	20 students with one instructor with computing coding experience
4.	Seattle	0 at the time of the proposal
5.	Providence	40 students split between two instructors. There was a hope to coach Internship Sales Managers to be fluent enough to sell for these internship seats. It was also noted by the National Academic Director that there were many opportunities in the greater Boston area and closer Massachusetts region. At the time of writing, the two instructors added great value, insight and expertise to the programming MOOC project for all sites to benefit from as a whole.
6.	New York	New York was dealing with a skills gap and change management at the time of the proposal.
7.	Boston	40 students divided between two instructors.
8.	Virginia	20 students with support from a national academic consultant

SOLUTIONS AND RECOMMENDATIONS

While the organization's feedback framework enhanced the accountability contract between students and teaching staff, regular instructor evaluations were incorporated into the program. The evaluation data offered instructors quicker insights into their practices and room for adjustments to differentiated learning styles. The accountability contract between students and staff in the MOOC initiative could be enhanced by the current instructor evaluation system to address, and to overcome challenges ranging from untrained teaching staff, the intense learning curve with high expectations, and a lack of teaching and learning transparency. The MOOC initiative also offered other valuable lessons for the national training organization, that ranged from:

- **Opportunities to Leverage Instructional Expertise Across the Organization:** The MOOC enabled NAT to leverage the unique talents of our instructors across the country, not just in any one city. With skills like scripting, programming, front end web development and software testing in high demand, NAT needs to be able to train students in these areas as per the demand from corporate clients. With the MOOC approach, we can have one Java or SQL expert instructor anywhere in the country and train hundreds of students across cities. It could enable the NAT to make the most of their relatively lean instructional workforce.
- **Providing More Support than a Traditional Classroom:** With an expert instructor online, on-the-ground support from a local instructor, and remote support from two academic fellows at NAT 's National office, students receive far more support from the MOOC than they do in a traditional classroom. In addition to the many forms of staff support, there was moreover a great deal of community support from peers across the country. Students share ideas and support around the clock and move together without lagging. The response time from both instructors and peers is faster and enables students to receive support from many different individuals with different perspectives and teaching styles.

If these opportunities are explored, these opportunities offer future research directions.

FUTURE RESEARCH DIRECTIONS

Other than leveraging instructional expertise across the country and providing more support than traditional classrooms, there needs to be research on the initial selection of scholarship courses which will be informed by labor market data and existing subject matter expertise (SME) pool of talent. While the MOOC service provider's courses will be offered as an elective to L&D students, it will be dependent on research that establishes students and training sites' readiness. Specific cities could then possibly start as early as at the start of their Learning and Development cycle.

In consultation with the involved instructional staff, a shorter and lower Introduction to Programming course was suggested as a better fit for trainees. Partly because of the lesser course content and time demands, but more so because of the intensive training demands of the national training organization and its program. Researching the impact of the Introduction to Programming course will yield valuable lessons, but also needs to be checked on whether that course would actually benefit trainees and position them for computer programming jobs with the requisite skills. The future potential that will also need to be researched and then explored, is whether to offer scholarships to graduates of the alumni program via their emerging Alumni Relations functions.

CONCLUSION

From the MOOC service provider 's perspective, a partnership with NAT will help advance their "completion agenda" (since most MOOCs have a 5% completion rate, and 95% drop-out rate) and position them well within then President Obama's "TechHire" initiative. For the telecommunications company, the partnership would be a highlight of a new corporate social responsibility (CSR) initiative focused on building out the nation's technology workforce and investing in youth training programs.

Their initial philanthropy with regards to this MOOC, lays additional groundwork for additional investments via grants, internships, and shadow training opportunities from the telecommunications company.

From an influence perspective, out MOOC initiative tangibly demonstrated our commitment to scaling innovative models for serving young adults with skills and knowledge to meet labor market demand opportunities. Whereas offering internship and employment opportunities, the partnership provided NAT with a valuable opportunity to advance our "pipeline of talent" discussions. Similarly, the scholarship program facilitated a major expansion of our students' skillsets and allowed NAT to increase agility in response to hiring needs in these software development areas.

ACKNOWLEDGMENT

My utmost appreciation to the virtual tutors, Kristin and Jeanny, for their dedication and support. Thank you to Luci Ponder, the on-the-ground instructor for her willingness to give up an hour of her lunch every week and to be brave enough to say to a student, "I don't know – let's search for an answer together". Ultimate appreciation to the virtual tutors for building an amazing ramp-up course and for responding to students' questions in the absence of MOOC system responses to students' questions. Gratitude to the CEO for his support in the face of the effort starting with no paid internships or job offers. At the end of the MOOC, the training organization who received 20 scholarships for the pilot phase, were awarded 500 scholarships over three years, and the organization was able to negotiate job shadow opportunities, 100 job opportunities, as well as paid internships.

REFERENCES

Andriessen, J., Baker, M., & Suthers, D. (Eds.). (2013). *Arguing to learn: Confronting cognitions in computer-supported collaborative learning environments* (Vol. 1). Springer Science & Business Media.

AT&T. (2014, September 14). *Att_and_MOOC service provider_launch_online_training_program_nanodegree*. Retrieved from http://about.att.com/story/att_and_MOOC service provider _launch_online_training_program_nanodegree.html

AT&T and Udacity to Offer Scholarships to Underserved Students for New Online Tech Training Program. (n.d.). *AT&T People | Planet | Possibilities.* Retrieved December 10, 2017, from http://about.att.com/content/csr/home/blog/2014/09/at_t_and_udacityto.html#sthash.U0r03eEW.dpuf

AT&T Aspire. (n.d.). *AT&T People | Planet | Possibilities.* Retrieved December 10, 2017, from http://about.att.com/content/csr/home/people/at-t-aspire.html

Beetham, H., & Sharpe, R. (2013). *Rethinking pedagogy for a digital age: Designing for 21st century learning*. Routledge.

Candy, P. C. (2004). *Linking thinking: Self-directed learning in the digital age.* Canberra, Australia: Department of Education, Science and Training.

Carter, T. J., & Adkins, B. (2017). Situated learning, communities of practice, and the social construction of knowledge. *Theory and Practice of Adult and Higher Education*, 113.

Dillahunt, T. R., Wang, B. Z., & Teasley, S. (2014). Democratizing higher education: Exploring MOOC use among those who cannot afford a formal education. *The International Review of Research in Open and Distributed Learning*, *15*(5). doi:10.19173/irrodl.v15i5.1841

Evans, B. J., & Baker, R. B. (2016). MOOCs and persistence: Definitions and predictors. *New Directions for Institutional Research*, *2015*(167), 69–85. doi:10.1002/ir.20155

Fischer, F., Kollar, I., Stegmann, K., & Wecker, C. (2013). Toward a script theory of guidance in computer-supported collaborative learning. *Educational Psychologist*, *48*(1), 56–66. doi:10.1080/00461520.2012.748005 PMID:23378679

Garrison, D. R., & Kanuka, H. (2004). Blended learning: Uncovering its transformative potential in higher education. *The Internet and Higher Education*, *7*(2), 95–105. doi:10.1016/j.iheduc.2004.02.001

Goodyear, P., Jones, C., & Thompson, K. (2014). Computer-supported collaborative learning: Instructional approaches, group processes and educational designs. In *Handbook of research on educational communications and technology* (pp. 439–451). Springer New York. doi:10.1007/978-1-4614-3185-5_35

Guzdial, M. (2014). Limitations of MOOCs for Computing Education-Addressing our needs: MOOCs and technology to advance learning and learning research (Ubiquity symposium). *Ubiquity*. doi:10.1145/2591683

Hsu, Y. C., Baldwin, S., & Ching, Y. H. (2017). Learning through making and maker education. *TechTrends*, 1–6.

Lave, J., & Wenger, E. (1998). *Communities of practice*. Academic Press.

Lazaroiu, G., Popescu, G. H., & Nica, E. (2016, July). The sustainability of udacity's business model of producing first-rate online content and incorporating interactive learning aspects into an online course. In *The International Scientific Conference eLearning and Software for Education* (*Vol. 3*, p. 40). "Carol I" National Defence University.

LeHong, H., Fenn, J., & Leeb-du Toit, R. (2013). *Hype cycle for emerging technologies*. Stamford, CT: Gartner Inc.

Loksa, D., Ko, A. J., Jernigan, W., Oleson, A., Mendez, C. J., & Burnett, M. M. (2016, May). Programming, Problem Solving, and Self-Awareness: Effects of Explicit Guidance. In *Proceedings of the 2016 CHI Conference on Human Factors in Computing Systems* (pp. 1449-1461). ACM. 10.1145/2858036.2858252

Majchrzak, A., Rice, R. E., King, N., Malhotra, A., & Ba, S. (2014). *Computer-mediated inter-organizational knowledge-sharing: Insights from a virtual team innovating using a collaborative tool*. Academic Press.

Meinel, C., Totschnig, M., & Willems, C. (2013). openHPI: Evolution of a MOOC platform from LMS to SOA. In *Proceedings of the 5th International Conference on Computer Supported Education (CSEDU), INSTICC, Aachen, Germany* (*Vol. 5*). Academic Press.

Mentor, D. (2016). EMxC3= e&mLearning Cultivating Connected Communities: Sustainable Workforce Talent Development. Handbook of Research on Mobile Learning in Contemporary Classrooms, 240-259.

Mentor, D. (2017, April 27). Cultivating Digital TLC Teaching and Learning Communities. In *ModSim 2017 International Congress on Modelling and Simulation - Modeling and Simulation in the Age of Data*. Retrieved December 24, 2017, from http://modsimworld.org/papers/2017/Cultivating_Digital_TLC_Teaching_and_Learning_Communities.pdf

Mercer, S., & Ryan, S. (2010). A mindset for EFL: Learners' beliefs about the role of natural talent. *ELT Journal*, *64*(4), 436–444. doi:10.1093/elt/ccp083

Montgomery, K. (n.d.). *Silicon Valley Now Selling Trade School Diplomas Called "Nanodegrees"*. Retrieved December 10, 2017, from http://valleywag.gawker.com/silicon-valley-now-selling-trade-school-diplomas-called-1638663780

Popenici, S. (2015). Deceptive Promises: The Meaning of MOOCs. *Macro-Level Learning Through Massive Open Online Courses (MOOCs): Strategies and Predictions for the Future*, 158-167.

Rai, L., & Chunrao, D. (2016). Influencing factors of success and failure in MOOC and general analysis of learner behavior. *International Journal of Information and Education Technology (IJIET)*, *6*(4), 262–268. doi:10.7763/IJIET.2016.V6.697

Sahin, F., & Mentor, D. (2017). Creating Teaching and Learning Accountabilities Through Data Analytic Feedback Loops. In *ModSim 2017 International Congress on Modelling and Simulation - Modeling and Simulation in the Age of Data*. Retrieved December 24, 2017 from http://modsimworld.org/papers/2017/Creating_Teaching_and_Learning_Accountabilities_Through_Data_Analytic_Feedback_Loops.pdf

Smith, S. (n.d.). *SVP HR Operations.* The Telecommunications Company. Retrieved from http://about.att.com/category/all_news.html

The Economist. (2017, January 12). *Established education providers v new contenders.* Retrieved December 10, 2017, from https://www.economist.com/news/special-report/21714173-alternative-providers-education-must-solve-problems-cost-and

Udacity. (2014). *Nanodegree Programs.* Retrieved December 10, 2017, from https://www.udacity.com/nanodegree

Walther, J. B., Hoter, E., Ganayem, A., & Shonfeld, M. (2015). Computer-mediated communication and the reduction of prejudice: A controlled longitudinal field experiment among Jews and Arabs in Israel. *Computers in Human Behavior*, *52*, 550–558. doi:10.1016/j.chb.2014.08.004

Wenger, E. (2010). Communities of practice and social learning systems: the career of a concept. *Social Learning Systems and Communities of Practice, 3*, 179-198.

Yousef, A. M. F., Chatti, M. A., Schroeder, U., & Wosnitza, M. (2014, July). What drives a successful MOOC? An empirical examination of criteria to assure design quality of MOOCs. In *Advanced Learning Technologies (ICALT), 2014 IEEE 14th International Conference on* (pp. 44-48). IEEE. 10.1109/ICALT.2014.23

Chapter 6
Blended Learning With a Virtual Mentoring Community:
Enhancing the Way We Mentor

Ruth Gotian
Weill Cornell Medicine, USA

ABSTRACT

Mentoring, the tutelage of one person by a more experienced one, is well documented to have a multitude of benefits. Mentoring, which can be traced back to Greek mythology and beyond, has seen its format evolve, especially with the advancement of technology. Traditional models of mentoring include in-person mentoring or more recently, online mentoring. For some underrepresented groups, mentors provide models for success. In 2014, capitalizing on technological advancements and the need for in-person dialogue with a larger constituency, Weill Cornell Medicine launched a science, technology, engineering, and math (STEM) blended learning curriculum during its summer program that prepares undergraduates for careers as physician-scientists. This initiative fuses the positive aspects of in-person mentoring along with the opportunities provided by virtual mentoring by a larger group of peer mentors. Blended learning, together with in-person and virtual mentoring, offers a newly charted multi-dimensional approach to fulfilling academic and career goals during a STEM summer program.

INTRODUCTION

This chapter will introduce an innovative blended Science, Technology, Engineering and Math (STEM) summer mentoring program, Gateways to the Laboratory Program (Gateways) at Weill Cornell Medicine (WCM). Gateways illustrates the manner in which technology, in conjunction with face to face interaction, can best develop, support and encompass the benefits of mentoring. This is exhibited through Gateways, a summer program which prepares underrepresented minority, disadvantaged college students, as well as those with disabilities, to qualify for a rigorous study for the combined medical (MD) and biomedical science (PhD) degrees, the premiere path for a student to become a physician-scientist. These groups were particularly chosen due to their underrepresentation in the physician-scientist workforce coupled

DOI: 10.4018/978-1-5225-4111-0.ch006

with the National Institutes of Health emphasis on recruiting these groups into physician-scientist training (MD-PhD) programs (National Institutes of Health, 2014).

The combined force of introduced virtual and face to face mentoring supports and challenges the participants. This program is vital as there is a national call for additional physician-scientists, especially those who are from underrepresented minority backgrounds (COMPETES Act, 2007; National Institutes of Health, 2014). This chapter is a case example, which illustrates one way in which technology, combined with in person mentoring can enhance the summer student's experience and possibly their success. The chapter will provide an overview of the history of mobile devices, blended learning and mentoring. It will provide a discussion grounded in literature on the need to improve the pipeline of underrepresented students into dual degree MD-PhD Programs, some background on the creation of Gateways, and a description of the Gateways curriculum and catalyst of the blended learning curriculum and virtual mentoring community. The chapter goes on to describe in some detail how to effectively mentor these students utilizing various tools, including the blended learning curriculum and provides suggestions on how to overcome fundamental challenges facing the pipeline. Finally, a description of the adult learning theories used to underscore this blended learning model are introduced as are the challenges faced while creating this innovative program. Recommendations for those who wish to take on the challenge of creating a blended learning curriculum with a virtual mentoring community close out the chapter.

MOBILE DEVICES AND BLENDED LEARNING

Personalized, portable and easily accessible technology is altering lives (Johnson, Adams Becker, Estrada, & Freeman, 2014). Through the use of mobile devices such as smartphones and tablets, people can communicate, collaborate, and share information from any place at any time. As such, learning can take place both virtually and in brick-and-mortar institutions (Evans & Johri, 2008). This allows for greater flexibility to modify content and adjust delivery channels.

This seamless learning is grounded in two main principles: personalization and mobility (Wong, 2013). The foundational origin of personalization is the understanding that the student is at the heart of the learning and everything—including intelligence, motivation, group dynamics, and feelings—must be considered (Motsching-Pitrik & Holzinger, 2002). Past research has been focused on either formal or informal learning, but little has been done on the bridging of the two learning modes (Looi et al., 2010), as is the case when virtual mentoring is used in conjunction with in-person mentoring.

Mobility emphasizes the ability to learn without restrictions. With the added efficiency of mobile devices such as smart phones and tablets, learning environments can be expanded as can social interactions, collaborations, and the potential for new formal and informal learning activities (Jones & Jo, 2004).

Recognizing that every student entered the institution with a mobile device, coupled with the software and the instructional knowledge to leverage the current learning management system (LMS), it was an opportune time to introduce blended learning at the institution. Gateways was the ideal platform for the trial.

The blended learning curriculum is most effective when there is active participation from the students. As such, the curriculum was developed with a multitude of opportunities for them to engage as they partake in everything active learning to reflective dialogue. Some of the learning strategies are passive such as watching video clips or reading, while others are more active such as writing scientific papers or delivering oral research presentations. Questions on the discussion boards are often open-ended without a clearly defined answer. This requires critical thinking and also offers the opportunity for multiple and varying responses. To properly engage, the students have to access various resources at their disposal. Every opportunity for learning is coupled with an opportunity to reflect on the learning.

Students learn from conducting experiments in the laboratory and giving research or journal club presentations, a model commonly known as "learning in action" (Heron, 1999; Schon, 1983; Yorks & Kasl, 2002). After the action takes place, there is a need to reflect on what happened, decipher why things worked or did not work, and understand the feelings that occurred during the learning (Heron, 1999; Schon, 1983; Yorks & Kasl, 2002). The discussion boards are the most popular technological tool for reflection in Gateways. People post questions address feelings they have experienced, thereby allowing others to understand that they are not alone. This also provides the students with multiple perspectives, with which they can better consider their own.

MENTORING

Mentoring, the guidance of one person by a more seasoned one is well recognized as a marker of success of the protégé (Buddeberg-Fischer & Herta, 2006; Gotian, 2016; Gotian, Raymore, Rhooms, Liberman, & Andersen, 2017; Kasworm, 2010). Proper mentoring is often considered a recruitment tool as students flock to those who will nurture and support their careers (Hathaway, Nagda, & Gregerman, 2002; Kardash, 2000; Lopatto, 2004; Nagda, Gregerman, Jonides, von Hippel, & Lerner, 1998; Seymour, Hunter, Laursen, & Deantoni, 2004). Research has shown that those who are routinely mentored outperform those who are not (Daloz, 1999; Eby, Allen, Evan, Ng, & DuBois, 2008; Sambunjak, 2006). Mentoring can come in multiple forms, both formal and informal. Gateways provides opportunities for both forms to occur.

The leadership of Gateways worked to incorporate a multitude of informal learning opportunities. In order to fully understand the benefits of mentoring, and before we consider its application, we must understand its roots. Mentoring has its origins traced to Greek mythology (Barondess, 1994). In Greek mythology, Mentor was a devoted friend and guide to Odysseus, the king of Ithaca. Odysseus' son Telemachus was raised by Mentor during the time Odysseus was away fighting the Trojan War. Mentor took on the role of Telemachus' teacher, counselor, coach, and defender. He worked diligently to create a bond based on friendship and trust.

The structure of mentoring has progressed over time, particularly with technological innovations. While in-person mentoring is still a staple of the relationship, it is often complemented by phone conversations, emails, FaceTime, Skype sessions, and other web based modalities (L. Bierema & Merriam, 2002; Martinez Witte & Wolf, 2003; Miller & Griffiths, 2005). Traditional models of mentoring include in-person mentoring or the more contemporary, online mentoring (Barondess, 1994; Daloz, 1986). While mentoring is advantageous for most, for some underrepresented groups, mentors are especially credited with developing students' potential to achieve (National Academy of Science, 2011), something which is openly acknowledged by the mentees. This is especially true for those who do not have role models from their chosen fields. Such is the case for many first-generation college students (Jack, 2014; Papadakis, 2017). Students often do not recognize what they do not know. Mentors aid in bringing vulnerabilities to light, putting them in context, and helping students develop strategies for overcoming them.

OVERHAULING THE PIPELINE

Traditional minority groups (African Americans, Hispanic Americans and Native Americans) continue to be severely underrepresented among physician-scientists. In 2012, the Association of American Medical Colleges (AAMC) reported that only 0.4% of medical school faculty members in the US, who had a MD-PhD or other health-related degree, were members of underrepresented minority (URM) groups (Association of American Medical Colleges, 2012).

It is important that we increase the number of physician-scientists who are members of URM groups or groups who are underrepresented for other reasons-social economic factors or disability, as they are likely to experience the many unique health challenges that affect their communities thus offering different perspectives on common problems (Gotian et al., 2017). This demographic, however, is severely underrepresented in the Science, Technology, Engineering and Math (STEM) workforce and almost undetectable among physician-scientists. This is of concern because URMs are the fastest growing sector of the US population (National Academy of Sciences, 2007). Not surprisingly, there is a national call to increase the number of URM physician-scientists (COMPETES Act, 2007). This is not an easy task as it demands those who find themselves in this diverse cohort to be something they have never actually seen modeled.

MD-PhD Programs are the most common avenue of training for physician-scientists (Brass LF et al., 2010). Nationally, there is an issue with the pipeline of underrepresented students entering the physician-scientists career track (Gotian et al., 2017; National Academy of Science, 2011; National Institutes of Health, 2014; Smedley, Stith, Colburn, & Evans, 2001). While Gateways helps to alleviate some of the problems, the issue has to be addressed. The pipeline is too narrow, so many students do not enter it. There are too many leaks, primarily because students give up on pursuing this career path when they do not receive the necessary encouragement at critical junctures in their training. As such, the number of underrepresented students pursuing a physician-scientist career is diminishing. The national demographics of physician-scientists do not, even closely, represent the national population (National Academy of Science, 2011). In turn, there are fewer physician-scientists studying clinical and scientific problems that are inherent in underrepresented minority communities (Gotian, 2017). Weill Cornell Medicine, one of the leading MD-PhD Programs in the country, decided to tackle these issues head on through its Gateways to the Laboratory Program. These problems can be traced to four major issues, all of which were addressed during Gateways:

1. The students have low expectations. They do not recognize their own existing fund of knowledge nor do they fully comprehend the need and their ability to challenge themselves.
2. The students are uninformed about career opportunities. Complicating this matter, they do not know how to approach investigating such options.
3. The students lack role models. There is an extreme scarcity of physician-scientists from diverse backgrounds. Therefore, the students do not see people whom they aspire to emulate.
4. The students maintain inadequate family and community support. Very often, the students are first-generation college students and do not have a support system that understands the conflicting demands of pursuing the dual degree.

CREATION OF A STEM-BASED SUMMER PROGRAM

In 1993, the Weill Cornell/Rockefeller/Sloan Kettering Tri-Institutional MD-PhD Program became the first MD-PhD program in the nation to take on the task of increasing and widening the pipeline of URM MD-PhD applicants and students. A training program was developed, Gateways to the Laboratory (Gateways), which encompassed everything one expects to see in a competitive MD-PhD applicant, particularly significant research experience and effective mentorship. The Gateways Program has evolved over time, modifying its class size, length of the program and introduction of numerous initiatives including a peer mentoring model and parental involvement.

From its inception and continuing today, Gateways focuses on the following four elements:

1. Research experience that will empower the student.
2. Effective and active mentoring focusing on the needs of the individual student.
3. Career development where the students learn fundamental skills.
4. Funding so that the student does not need to choose between supporting themselves and furthering their career.

Figure 1. Evolution of the gateways to the laboratory program

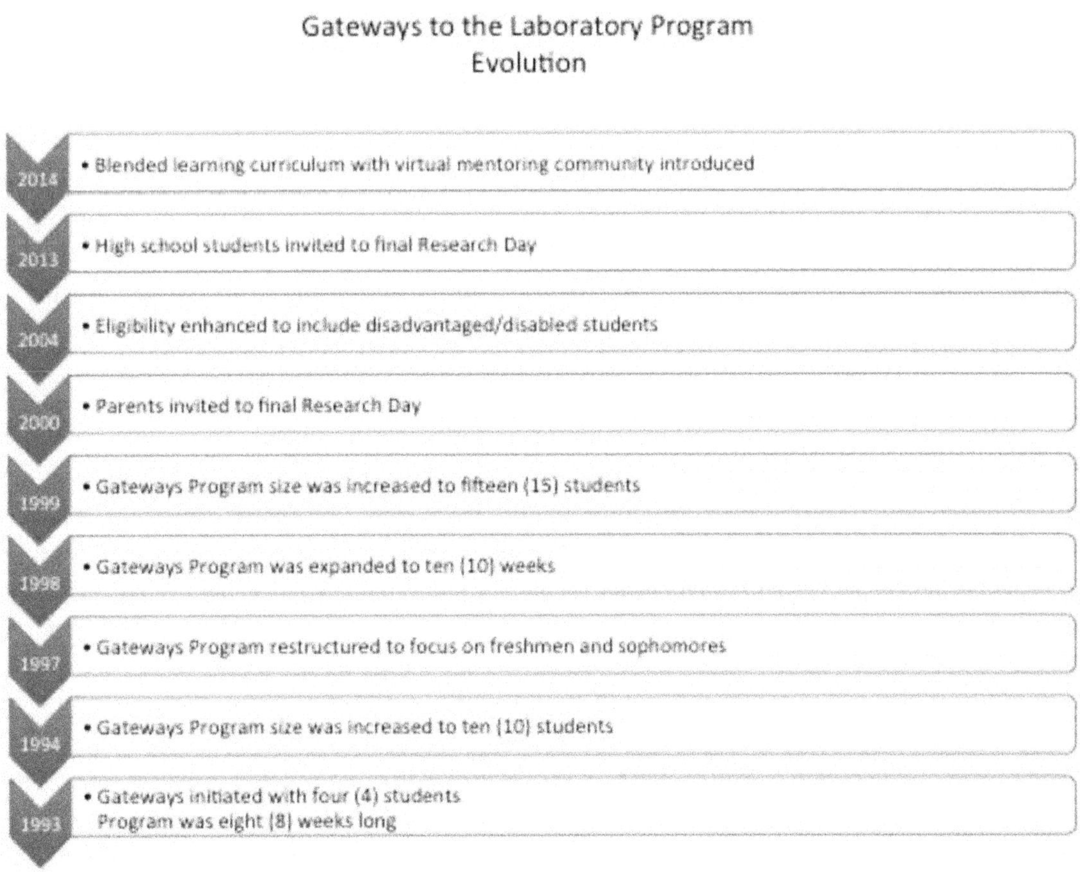

These elements are similar to those subsequently deemed to be characteristics of summer programs that have a positive impact on student decisions to pursue STEM careers (National Research Council, 2005).

Like all summer programs focused on STEM fields, Gateways students enjoy enriching research opportunities in the laboratory. They conduct hypothesis-driven research under the guidance of a faculty member, post-doctoral fellow, or graduate student. The laboratory time combined with weekly professional development workshops leave the Gateways students well equipped with the tools they need to gain acceptance into coveted physician-scientist programs, the structured programs that produce the highest caliber of physician-scientists nationwide (Brass LF et al., 2010; Gotian, 2017). Having Gateways students shadow in clinical settings, such as scrubbing into surgeries, and participate and present in their own weekly journal club rounds out the physician-scientist experience. They take a mock MCAT exam, the professional exam needed to gain admittance into medical school, as well as a mock interview. Finally, they have someone who reviews their presentation with them to help ensure it is successful.

In 2014, capitalizing on advancements in technology and pedagogy while recognizing the need for in-person dialogue with a larger constituency became the cornerstone guidelines for an innovative learning system delivery for college students. Weill Cornell Medicine, through its combined MD-PhD Program, launched a Science, Technology, Engineering and Math (STEM) curriculum for its summer program. This initiative melds the positive features of in-person mentoring with the opportunities provided by virtual mentoring. Blended learning, combined with in-person and virtual mentoring meld to provide a multidimensional approach for achieving academic and career objectives in a STEM summer program.

Studies have shown that the quality of the undergraduate research experience improves with the insertion of mentors into the training (Eby et al., 2008; Pfund, Pribbenow, Branchaw, Lauffer, & Handelsmann, 2006). Mentors are also critical to attracting students in general, but especially for URMs seeking careers in STEM fields (Hathaway et al., 2002; Kardash, 2000; Lopatto, 2004; Nagda et al., 1998; Seymour et al., 2004). Personal mentoring was introduced as a compliment to the formal group wide developmental activities. The need for a peer mentoring system in Gateways was recognized in 1997 with the establishment of the Big Brothers/Sisters initiative, commonly referred to as 'Big Sibs'. Graduate students were selected to provide one-on-one mentoring to the summer students. There were two major goals of the Big Sib mentors. The first was to offer summer students role models who look like them, specifically the same ethnicity, and were close to them in age. The second was to recruit mentors who were knowledgeable about the challenges of dual-degree training. The peer mentoring resource was developed in the late 1990s, at a time when most students did not have email addresses or mobile phones. Websites within higher education were just being considered and learning management systems did not yet exist. As such, everything was done in person. The challenge with this model was that the summer students only received mentoring on a regular basis from one person, and these interactions were based on schedule availability. Therefore, each student only obtained the perspective and tutelage of one person. This completely limits their assessments of current challenges and how to approach new situations. By having multiple perspectives the students are afforded the opportunity to critically think about the various options and opinions presented before them and make their own informed decisions. The mentoring model had to be transformed in order to enhance the experience, increase the underrepresented minority applicant pool, and improve retention.

Gateways to the Laboratory focuses directly on the issues that cause URM to leave the physician-scientist pipeline, namely low expectations, lack of information about career opportunities, lack of role models and support system (). Gateways is designed to address they key problems aptly identified in the literature by identifying and raising expectations, clearly communicating about career opportunities, introducing role models through the Big Sib Program, and involving their families.

Identifying and Raising Expectations

What sets Gateways apart is the emphasis on preparing students for combined physician-scientist training, as opposed to solely one degree or the other. The underlying principle is that the students are held to high standards—"good enough is not good enough"—from the first day of Gateways. While expectations are made clear to students at the beginning of the program, they are expertly modeled in an environment where everyone sets and meets high expectations. They are also given examples throughout the summer of others who have set and met high expectations. Sentences from actual letters of recommendation are read to them so that they can appreciate how admissions committees are expecting them to be perceived by their letter writers. The Gateways structure helps the student learn how to set and meet goals and deadlines. They are firmly told that deadlines are not recommendations. They have milestones to meet throughout Gateways, and there is regular follow-up to make sure they are on target to meet, if not exceed, all expectations.

Clear Communication About Career Opportunities

The goal of Gateways is to prepare its students to become competitive dual-degree applicants and successful physician-scientists. The students need to understand what a combined program is, what types of people pursue this course of study, and what a graduate does once they complete their training. This is done both formally, through scheduled programs such as panels and seminars, as well as organically through informal networking events held throughout the summer. They are consistently exposed to people who are enrolled in, are graduates of, or lead physician-scientist programs. By the end of the summer, Gateways students typically remark that they feel knowledgeable and comfortable about the dual program application process. Being exposed to so many physician-scientists reduces the levels of imposter syndrome they may have experienced, while increasing their self-efficacy.

Introduction of Role Models

According to a National Research Council assessment (National Research Council, 2005), undergraduates see positive mentoring as the most valuable feature of a training program, second only to significant research experience. The foundation of Gateways is consistent mentoring made up of multiple professional, formal, in person development meetings held weekly both in groups and one on one sessions. These mentoring sessions cover an array of topics ranging from competitive admissions applications to recognizing and overcoming imposter syndrome.

A multitude of people make up the Gateways mentoring program including the program director, administrative director, principal investigator, members of the lab, the previously described Big Sibs, journal club leaders and moderators, mock interviewers, oral presentation reviewers, and students in the physician-scientist program. During the summer, Gateways students are given the opportunity to and taught how to create and expand their professional network. Having a diverse group of mentors provides multiple advantages. The summer students hear multiple perspectives, gain professional contacts, attend seminars, and learn about different mentoring styles to best understand which mentoring style works for them.

Involving Family and Community

After ten weeks of dedicated time spent on research, the students are finally able to present preliminary results and explain the purpose and importance of their research projects. Prior to the final day of the program, Where Gateways students present their research in front of the academic community, the Gateways students review their presentation with a panel of peer mentors. They also are given the opportunity to present their research at a local conference highlighting summer projects of other undergraduates.

The parents are invited to the final Research Day, which allows families to meet the program's administration and get answers to questions they may have regarding the pursuit of the combined physician-scientist degree. The program provides travel support and housing, and all students have one or more family members in attendance. The family members are always in awe of the high-quality presentations as well as the large crowds of students and faculty who come to listen to each student's presentation. Families are inspired and marvel at what the students are able to accomplish in just ten weeks and then to imagine the heights to which they can soar once they have their doctorates.

CREATION OF THE GATEWAYS BLENDED LEARNING CURRICULUM

Nationally, an administrative office within a medical or graduate school usually runs STEM summer programs for undergraduate college students (Association of American Medical Colleges, 2017). These offices generally have a multitude of other administrative responsibilities related to their year round students and are not equipped to effectively develop curriculum, especially using various technological instruments. The staff members who run these summer programs also have competing demands of their time. Doing things as they have always been done is sometimes easier than conceiving, planning and executing something new. Producing an effective blended learning curriculum requires some instructional technology background and understanding. In higher education, especially at the graduate level, most faculty and staff are enthusiastic to meet the changing needs of their students but do not always have the instructional technology and experiential knowledge to do so. Unfortunately, too many faculty members simply upload videos of lectures and corresponding readings thinking this equates to effective use of the blended learning technique. This uploading of lectures and readings is similar to a data dump and severely limits active learning.

As previously noted, the in-person delivery medium limited the perspective shared to only one point of view. With the rapid development and reliance on technology, there was merit in contemplating alternative learning methods. At the time of the discussions to innovate the delivery medium for Gateways, the medical curriculum at Weill Cornell Medicine, was using Canvas, a Learning Management System (LMS), for its medical curriculum. Capitalizing on this existing technology, Canvas was used as the technological tool to fuse the new blended learning and virtual mentoring techniques for Gateways. The new blended learning format featured a multitude of static pages with valuable information. In order to provide a space for critical reflection, discussion boards were actively used and moderated by the Big Sibs, program leadership, and journal club leaders. Every page had an accompanying discussion board, which allowed for both passive and active practice and discussion with peers. Some of the discussion topics had deadlines, but most conversations stayed open and active. Allowing the discussions to be freely form afforded students the opportunity to digest the reflections of others, make connections, and think critically about what was written. It was especially advantageous for the students who were more introverted in nature, as it offered them sufficient time to gather their thoughts and articulate them (Cranton, 2006).

HOW TO MENTOR GATEWAYS STUDENTS

As they haven not been readily exposed in their lives to physician-scientists, the Gateways students often lack role models in their lives. Furthermore, they do not have people who can motivate and advise them. During their brief summer stay with us, it is the mentor's role to be that role model for them who can appropriately prepare them for the rigors of applying to physician-scientist training programs and increase their chances of becoming competitive applicants.

When assessing the success of mentorships, there are conflicting viewpoints as to who makes the most effective mentors. Atkinson (1991) reports that both mentors and mentees alike deemed the relationships to be more rewarding when they were matched for race or ethnicity. This is in fact, what the Big Sib Program was like in the early years of Gateways. This viewpoint is concurred by the National Academy of Science (2011) disagrees with the aforementioned notion of selecting mentors based on race or ethnicity. Weinberg states that as a positive mentorship develops, gender difference becomes less crucial. Likewise, there may be benefits to not matching mentors based on age, culture, gender, and area of interest (Shedletsky, 2006). The latter point of view was the catalyst for building the first STEM summer program virtual mentoring community coupled with in-person one-on-one mentoring. Overall, it is the mentoring mentality, including active listening, communication style and sponsorship that is pivotal.

In this specific mentoring situation, the Big Sibs need to remember they are dealing with unique students. A "one size fits all" mentoring model does not suffice. Instead, a combination of face-to- face mentoring, virtual mentoring via the discussion boards, and blended learning with links and videos is optimal. That said, there are time limitations and numerous demands placed on the students, so the expectations involved with being a successful mentee need to be clear.

Big Sibs undergo mentor training to receive coaching on being effective in their role. Some of the basic mentoring skills taught and reinforced include:

1. **Listening With Curiosity:** Find out each student's unique story. Why do they want this advanced combined degree? What motivates them? What do they perceive to be their biggest challenge? Do they have a place at their home institution that they can implement the strategies we recommend? Do they have a support system? Have they done research before? The introductions thee Gateways students submit online prior to their arrival answer many of these questions and prompt future discussions.
2. **Retaining Information about Mentees:** Answers to the aforementioned questions will be different for each student. This is why a standard mentoring plan would not be advisable. Mentors need to tailor their advice and sponsorship based on the students' access and abilities.
3. **Reflecting With Accuracy.** This is critical. Everyone's ultimate goal, of getting into a physician-scientist training program is identical, but each journey to achievement will be vastly different. It is imperative that mentors can accurately advise on mentees' visions and perceptions in a personalized

manner. They will have the opportunity to do so in person and online. Accurately understanding the motivation, concerns, abilities and access of the Gateways students will allow the mentors to provide the most personalized guidance.

4. **Questioning to Explore:** Every student's journey to the same destination will be different, as each student comes from a different background and support system. While telling the mentee how to solve a problem is a quick fix, it is useless in the long term. Teaching the mentee to ask pivotal questions will help them move forward and teach them how to help themselves. As such, it is imperative mentors give mentees questions to ponder while reflecting on their experiences and those of others.
5. **Providing Feedback for Development:** Based on each individual story (some of which are shared in their online introductions) and answers to questions previously addressed, planning of next steps can be initiated. It is at this point we not only consider their future direction, but what they have done to date and how they might repeat these efforts more effectively and efficiently. With this feedback, the Gateways student can move forward with tips for success which become their toolbox which can be used as they face future challenges.

ROLE OF A GATEWAYS MENTOR

The success of the Gateways program is based first and foremost on mentorship. The backbone of mentorship is to help students organize their goals and formulate an action plan for meeting those goals. Offering feedback is a critical component of mentorship. These students will need guidance along the way. They will also make some mistakes. That is why it is imperative for the mentors to be available both physically and virtually to help the students figure out what went wrong and how to prevent such errors from recurring. By reflecting on an experience with members of a virtual community, others may learn from both the students' and mentors' experiences, thereby increasing the number of people impacted by the learning.

Along with the Big Sibs and the journal club directors, the program director and administrative director act as Gateways student advisors, instructors, cheerleaders, and even political sponsors. These are four types of roles enjoyed by many a successful coach (Bacharach & Bacharach, 2012).

1. **Counselor/Advisor:** The online forum gives the Gateways students a platform to ask any and all questions, even those they are afraid to ask. The goal is to support a virtual mentoring community that will be a protected place for the Gateways students to voice concerns. This place is a judgment-free zone concerned solely with the development of the student. This safe space is modeled and reinforced from their initial introductory post. Students and Big Sibs are told verbally and on-line that for true learning to occur, comments should be discussed for clarification, but the author and their intentions, should never be judged.
2. **Instructor:** In addition to the various in-person career development workshops, there is supplemental material including links to articles, videos, tips, and discussions. Popular career development discussion topics include the program application process, study tips, interview tips, professional email etiquette, research presentation skills, time management skills, and the like. The mentors provide real examples and offer reflection on various experiences. The virtual community affords multiple mentors to share their experiences while allowing the students to ask clarifying questions.

3. **Ultimate Cheerleader:** When experiments in the laboratory are not working and the students are not reporting data, the mentor as cheerleader must motivate them to dig deeper and keep going. With support via the discussion forum, the Gateways students can see that they are not alone in facing these challenges and are not the only ones feeling these anxieties. It is critical that the students realize the pressure they feel is not unique to them and is, in fact, shared by most of the students in the group. By realizing this, they will feel less isolated (National Academy of Science, 2011).
4. **Political Sponsor:** With mentoring via the online portal, the Gateways students broaden their networks. It is not unusual for a Big Sib to pick up the phone and call a faculty member or senior ranking official at their alma mater to see if that person can try and arrange a research opportunity for a Gateways student. It might be difficult for this junior student to arrange such an opportunity on their own, but when someone picks up the phone on their behalf, things seem to move at a rapid pace (Hewlett, 2013).

CREATION OF A VIRTUAL MENTORING COMMUNITY

General Benefits of Virtual Mentoring

Virtual mentoring is seamless—transcending time, place, and delivery modes. It supports lifelong learning anywhere and anytime (Marin et al., 2016). Technology has provided access to infinite knowledge and new ways of learning, thinking about, and processing information (Kereluik, Mishra, Fahnoe, & Terry, 2013). The amount of information that is now available is growing exponentially. It is to the mentor's benefit to take advantage of this and adjust their mentoring philosophy and delivery.

Specific Features of the Gateways Virtual Mentoring

Virtual mentoring has a multitude of benefits not least of which is flexibility (L. L. Bierema & Hill, 2005). However, Bierema examines a one-on-one virtual mentoring model while the Gateways virtual mentoring community is a many-to-many mentoring model further enhanced by an in-person one-to-one mentoring model. When developing the Gateways program, the leaders recognized that the two aspects of mentoring should not be isolated from one another. An association built on trust can be developed in person and the virtual mentoring builds upon that trust. As all posts on line are made using the person's real name and the fact that all of the students and Big Sibs know each other, increases the honest and positive tone used in the posts. Everyone knows that within days, they will see each other in person. This provides a further incentive to avoid negative, judgmental posts and confrontations. Without having the in-person experience as an integral part of the mentoring relationship, the virtual mentoring takes on a serious risk of becoming analogous to various online chat rooms where guidance can be provided by anyone, often anonymous and often without the benefit of mentor training.

To our knowledge, it is the first STEM summer program of its kind, run by a medical or graduate school, to use a blended learning curriculum linked with a virtual mentoring community. With the use of the learning management system Canvas, a self-directed and blended learning virtual classroom was generated and immediately exhibited a multitude of opportunities improving the functions of the curriculum (including pre-work, journal club, calendar, assignments and evaluations), virtual mentoring, career and scientific development and a sense of community.

1. **Pre-Work:** Getting all students to the same starting line is critical to the success of Gateways. Pre-work is a great way to create a culture of excellence and get all participants to the same starting point on the first day (Yiu & Parker, 2000). The three parts to the summer program's pre-work program consisted of (a) online introductions; (b) online assessments ;and (c) lab-related reading materials. Introductions, which are submitted prior to the official commencement of the program, launch a feeling of community and allow the summer students to express themselves creatively. The introductions created a social connectedness. It allowed for the students to create relationships with people beyond their social circles. Ultimately, these connections can help propel careers. Introductions often include a written narrative, video, or photo. The students have enough time to leave feedback for their classmates, ask each other questions, decide in advance who will be bringing some of the basic living essentials—coffee pot, iron, etc. The online assessments are an opportunity for the students to familiarize themselves with the LMS and allow the administration to gauge expectations, experience levels, etc. Lab readings sent in advance to the students provided a vehicle for self-directedness, as they allowed the students to read about the research being done in the lab prior to their arrival. The pre-work saves time, helps alleviate anxieties, sets a high standard for preparation, and gets each reader used to the idea of regularly using the LMS.
2. **Journal Clubs:** The Gateways students meet weekly with a MD-PhD student moderator. Every week, two Gateways students each present an article, which are read by the entire class. The articles are discussed, issues are raised and figures are clarified. The journal club discussion board is used before and after the weekly journal clubs to discuss and answer questions related to the techniques used in the articles. Two journal club student leaders moderate this discussion board. The discussion board page links directly to the journal club articles. The document is broken down by week, and each article has introductory slides that introduce the techniques and critical thinking questions. This ensures that the discussion continues after the journal club session concludes. In addition, it provides an avenue to ask questions which may come up after the journal club session.
3. **Calendar:** Summer students often hear the program leadership state that "Deadlines are not recommendations." In order to jumpstart their organizational and time management skills, technology provides one possible solution. The fact that the LMS has a mobile application version was an added benefit. The calendar feature in the LMS lists all of the deadlines and can be synced with mobile phone and desktop calendars, so that the students can always be aware of upcoming deadlines.
4. **Assignments:** Assignments used to be submitted in a multitude of formats including emails and printed assignments. Tracking the submissions became tedious and an administrative burden for the staff. With the blended learning format, assignments can now be submitted online using the LMS and they are all date stamped. The program leadership can now access the assignments in bulk from any computer, make comments, and upload them directly to the LMS. The documents are archived in perpetuity in the LMS.
5. **Evaluations:** Evaluations, a key source of continuous feedback, were submitted electronically using the LMS. This feature saved staff time in distributing, collecting, and analyzing evaluations. Furthermore, by submitting feedback via the LMS, results could be compared to each other and presented in aggregated formats as needed. There are four evaluations throughout the summer. The first assessment, part of their previously described pre-work, is completed prior to the students' arrival and start of the program. Here they outline their expectations, provide a brief backdrop of their scientific background, inform us of the scientific journals they read, and let us know if they have ever been officially mentored. The second evaluation is a check in after two weeks, whereby

the program leadership is able to assess how things are progressing. This is a pivotal juncture in the program, as there is still time to make high impact changes to the program which could positively impact the experience for the students. The midpoint evaluation lets us know if the students are still on track, and the final evaluation lets the program leadership understand if we have met their expectations.

6. **Virtual Mentoring:** Online discussion board usage allowed the Big Sibs to mentor all of the Gateways students in a virtual forum in addition to having face-to-face meetings with each individual. This provided each summer student access to the guidance of numerous mentors without being inhibited by place or time. Furthermore, the face-to-face meetings generated a level of trust so that the summer students did not feel as if they were receiving on line advice from anonymous sources.
7. **Career and Scientific Development:** The summer is rich with career development opportunities and workshops for the students. Links to the workshop handouts and articles of interest are posted on the LMS. Reflective questions are posted on the discussion forum regarding each career development topic. In order to provide a strong foundation for scientific techniques, a pictorial guide was produced to highlight every common scientific technique that might be introduced over the summer.
8. **Creating a Community:** Being part of a community of peers is essential if the students are to overcome feelings of isolation (Elam et al., 2009; Odom et al., 2007). The online discussion forum aids in building a community by providing a setting for students to reflect, post discussion questions, share information on conferences, and more. This starts with the introductory posts. Students report that when they read that another fellow student enjoys the same sport, movies, music or books as them, they tend to feel a connection toward the other person and therefore feel less isolated.

CHALLENGES

There are a multitude of challenges to face when developing the blended learning curriculum with the virtual mentoring community:

1. There is a great deal of work for the Gateways students to accomplish in an abbreviated timeframe. In order to meet the goals, it is imperative to set milestones along the way. It is the Big Sibs' and the program administrators' responsibilities to follow up with the Gateways students to make sure milestones are being met in a timely fashion. Discussions about any challenges or successes that may have occurred on the journey to reach each milestone take place on a regular basis.
2. The Gateways students have many demands on their time. When people are very busy and up against deadlines, it is often difficult to understand the importance of meeting with a mentor. For this reason, the students attend regularly scheduled lunch meetings with the administrative director. Lunch is included and the meetings are part of their routine, so other demands on their time are scheduled around it. Furthermore, each Gateways student has an assigned Big Sib, so that there is always someone who can also meet with him or her if the administrative director is not available or if they seek another perspective. The Big Sib, a current MD-PhD student, can also provide technical assistance.

3. As with any technological product, there are various associated challenges. While Canvas as a learning management system is rather self-explanatory, there were some issues related to non-student access of the system. For example, everyone appears on the class roster, including the staff. The roles assigned through the LMS for access to the course are either too lenient or too stringent. As a significant number of students were in the roles of mentors, there was only a need to provide them with limited access. The only available role in the LMS gave them access to grades and all discussion boards, rather than specific boards. They also had the ability to add and remove people from the class. As the only way to move forward was to give them the full access, the access came with explicit directions and a high level of trust. As Canvas is the LMS being used at Weill Cornell Medicine and the only one with institutional technological support, it was the only option available in order to build this type of course.

ADULT LEARNING THEORY AS A FOUNDATION TO A BLENDED-LEARNING MODEL

While blended learning has become a staple in higher education, it is an unheard of phenomenon in STEM summer programs. Nearly every medical and graduate school in the country runs a STEM summer program (Association of American Medical Colleges, 2017; The Leadership Alliance, 2017). Running the summer program is one of the many competing demands of the staff of these offices. Having been involved in the running of summer programs for two decades and being an active member of its community of practice, I am told repeatedly that the staff often do not have the knowledge and/or bandwidth to create a blended learning curriculum with a virtual mentoring community. The staff are experience din organizational and change management, not curriculum and program development. In addition, none of the summer programs market a blended learning curriculum or a virtual mentoring community.

The time for developing such a model was right. Adult learning theories and technological advances were used as the groundwork for this new learning model. Kolb's Theory of Experiential Learning (DA Kolb, 1984) which emphasizes the process of learning and York-Barr's Theory of Reflective Practice (Merriam, Caffarella, & Bamgartner, 2007) which underscores the process of thought, were used as the foundation for this blended learning delivery medium.

Kolb's Theory of Experiential Learning

The groundwork of Kolb's Theory of Experiential Learning is that the act of learning is cyclical and includes four elements: concrete experiences (feeling), reflective observation (watching), abstract conceptualization (thinking) and active experimentation (doing). This sequence can repeat as often as is necessary and can begin with any of the four elements, although it usually starts with a concrete experience (DA Kolb, 1984).

This results in four types of learners: diverger (who ask "what is it?"), assimilator (who ask "what does it mean?"), converger (who ask "what can I do?") and accommodator (who ask "how can I do it?") (D Kolb, 1984; D. Kolb, 1984).

In Gateways there are multiple tools available to support and encourage every type of learner. Realizing this, Gateways is a potpourri of facilitation styles ranging from lectures, small group discussions, model building, visual learning, papers, presentations and on-line forums. The Gateways students are

Figure 2. Kolb's Experiential Learning Model

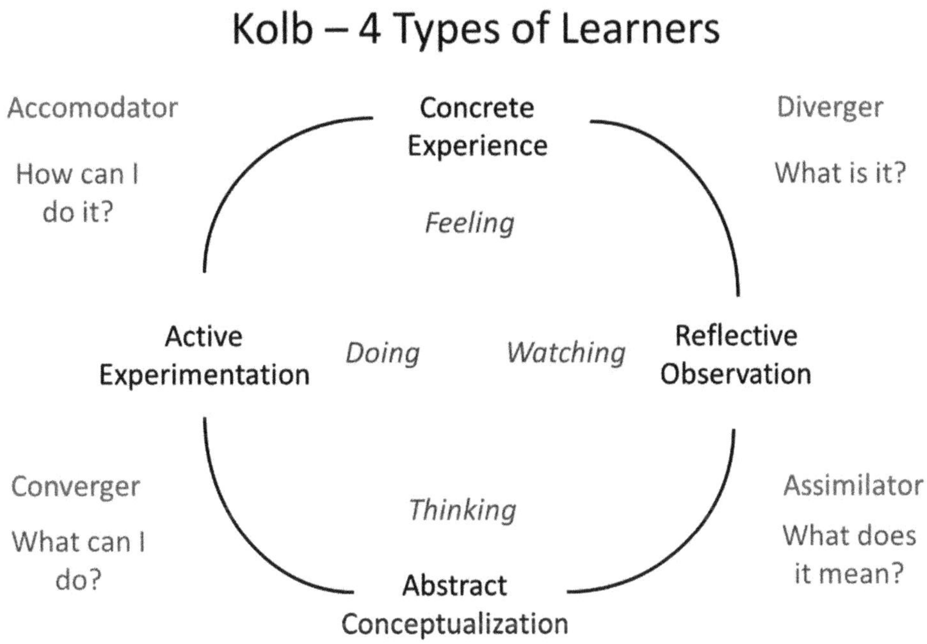

also put into positions where they need to work with learners who are different from them. This teaches them to learn from their peers and in manners they may not be accustomed to.

Self-Directed Learning

No blended learning curriculum would be complete without acknowledging the role of self-directed learning. Self-directed learning focuses on the goals, processes, and attributes of learning, as recognized by the learner in question. Tough (1967) along with (Knowles, 1970, 1984), laid the foundation for self-directed learning (SDL) in adult education. Merriam and Cafarella describe three uses for SDL (Merriam et al., 2007): (a) SDL as a goal; (b) SDL as a process; and (c) SDL as a characteristic (Candy, 1991). As a goal, SDL is used to increase self-directedness. It is focused on personal growth and autonomy. As a process, SDL describes the delivery mode. As a characteristic, SDL describes a learner's attribute.

All three aspects of self-directed learning are inculcated into the fabric of Gateways. As a goal, Gateways students are routinely encouraged to ask questions especially those that start with *why*. They can do this in person and in the on-line forums. As a process, Gateways students are given a multitude of tools to search for answers to their myriad of questions. Some of these strategies include on line learning modules, reading on their own and Internet searches. Finally, when they note the expressed expectations of the Gateways program, the bar is raised and they are continuously working to meet those newfound expectations.

While self-directed learning is often criticized for disregarding cultural perspectives (Brookfield, 1993), Gateways embraces cultural experience. The Garrison (1985) model of self-directed learning which includes motivation (entering of the task), self-management (control), and self-monitoring (responsibility) (Merriam et al., 2007). This model is more than appropriate to describe the self-directed learning experiences of the Gateways students.

Figure 3. Garrison's Model of Self Directed Learning

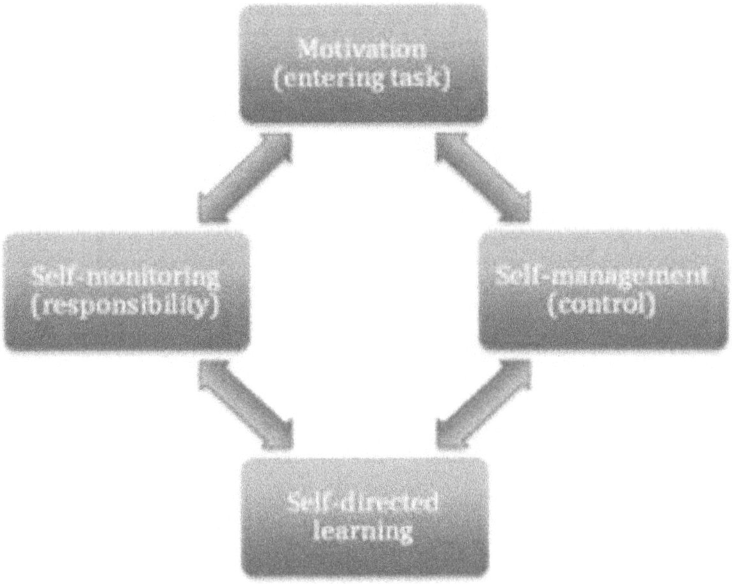

RECOMMENDATIONS

There are a number of recommendations that can be offered in order to successfully implement a similar STEM blended learning curriculum with a virtual mentoring community. These recommendations are based on the feedback of students, faculty, and staff as well as the author's personal experience developing this program.

In order to effectively use the LMS, students often need prompts. As such, weekly questions to reflect upon should be introduced at the culmination of each week. They will provide an opportunity for students to appropriately reflect on their actions and observations. The teacher leads by example. If the teacher responds in an authoritative manner, the students are more likely to engage with each other on the discussion board. By responding directly to most/all of the initial posts, the students will understand the expectation.

Articles of interest should be routinely uploaded onto the LMS in conjunction with discussion questions. This will allow students to actively keep abreast of and engage with the latest scientific findings while understanding the critical relevance of communicating scientific findings. Engaging with such materials in a safe virtual environment will make it easier for students to write about their own research findings later.

Profiles of physician-scientists should be added to the discussion board, preferably those who are of the same genders and ethnicities as the current students. This will provide a visual presentation of role models and hopefully decrease any feelings of inadequacy.

Finally, there are numerous ethical dilemmas that can arise in research and clinical care, as there are in many industries. It is critical that the students get familiar with the feeling of being uncomfortable. These moderated discussions should provide an opportunity for students to be open and honest with themselves and others as they discuss ethical dilemmas and consider how they might approach such quandaries.

CONCLUSION

There is paucity in the number of physician-scientists in the United States. This number is especially low for those who are underrepresented minority. This chapter describes a case example of a STEM summer program designed to increase the pipeline of competitive URM applicants. In this context of a summer preprogram within a medical school, technology coupled with in-person interaction helped fill a void where a "one size fits all" mentoring model would not suffice. This is why a combination of virtual mentoring, face-to-face mentoring, and blended learning with links and videos is ideal.

Providing feedback along the way is a critical component of mentorship as is reflecting on experiences with a virtual community. There are numerous avenues to give and receive feedback, information and questions in this program, including formal and informal interventions, in person and virtual modes. There are numerous people who make up the mentoring team, chief of which are the program leadership and Big Sib peer mentors. The various roles these people take on include advisor, instructor, cheerleader and sponsor. Every class is different and has varying learning styles among its students. It is critical that the curriculum be consistently monitored and modified, as needed, in real time.

REFERENCES

Association of American Medical Colleges. (2012). *Faculty Roster, Table 5*. Retrieved from https://www.aamc.org/data/facultyroster/reports/

Association of American Medical Colleges. (2017). *MD-PhD Summer Undergraduate Research Programs*. Retrieved from https://students-residents.aamc.org/applying-medical-school/article/md-phd-summer-undergrad-research-programs/

Atkinson, D., Neville, H., & Casas, A. (1991). The mentorship of ethnic minorities in professional psychology. *Professional Psychology, Research and Practice*, *22*(4), 336–338. doi:10.1037/0735-7028.22.4.336

Bacharach, S., & Bacharach, Y. (2012). The Coaching Mindset for Engaging and Developing Others. *eCornell.*

Barondess, J. (1994). A brief history of mentoring. *Transactions of the American Clinical and Climatological Association*, *106*, 1–24. PMID:7483164

Bierema, L., & Merriam, S. (2002). E-mentoring: Using computer mediated communciation to enhance the mentoring process. *Innovative Higher Education*, *26*(3), 211–227. doi:10.1023/A:1017921023103

Bierema, L. L., & Hill, J. R. (2005). Virtual Mentoring and HRD. *Advances in Developing Human Resources*, *7*(4), 556–568. doi:10.1177/1523422305279688

Brass,, L.F., Akabas,, M.H., Burnley,, L.D., Engman,, D.M., & Wilsey,, C.A., & Os, A. (2010). Are MD-PhD programs meeting their goals? An analysis of career choices made by graduates of 24 MD-PhD programs. *Academic Medicine*, *85*, 692–701. PMID:20186033

Brookfield, S. (1993). Self-directed learning, political clarity, and the critical practice of adult education. *Adult Education Quarterly*, *43*(4), 227–242. doi:10.1177/0741713693043004002

Buddeberg-Fischer, B., & Herta, K. (2006). Formal mentoring programmes for medical students and doctors-a review of the Medline literature. *Medical Teacher*, *28*(3), 248–257. doi:10.1080/01421590500313043 PMID:16753724

Candy, P. (1991). Four Dimensions of Self-Direction. In Self-direction for lifelong learning (pp. 97-126). Jossey-Bass.

Cranton, P. (2006). *Understanding and Promoting Transformative Learning: A Guide for Educators of Adults* (2nd ed.). San Francisco, CA: Jossey-Bass.

Daloz, L. A. (1986). *Effective Teaching and Mentoring* (A. B. Knox, Ed.). San Francisco, CA: Jossey-Bass.

Daloz, L. A. (1999). Mentor (2nd ed.). San Francisco: Jossey-Bass.

Eby, L. T., Allen, T. D., Evan, S. C., Ng, T., & DuBois, D. (2008). Does Mentoring Matter? A Multi-diciplinary Meta-Analysis Comparing Mentored and Non-Mentored Individuals. *Journal of Vocational Behavior*, *72*(2), 254–267. doi:10.1016/j.jvb.2007.04.005 PMID:19343074

Elam, C. L., Stratton, T. D., Hafferty, F. W., & Haidet, P. (2009). Identity, Social Networks, and Relationships: Theoretical Underpinnings of Critical Mass and Diversity. *Academic Medicine*, *84*(10Supplement), S135–S140. doi:10.1097/ACM.0b013e3181b370ad PMID:19907377

Evans, M. S., & Johri, A. (2008). Facilitating guided participation through mobile technologies: Designing creative learning environments for self and others. *Journal of Computing in Higher Education*, *20*(2), 92–105. doi:10.100712528-008-9004-1

Garrison, H. H. B., & Prudence, W. (1985). *Minority Access to Research Careers: An Evaluation of the Honors Undergraduate Research Training Program*. National Academies Press.

Gotian, R. (2016). Mentoring the Mentors: Just Because You Have the Title Doesn't Mean You Know What You Are Doing. *College Student Journal*, *50*(1), 1–4.

Gotian, R. (2017). *Optimizing Success of Physician-Scientists* (EdD). Teachers College Columbia University. ProQuest database. (10255977)

Gotian, R., Raymore, J., Rhooms, S.-K., Liberman, L., & Andersen, O. S. (2017). Gateways to the Laboratory: How an MD-PhD Program Increased the Number of Minority Physician-Scientists. *Academic Medicine*, *92*(5), 628–634. doi:10.1097/ACM.0000000000001478 PMID:28441673

Hathaway, R., Nagda, B., & Gregerman, S. (2002). The Relationship of Undergraduate Research Participation to Graduate and Professional Education Pursuit: An Empirical Study. *Journal of College Student Development*, *43*(5), 614–630.

Heron, J. (1999). *The Complete Facilitator's Handbook*. Sterling, VA: Kogan Page.

Hewlett, S. A. (2013). *Forget a Mentor: Find a Sponsor (Thew New Way to Fast Track Your Career)*. Boston, MA: Harvard Business Review Press.

Jack, A. A. (2014). Culture Shock Revisited: The Social and Cultural Contingencies to Class Marginality. *Sociological Forum*, *29*(2), 453–475. doi:10.1111ocf.12092

Johnson, L., Adams Becker, S., Estrada, V., & Freeman, A. (2014). *The NMC Horizon Report: 2014 Higher Education Edition*. Retrieved from http://cdn.nmc.org/media/2014-nmc-horizon-report-he-EN-SC.pdf

Jones, V., & Jo, J. H. (2004). Ubiqutous learning environment: An adaptive teaching system using ubiquitous technology. In R. Atkinson, C. McBeath, D. Jonas-Dwyer, & R. Phillips (Eds.), *Beyond the comfort zone: Proceedings of the 21st ASCILITE Conference* (pp. 468-474). Perth, Australia: ASCILITE.

Kardash, C. M. (2000). Evaluation of an Undergraduate Research Experience: Perceptions of Undergraduate Interns and Their Faculty Mentors. *Journal of Educational Psychology*, *92*(1), 191–201. doi:10.1037/0022-0663.92.1.191

Kasworm, C., Rose, A. D., & Ross-Gordon, J. M. (2010). *Handbook of Adult and Continuing Education* (2010 ed.). SAGE Publications, Inc.

Kereluik, K., Mishra, P., Fahnoe, C., & Terry, L. (2013). What Knowledge Is of Most Worth: Teacher Knowledge for 21st Century Learning. *Journal of Digital Learning in Teacher Education*, *29*(4), 127–140. doi:10.1080/21532974.2013.10784716

Knowles, M. S. (1970). *The modern practice of adult education* (Vol. 41). New York, New York: Association Press.

Knowles, M. S. (1984). *The adult learner: A neglected species* (3rd ed.). Houston, TX: Gulf.

Kolb, D. (1984). *Experiential Learning: Experience as the Source of Learning and Development*. Englewood Cliffs, NJ: Prentice-Hall, Inc.

Kolb, D. (1984). The Process of Experiential Learning. In Experiential learning: Experience as the source of learning and development (pp. 20-38). Englewood Cliffs, NJ: Prentice Hall.

Kolb, D. (1984). Structural Foundations of the Learning Process. In Experiential learning: Experience as the source of learning and development. Englewood Cliffs, NJ: Prentice Hall.

Looi, C.-K., Seow, P., Zhang, B., So, H.-J., Chen, W., & Wong, L.-H. (2010). Leveraging mobile technology for sustainable seamless learning: A research agenda. *British Journal of Educational Technology*, *41*(2), 154–169. doi:10.1111/j.1467-8535.2008.00912.x

Lopatto, D. (2004). Survey of Undergraduate Research Experiences (SURE): First Findings. *Cell Biology Education*, *3*(4), 270–277. doi:10.1187/cbe.04-07-0045 PMID:15592600

Marin, V. I., Jaakela, P., Hakkinen, P., & Juntunen, M. (2016). Seamless Learning Environments in Higher Education with Mobile Devices and Examples. *International Journal of Mobile and Blended Learning*, *8*(1), 51–68. doi:10.4018/IJMBL.2016010104

Martinez Witte, M., & Wolf, S. E. (2003). Infusing Mentoring and Technology Within Graduate Courses: Reflections in Practice. *Mentoring & Tutoring: Partnership in Learning, 11*(1).

Merriam, S., Caffarella, R., & Bamgartner, L. (2007). *Learning in Adulthood*. San Francisco, CA: Jossey-Bass.

Miller, H., & Griffiths, M. (2005). E-mentoring. In D. L. DuBois & M. J. Karcher (Eds.), *Handbook of Youth Mentoring*. Thousand Oaks, CA: Sage Publications. doi:10.4135/9781412976664.n20

Motsching-Pitrik, R., & Holzinger, A. (2002). Student-Centered Teaching Meets New Media: Concept and Case Study. *Journal of Educational Technology & Society*, *5*(4).

Nagda, B. A., Gregerman, S. R., Jonides, J., von Hippel, W., & Lerner, J. S. (1998). Undergraduate Student-Faculty Research Partnerships Affect Student Retention. *The Review of Higher Education*, *22*(1).

National Academy of Science, National Academy of Engineering, and Institute of Medicine. (2011). Expanding Underrepresented Minority Participation: America's Science and Technology Talent at the Crossroads. Washington, DC: The National Academies Press.

National Academy of Sciences. (2007). *Understanding Interventions that Encourage Minorities to Pursue Research Careers. Summary of a Workshop*. Washington, DC: Author.

National Institutes of Health. (2014). *Physician-Scientist Workforce Working Group Report*. Retrieved from http://acd.od.nih.gov/reports/PSW_Report_ACD_06042014.pdf

National Research Council. (2005). *Assessment of NIH Minority Research Training Programs: Phase 3*. Washington, DC: National Academies Press.

Odom, K. L., Roberts, L. M., Johnson, R. L., & Cooper, L. A. (2007). Exploring Obstacles to and Opportunities for Professional Success Among Ethnic Minority Medical Students. *Academic Medicine*, *82*(2), 146–153. doi:10.1097/ACM.0b013e31802d8f2c PMID:17264692

Papadakis, M. (2017). *Experiences of first-generation students at one ivy league campus* (EdD). Teachers College Columbia University, New York, NY. Retrieved from https://search.proquest.com/docview/1906740375?accountid=13314 (1906740375)

Pfund, C., Pribbenow, C. M., Branchaw, J., Lauffer, S. M., & Handelsmann, J. (2006). The Merits of Training Mentors. *Sciencemag.org, 311*, 473-474.

Retrieved November 19, 2017

Sambunjak, D., Straus, S. E., & Marusic, A. (2006). Mentoring in Academic Medicine: A Systematic Review. *Journal of the American Medical Association*, *296*(9), 1103. doi:10.1001/jama.296.9.1103 PMID:16954490

Schon, D. A. (1983). *The reflective practitioner: How professionals think in action*. New York: Basic Books.

Seymour, E., Hunter, A.-B., Laursen, S. L., & Deantoni, T. (2004). Establishing the Benefits of Research Experiences for Undergraduates in the Sciences: First Findings from a Three-Year Study. *Science Education*, *88*(4), 493–534. doi:10.1002ce.10131

Shedletsky, L. (2006). Internet training for older adult learners: An intergenerational mentoring approach. *The LLI Review*, *1*, 34–43.

Smedley, B. D., Stith, A. Y., Colburn, L., & Evans, C. H. (2001). *The Right Thing to Do, The Smart Thing to Do: Enhancing Diversity in the Health Professions*. Washington, DC: National Academy Press.

The Leadership Alliance. (2017). *Summer Research - Early Identification Program (SR-EIP)*. Retrieved from http://theleadershipalliance.org/programs/summer-research

Tough, A. (1967). *Learning without a teacher*. Toronto: Ontario Institue for Studies in Education.

Weinberg, F. J., & Lankau Melanie, J. (2011). Formal Mentoring Programs: A Mentor-Centric and Longitudinal Analysis. *Journal of Management*, *37*(6), 1527–1557. doi:10.1177/0149206309349310

Wong, L.-H. (2013). Analysis of Students' After-School Mobile-Assisted Artifact Creation Process in a Seamless Language Learning Environment. *Journal of Educational Technology & Society*, *16*(2), 198–211.

Yiu, L., & Parker, J. (2000). *Action Learning in Cyberspace: A Case Example of a Technology-Supported Laboratory Management Development Program.* Paper presented at the Transformative Learning Conference, New York, NY.

Yorks, L., & Kasl, E. (2002). Towards a theory and practice for whole person learning: Reconceptualizing experience and the role of affect. *Adult Education Quarterly*, *52*(3), 176–192. doi:10.1177/07417136020523002

Chapter 7
Digital Narratives of Immigrant Youth:
Cultivating Mobile Activism and Mobile Journalism Skills

Regina Casale
Teachers College, Columbia University, USA

Dominic Mentor
Teachers College, Columbia University, USA

ABSTRACT

This chapter focused on cultivating mobile activism mobile journalism with middle and high schoolers of a town in Long Island. The youth film production effort was in response to a hate crime. An immigrant was attacked and killed by a group of young males after a suspected spree of other attacks that same night. After the murderous incident, immigrant parent and students of the local schools feared for their lives. Working towards the goals, the organizers set out to teach students how to use mobile and computer technologies for filmmaking. Using themes of human rights, they also focused on responding to hate crimes and immigration issues. This chapter offers key discoveries and lessons. The short intensive program provided academic and workforce development skills as well as how to use computer technology for digitizing personal narratives. The program also offered informal academic purposes, along with observations, opportunities, and recommendations from the findings for other K-12 digital video filmmaking endeavors.

INTRODUCTION

In 2011, a summer media workshop was implemented to cultivate mobile activism and develop mobile journalistic instincts with 1.5 generation Ecuadorian middle and high school students from Patchogue, New York. To start working towards the goals of mobile activism and mobile journalism, the organizer and her partners arranged a hands-on, project based, short term engagement. The summer workshop was set up to teach students how to leverage technology for digital narratives with the theme of human rights.

DOI: 10.4018/978-1-5225-4111-0.ch007

The 2011 summer human rights media workshop was the second pilot of the effort and employed a professional videographer. The goals were to enhance 21st century academic skills, promote civic engagement, encourage intercultural dialogue and empower youth. This chapter will focus on the 2011 summer media workshop, explore the process and highlight the lessons learned as well as offer some examples from around the world and how these video projects were used to connect young students from around the world.

This chapter also focuses on the broad domains of practice, in terms of youths' digital narratives and video production that foster a host of critical thinking, hard and soft skills, agency as well as academic enthusiasm. Invoking enthusiasm was also achieved by utilizing multimodal stimulation and active engagements. The review will also bring attention to how mobile messaging has been leveraged by formal organizations and educational institutions. This literature review has the potential to aid international students or offices who might consider leveraging the ever growing mobile messaging services to help students acculturate to a new environment. Simple information sharing with these students could aid connecting them to their home support networks or by forming new virtual and/or physical networks.

This chapter will offer key discoveries and lessons from a summer media and human rights program. The short intensive summer program provided access, use, as well as highlighting the value of computer technology for digitizing personal narratives of understanding. The program also offered informal academic purposes, along with observations, opportunities, and recommendations from the findings for other middle and high school digital video narrative and filmmaking endeavors. The program inadvertently also empowered and provided young students with new skills that they could explore for future studies, or career tracks. Whether they focus on digital autobiographies, family stories, capturing oral histories or responding to hate crimes, immigration, or protests and activism.

BACKGROUND

The summer media and human rights workshop was in response to a hate crime in 2008 when Marcelo Lucero, an Ecuadorian immigrant was attacked and killed by a group of young males who were out "beaner hopping". A derogatory term used to describe the despicable act of group attacks on people perceived as undocumented immigrants. After the murderous incident, the Southern Poverty Law Center released "Climate of Fear", detailing other incidents of violence against Latino immigrants in the community over the last decade. The finding of this report triggered an investigation by the Department of Justice into the Suffolk County Police Department. Immigrant students of the local high schools feared for their lives and parents and teachers concerned for the safety of their children suggested that their children never walk alone and always walk in a group, to and from school.

LIFE through My Eyes, a human rights media project was a result of the tragedy. It was initially funded by the WK Kellogg Foundation through their America Healing initiative and co-sponsored by Film and Educational Research Academy (FERA) as well as the Long Island Teachers Association (LILTA) as a summer workshop for immigrant Ecuadorian youth. Apart from educating students on human rights and how bullying violates those rights, the program goals also focused on empowering youth with 21st century academic skills, promoting civic engagement as well as encouraging intercultural dialogue. Computer and information technology is interwoven throughout students' lives and does not run separately from their daily interactions. Be it their mobile phones, social media or other computer technology, students traverse their lives in an ever-growing computer-mediated manner, that demands learning how to use and leverage the technology as producers and not just as passive consumers of media. The 21st-century academic skills involved computer-mediated educational engagement and taught students how to create digital video, giving students a chance to tell their stories, show their understanding and have their voices later shared in several venues throughout the community. Allowing the students to further develop job and life skills by not only learning how to conduct public speaking, but also an introduction to media literacy and the importance of decoding messages.

Real life connections were further enhanced through field trips addressing human right themes that coincided with the summer workshop. Connections were made through visits to different religious and cultural institutions, luncheons with elected officials, university campus tours and participating in community based projects. These experiences lead to a specific topic on which to base their digital narratives. Students created storyboards, researched information, incorporated different camera angles, write voice over scripts, and learned how to take notes in an interview setting with their mobile devices. After the editing process was complete students shared their digital narratives in conjunction with the United Nations Plural + film festival screened at the public library, cultural centers and community theaters. Giving voice and developing agency with these students, also shed light on the concept of immigration, a natural, global phenomena.

IMMIGRATION ISSUES

The decision to immigrate is more than an economic one. Immigrant families interviewed for the Longitudinal Immigrant Student Adaptation (LISA) study revealed, that 70% of them came to the US to provide better opportunities for their children and 18% emphasized the opportunity for a better education (Suarez-Orozco, Suarez-Orozco, & Todorova, 2008).

The US immigrant population has increased considerably since 2000. By 2008, the Ecuadorian population was double that of 1990, becoming the fourth largest Latino population in NYC, following Puerto Ricans, Dominicans, and Mexicans. More than two-thirds of these Ecuadorians reside in Queens, NY (Caro-Lopez, 2010). In the top ten US counties with the largest Hispanic population, Ecuadorians rank first in Queens and tenth in Suffolk (Pew Hispanic, 2012). In Suffolk County the 2010 US Census reported that there was a 64% increase in the Latino population from 2000-2010. The village of Patchogue is home to the largest population of Ecuadorians on Long Island. In Patchogue, as for other schools on Long Island and across the country, immigrant communities have changed the faces of the students in the classroom.

Migration is not just a country specific issue, nor is it restricted to only one segment of society. Countries around the world have seen the migration of people due to cheap labor to increase business profits, conflict, natural disasters or even to pursue higher education. Universities, international officers as well as business organizations has seen that the world has become a global village with human migration happening at a rapid pace. This change is evident in the number of international students at various universities as can be seen in Table 1 from the *Open Doors 2010 Report on International Exchange* which shows the top 20 universities in the USA with the highest number of international students.

According to the Institute of International Education *Open Doors* 2012 *Report on International Educational Exchange*, the number of international students at colleges and universities in the United States increased by six percent to a record high of 764,495 in the 2011/12 academic year. With that increase of international students, comes a variety of business advantages, and cultural knowledge sharing. The

Table 1. Top USA institutions with international students for 2009 and 2010

Institute	City	State	International Total
University of Southern California	Los Angeles	CA	7,987
University of Illinois at Urbana-Champaign	Champaign	IL	7,287
New York University	New York	NY	7,276
Purdue University – Main Campus	West Lafayette	IN	6,903
Columbia University	New York	NY	6,833
University of Michigan – Ann Harbor	Ann Harbor	MI	6,095
University of California – Los Angeles	Los Angeles	CA	5,685
Michigan State University	East Lansing	MI	5,358
University of Texas – Austin	Austin	TX	5,265
Boston University	Boston	MA	5,172
University of Florida	Gainesville	FL	4,920
SUNY University at Buffalo	Buffalo	NY	4,911
Harvard University	Cambridge	MA	4,867
Indiana University – Bloomington	Bloomington	IN	4,819
Ohio State University – Main Campus	Columbus	OH	4,796
University of Minnesota – Twin Cities	Minneapolis	MN	4,665
Texas A&M University	College Station	TX	4,611
Penn State University – University Park	University Park	PA	4,561
University of Pennsylvania	Philadelphia	PA	4,522
Arizona State University	Tempe	AZ	4,483

Data source: 2012 Open Doors Report on International Educational Exchange

tools and practices in the Higher Education space are a lot more robust to help acculturate students to their new study and living environments as they familiarize themselves to a new environment, stay in touch with family and friends back home and make new found ones in the USA. 76% of international students surveyed in 2016 at a prominent New York university indicated that they feel very positive after connecting with friends and family from back home, but K- 12 school employment, curricula, tools, practices and educators are still working to reflect those changes in their classroom practices. In addition to cultural differences, educators must also be sensitive to immigration status of their students and family members.

This raises the question of cultural competency in K- 12 teaching, as in the need to have an awareness of one's own cultural identity and views about similarities and difference, and the ability to learn and build on the varying cultural and community norms of students and their families (NEA.org, 2015). In education cultural competency is being given a more critical role due to the acknowledgement that the linguistic, racial, cultural and class differences between students and teachers have been documented as playing significant roles in the achievement gap (Boykin & Noguera, 2011).

NARRATIVES

Culturally relevant pedagogy (CRP) has been suggested by Ladson-Billings as a way to improve teacher's receptivity to our increasingly diverse student populations. She proposes that pedagogy focusing on student achievement, acceptance and affirmation of cultural identity, as well as the development of critical consciousness, can challenge inequities in education and society (p.469). She documents how CRP provides a way for students to achieve academic success while maintaining cultural integrity. As part of the process, she proposes that students need to identify and critique social inequities which plague their communities. CRP aims to develop a learning environment in which the following are constitutive elements:

(a) *Students must experience academic success;*
(b) *Students must develop and/or maintain cultural competence; and*
(c) *Students must develop a critical consciousness through which they challenge the status quo of the current social order. (Ladson-Billings, 1995b, p. 160)*

The most current CRP has been framed within a critical race theoretical (CRT) framework and addresses issues of relevance to education through curriculum, instruction, assessment, funding and desegregation. According to Ladson-Billings, school curriculum is a culturally specific artifact that can sometimes downplay the experiences of students of color, although extensive research shows that including students' culture, language, and experiences can lead to academic success (Nieto, 2002). Many stories of youth marginalized by race and class are, however, reconstructed as acculturize (Delgado, 1989; Solorzano and Yosso, 2002) as CRT supports and encourages the retelling of stories by those people whose experiences have not been told from the perspectives of those who have lived the discrimination directly. Retelling stories is used as a tool to analyze, expose and challenge the stories of racial privilege and racial discrimination found in the dominant discourse. It is from this tradition that many scholars have used the genres of storytelling, biography, and autobiography and other methods to challenge dominant stereotypes (Yosso, Villalpando, Delgado Bernal, Sol6rzano, 2001). Vasudevan (2004) has named counter-storytelling as a site of discursive possibilities for learning how youth experience, live, mediate, and embody race and other subjectivities across their everyday social practices.

LIFE THROUGH MY EYES

LIFE through My Eyes, is a student led video project focusing on film and human rights that was first implemented shortly after the death of Marcelo Lucero, an Ecuadorean immigrant,, who was victim of a hate crime in 2008 in Patchogue, New York. The initial funding for this project was received through Kellogg Foundation America Healing Initiative grant from 2010-2013. The program has continued over the years through additional grants from Office of Juvenile Justice and Delinquency Prevention, NYSUT Retiree Council and Target. In the last two years the program was conducted in the Longwood Junior High School as part of the Spanish heritage learners program.

The participating families from 2010-2014 were originally from Azuay, Ecuador, which is the largest province in Ecuador experiencing an exodus of emigrants. Today they reside in Patchogue, a suburb of Long Island located 60 miles east of New York City where the immigrant population has doubled in the last ten years. Their children, the 1.5 generation Ecuadorian youth, also called DREAMERS, were currently or previously enrolled in an ESL program. According to Immigration Policy, dreamers are young immigrants that meet the criteria for the Development, Relief, and Education for Alien Minor Act (DREAM). These individuals are under the age of 31 who have arrived in the US under 16 years of age and continuously lived 5 year or more in the US. These individuals also have not been convicted of a felony, or misdemeanor, are studying in school or college, received a GED or have served in the military. Many of these young people speak Spanish at home but English is the language of instruction at school and socializing with friends. Although some of them scored below average on the New York State ELA test, their reading and writing skills in English are stronger than in their mother tongue. Across the board, many of these young people have experienced struggles with different types of discrimination in their community and schools with limited political and social capital needed to help to address the social maladies they encounter. Mills (1959) proposed that the first step in social change is to mobilize communities with similar issues, which is one reason why the project LIFE through My Eyes was implemented.

This innovative educational project was first implemented in 2010 in the Patchogue community as an after school program, summer program and parent workshops, and its success is attributed to its support from principals and staff, co-facilitators; Film and Educational Research Academy at Teachers College, Long Island Teachers Association, and most importantly parental support and collaboration. Over the course of three years we nurtured the relationship with each family through dinners, graduation and confirmation parties, group excursions, soccer tournaments and LIFE through My Eyes activities. As the relationships strengthened through community building activities we saw an increase in students' confidence as well during the filmmaking workshops. As shown in figures 1 and 2 students did not hesitate to approach strangers in the streets to interview or investigate concerns directly affecting their lives. "I got to say my opinion and how I feel", a student shared, "and I actually got to speak up".[1]

Methodology

The Patchogue participants attended several filmmaking workshops held at the public library after school, on the weekends, and in the summer. The goal was to capture the collective stories of these youth on video. These digital narratives were inspired by human rights addressing religion, language, culture, family, education, and immigration. All aspects that impacted them that they did not necessarily know were rights that they had, or rights that would also apply and serve them well in the workplace. For the first few years we contracted facilitators from FERA at Teachers College and Educational Video Center (EVC). An example of the career interests and job skills cultivated through the Video Production program was that eventually we were able to recruit one of our alumni Estefania, a young filmmaker, who studies television and radio broadcasting to facilitate our workshops.

First, participants were encouraged to share topics that affected them most, either in the community, school or at home. Many of the common themes included; discrimination, racism, hunger, bullying and keeping the community clean. At the EVC workshop, students were directed to choose an article from the universal declaration of human rights and complete the writing prompts as shown in figures 3 and 4 such as; my third eye sees…my heart feels…, if I ruled the world I would….

Figure 1. Child labor on the streets of Patchogue resident is being interviewed about child labor

Figure 2.

After these activities students were able to organize themselves into groups with common themes to write the script, and questions.

Secondly, once the script is completed each group works on their storyboard and discuss necessary shots and photos, match script and quotes with each image, and choose music. Before going out on camera shooting, groups must assign the following roles; soundman, cameraman, interviewer, director, and someone to scout the scene. After each interview, students switched so everyone gets an opportunity to experience each role. Third, after filming was completed groups began the editing session and again shared responsibilities. As shown in figures 5, and 6 editing included hard skills like video editing, adding

Figure 3. Writing prompt, freedom of movement

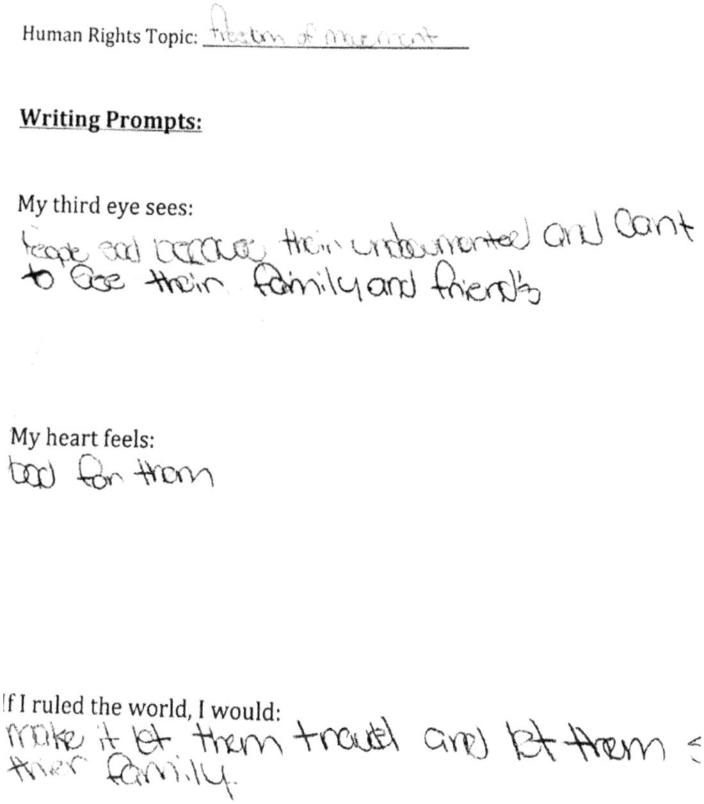
Human Rights Topic: Freedom of movement

Writing Prompts:

My third eye sees:
People sad because their undocumented and cant to see their family and friend's

My heart feels:
bad for them

If I ruled the world, I would:
make it let them travel and let them s thier family.

sound or images and also took the critical thinking development further with scene selection, soundbites to include, appropriate music and font of subtitles, to mention but a few actions.

Several studies have documented how participatory media can be used as a tool to empower children's voices. Other examples from around the world are New York's Educational Video Center (EVC) offers documentary workshops to students and teachers at its own facility and at schools. Its founder Steve Goodman (2010) strongly defends the use of digital media as a legitimate and productive pedagogy, especially for impoverished districts, to foster critical media literacy and civic engagement, supporting the New York State Common Core standards. He conducted a case study in a documentary workshop, inviting a team of students to engage in a sustained video –based inquiry into a social issue within their community. The short video 'Young Gunz', highlighting gun violence in the neighborhood, allowed students to voice ideas, questions and possible solutions regarding issues that were affecting them personally or at the systemic level.

Children in Immigration about Migration (CHICAM), was an action research project funded by the European Commission from 2001-2004, that involved young people in cinema clubs in Sweden, Netherlands, Germany, Italy, Greece and the UK. It was created to give children a voice and to promote intercultural communication through filmmaking (de Block, L., Buckingham, D., & Banaji, S, 2005). The various clubs were comprised of refugee and migrant students ages 10-14, one researcher and one facilitator. Not only did participants create short films but the films were then shared with their international counterparts via the Internet.

Figure 4. Writing prompt, describing problems

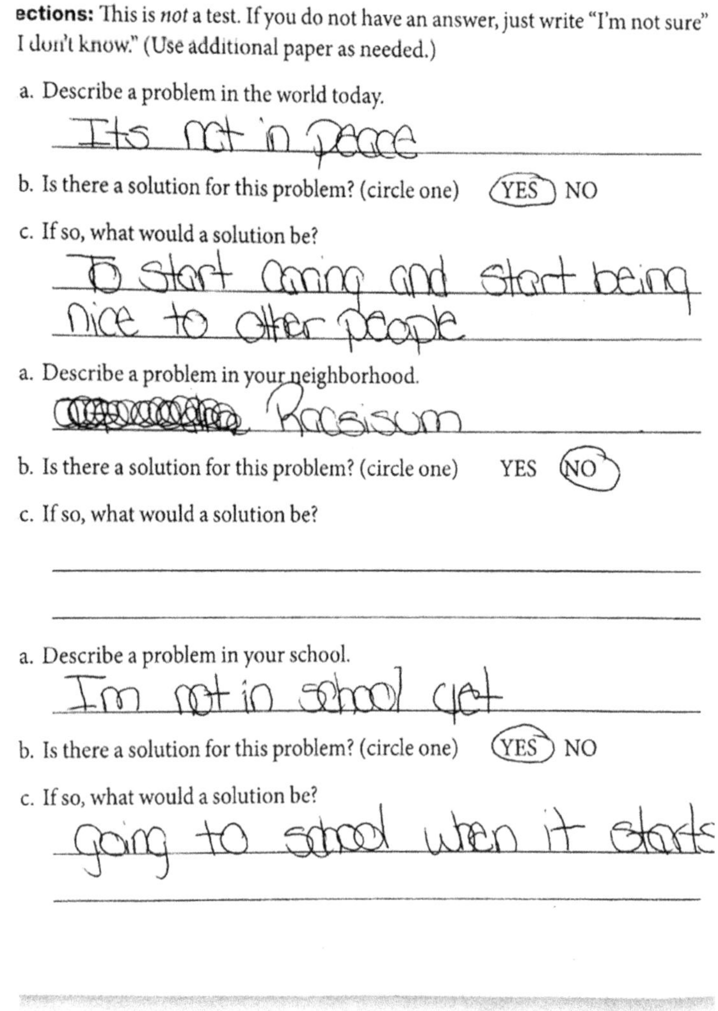

ections: This is *not* a test. If you do not have an answer, just write "I'm not sure" I don't know." (Use additional paper as needed.)

a. Describe a problem in the world today.

Its not in peace

b. Is there a solution for this problem? (circle one) (YES) NO

c. If so, what would a solution be?

To start caring and start being nice to other people

a. Describe a problem in your neighborhood.

Racisum

b. Is there a solution for this problem? (circle one) YES (NO)

c. If so, what would a solution be?

a. Describe a problem in your school.

I'm not in school yet

b. Is there a solution for this problem? (circle one) (YES) NO

c. If so, what would a solution be?

going to school when it starts

The ethnographic research project, Video Culture was conceived and first implemented in Ludwig, Germany in 1997, and explored how audiovisual media production can be used to communicate between young people in different regions. This research focused on how the form of transcultural symbolic language can overcome cultural and linguistic differences. The project included groups of young people ages 14-19, from different socioeconomic backgrounds in Germany, England, the Czech Republic, Hungary and the United States. These students produced, exchanged and interpreted thematically oriented videos based on work from different educational settings. Selected students had no previous media production experience and were given five days to complete a three-minute short film based on a specific theme (Niesyto, Buckingham & Fisherkeller, 2003).

Students developed individual leadership and group work skills. From the outset, we attempted not to frame our approach in formal methodology, which afforded many pros and cons. Advantages of operating from a flexible framework meant that the facilitators, support staff and students were not restricted to what they could envision or design within the framework laid out previously. Disadvantages to this

Figure 5. Editing workshop with FERA at Pat-Med HS using I-Movie

Figure 6. Editing workshop with EVC at South Ocean MS using I-Movie

flexible approach, at times, did not fit well with the formal approach or the academic modus operandi of K -12 teaching settings. Furthermore, we also needed to operate within the confines of the learning centers digital tools the limited after school and summer program time and their own personal audience's identified goals. In addition to requiring the students to conduct a needs assessment of their assigned themes or interests, we also asked students to get a baseline of what they currently understood from one of the UN Human Rights' themes. We also had to ascertain what digital tools we would have access to at each digital center in terms of technology (computers, mobile phones, WiFi or what their knowledge or use of producing video was, as well their computer and media literacy skills. Our informal needs assessments were conducted as part of our program, and also looked at what technology the student were currently using or had access to from a computer technology hardware and software perspective. In keeping with our learning theories that informed our methodology and the reality of time frames and time limitations, the idea was to leverage what knowledge and levels of computer proficiency they currently have, and to build on what they identified as needing to produce their digital video narrative. Ultimately, we increased their computer technological literacy and life skills.

CONCLUSION

Over the past few years we have seen how social media, specifically video, has been used successfully as a tool in mobilizing communities. Examples include the following: transmitting the democratic uprising of the Arab Spring; documenting verbal abuse toward an Uber driver which led to the resignation of a NYC officer; the recorded death of Eric Gardner resulting from an illegal police choke-hold in Staten Island and the recordings of a Suffolk County police officer caught stealing from undocumented immigrants during unauthorized stops. These recorded actions have all lead to a call for action once viewed by the public.

The process of working up to the video production, as well as the post public presentation of the produced videos, fostered positive influence on their lives, their families and the communities. The analogue and digital learning journey also reflected their continuous and affective interactions and experiences and aided their social emotional well-being, which in turn can have a positive effect on their personal lives as well as their academic progress and success.

Observations indicate that a collaborative project of this nature can be highly beneficial for all stakeholders as it delivers a layered work that reveals the hosting of multiple perspectives and multi-directional memory. The students depiction of the Universal Declaration of Human Rights (UDHR) and other immigration issues posits a post-colonial tension within different contexts. The digital video production also navigates the inner thinking and at times conflicting ideas of prescribed and self-actualized identity. And does so through the promotion of agency in resisting racial policies and practices. Echoing Hall's (1972) deconstruction of race as a "floating signifier", some of the students' videos initially depicts the themes through an idealized and romanticized lens but at the same time offering multiple perspectives of a harsh politically imposed reality. The multi-dimensionality of identity and power of agency amidst a forced upon dominant culture was also explored through their production of the digital video narratives.

The video production processes were initiated in ways that increased collaborative efforts, online and offline. Using applied theoretical knowledge and experiential teaching systems, we not only strengthened the capacities of the current young literacy digital narrative programs, but also empowered the learners to draw on past experiences and individual expertise. Thus, they became active collaborative agents who had the ability to use technology to build and contribute to the film, real and online world. Youth tend to enter settings ready to learn and the program capitalized on those aspects to engage them in video production and digital narrative projects that, while they focused on themes, community needs or their group problem solving techniques, they were further motivated by intrinsic factors. The systems and instruction was designed to meet the self-identified goals and objectives of the learners, and the learners' use of the digital media tools ultimately developed them in ways that were difficult to formally assess. In general, current computer literacy, digital production and purposeful internet usage in NYC's K-12 education programs consists predominantly of desktop word processing, and unstructured Internet surfing with search engines. There were only introductory instruction on, or with digital media tools to produce the video, but students were provided space to build on the introductory foundation.

Facilitators realized there were different meta-cognitive processes at play for each student and groups of students. This was evident from the devices they used and how they employed the tools. Tools used by students showed careful consideration developed by them prior to, during and after the video production process. Teaching support staff had the opportunity to see the underlying and overarching pedagogical principles at play in incorporating digital video production in a program and as a practice in general. Shifting some of the teaching assistants' ideas, who were long term serving teachers where passive receiving of education is the order of the day, offered new disciplines. While receiving support from the teaching assistants was challenging, it was not a deterrent to the students using the digital tools.

Success also came in the form of introducing the young learners to the numerous digital tools that went into video production tools and that was available to them. Universal user interface offered familiarity with the tools and their quick uptake of the digital tools offered a further functionality, which aided in the success of the initiatives, by anchoring the overarching goals to an objective or level of proficiency that the learner was comfortable with. Learners were reminded that success was defined by a digital narrative project that was authentic and important to them. Lessons learned lasted beyond the piloted time frame of the LIFE digital video narrative initiative. Examples of success ranged from students' own personal growth, proficiency, as well as transformative learning experience for students and for some of the full-time teachers helping out as teaching assistants

It is difficult to say which learning activities or initiatives were most appealing, as well as to have measured the students learning enjoyment. The creation of a digital narrative and a collective space for past, present, and future films produced in the program, showed a body of work that they could be proud of and also showcased the skills they picked up along the way. The past and new program designs received positive reviews. These youth produced films have been screened in film festivals at universities, local theaters, libraries, cultural centers and recently in conjunction with the United Nations Plural + film festival. In 2015 and 2016 the Suffolk County Inter-Faith Anti-Bias Taskforce recognized the program, its facilitators and *Immigration and Diversity*, a short film by a Estefania and *Speak Up Words Matter* filmed by Longwood JHS Hispanic Heritage Learners as scene in figure 7.

Although many of the digital media tools still need to be integrated into their formal schooling, students now have the knowledge and skill to present any of their schoolwork or personal stories, in a well-crafted digital film manner. Parents and staff also provided positive feedback which resulted in an

Figure 7. Words Matter Speak Up anti discrimination film recognized at SC Interfaith Anti-bias task-force 2016

increase of students participating in the program over the years and an increase in the number of digital narratives produced.

In the first 34 days of the 2017 USA presidential administration there were 1,094 bias incidents reported. According to the Southern Poverty Law Center, there has been an increase to 917 with an increase of 197% in anti- Muslim hate groups. Empathy and desire for social change begin with an understanding of others; to listen to their stories and share one's own. In today's current climate it's most important for educators and community leaders to include the diverse perspectives our students bring to the classroom. Many of our immigrant youth are recognized as digital natives, capable of sharing narratives about how life is experienced through their eyes using computer technology. I encourage my colleagues to support the use of film as a pedagogical tool to foster cross-cultural dialogue between our educators and newcomers. I hope through the lenses of critical pedagogy and critical race theory more teachers will explore how the process of creating digital narratives nurture students' voices and increase civic engagement in a target population whose voice is limited by their immigration status.

Through our collaborative efforts, the middle and high school students emerged with a sense of innovation that catapulted them to the forefront of being able to traverse and take ownership of their physical and digital world. With the team leadership of the program organizer, the young learners increased their confidence in navigating and leveraging their new digital video production skills. The efforts by the LIFE through their eyes program, had the effect of making some of the digital learning centers more visible, not only in terms of use, but also in terms an community presence. The digital video projects that were produced also instilled the potential for these young learners and their affiliated learning organizations to not only connect their youth with one another but also with societal issues and enhancing further support for their cause of highlighting Marcelo's unfortunate death as inspiration for the empowering program. This social community digital video project, which was started for the betterment of young learners in New York City, put a strategy in motion for the Long Island Communities, New York City, and their similar organizations from around the world. Valuable lessons were learned from the implemented digital media projects and key issues were revealed that could also be applied to other digital video production centers that are pushing for more digital narrative initiatives of marginalized youth.

The initial short term implementation delivered not only lessons, but also built momentum for future projects within the program, the digital learning centers, and other non-for profit organizations that focus on informal digital video production learning opportunities.

RECOMMENDATIONS

Our immediate recommendations for a second iteration of this project encompassed the changing of digital center venues, leveraging other video production persons and organizations, as well as increasing the time frame of engagement to allow students more time to develop and hone their video narrative skills. The recommendations were multifold and interrelated with all stakeholders concerned and addressed the previous challenges faced by the students and at the learning centers. The overseeing committee also looked into how to enhance the technical and support infrastructure of the program, and either increase or reduce the involvement of the full time teaching assistants. While the first runnings of the program was done in short pilot format bursts, the entire process could have benefited from a longer time span where students could possibly have worked on more than one digital video project over a semester. A recommendation for the program organizers was to train and empower a select group of full time teachers as assistants and to utilize technically gifted learners to help manage the program, the center and the supportive adult group of parents and other stakeholders. Learner's social media could also have been explored to add further ownership, pride as well as to market the program objectives and their work. Utilizing the social media of the learners would offer further ownership of the projects and offer the students opportunities to revisit and hone their technical, digital video production skills, and leadership skills.

Consultants contacted could have met earlier, virtually and longer with all the major stakeholders, ranging from the program organizers, teaching assistants as well as the learners. So that the consultants on the project could also have gained deeper insight in to the students' digital literacy levels, their likes, learning styles and various other social media initiatives they embraced. This could be addressed by embedding the consultants within one or two of the projects during the semester or before the summer part of the program. They could act as passive observers and help in identifying a project management sharing tool, to provide cognitive surplus through collaborative tools and build out the infrastructure of knowledge management, and to act as a virtual hub and reference point for the digital video projects. All of which, would have been beneficial to all stakeholders who were mostly full time employed with other competing demands. An online central location could also be identified for the project to be used to share resources among all involved, where everyone could document and ask question of the program organizers, and cultivate collaboration amongst themselves with virtual tools. The online tools could offer constant access and opportunity for feedback when it was convenient for all agents involved in the project rather than confining the sharing to just before the start of the summer program. All involved would then have had greater cognitive resources to share amongst themselves, cater for ongoing feedback, and everyone could have had a living document to reference during the semester running of the afterschool portion of the program. The students, who operated in teams of two or three, could also have had their own online spaces that they could have used to keep track of project progress, offered additional reflection and feedback and could have done so in that virtual space with different support staff who could have offered affirmations or ideas.

The project wouldn't have been successful without the support of the parents. Their trust is critical for educators to successfully advocate for students. As parental approval increased so did the desire for community organizations to become more involved. From the beginning parents and students had input on how to organize the program and over the years they have taken on leadership roles. To run a successful program it is crucial to empower parents to become community leaders and to identify globally aware, competent educators to serve as liaisons between the community and the school. The ability to receive independent funding also allows educators more freedom to implement programs through a grassroots framework.

FUTURE RESEARCH DIRECTIONS

Since the inauguration of LIFE through My Eyes in Patchogue in 2010 there have been several success stories, students pursuing degree in the visual arts, film and television. Alumni have taken on leadership roles as facilitators and graphic designers. Many have been encouraged to submit work to film festivals or other competitions and some have been recognized for their talents as recipients of the Marcelo Lucero Award as well by local legislators and organizations. There have been a total of 19 youth produced digital narratives and over 150 participants at four different schools.

There is still a need to grow, teach and expose the unique advantages of interacting with as well as producing digital film and for a wide variety of audiences. There is an additional need, not just within this program, but also other youth programs of this nature to teach digital engagement and production of digital video with and via hypermedia concepts and means. Using a blended learning approach, the computer technology's hypermedia capabilities, can be used teach at all levels and can actively promote personal reflection, participation, the collation of multiple perspectives and provide opportunities to learn from others youths' previous digital film examples that could be interacted with via availability online.

The teaching of digital film via hypermedia offers collated means and incorporates the use of online environments for the real-time teaching and learning of films. With the 'inter-web' paradigms shifting from end-users to active participators, the milieu of the web have increased human ICT use and interaction. The hypermedia framework is set up in such a didactic manner that teachers and students of this program and others of the same nature can be exposed to an online learning environment that is hopefully more than a repository for film clips, but rather an interactive and dynamic learning environment that offers book-marking within the film clips, space for saved reflective notes as well as asynchronous interaction with other class participants on the aforementioned points. All of which, would not only activate reflection, but provide valuable tips, tricks, styles and new ideas of how the new students in the program could approach their digital video production. Hence, instead of starting from complete scratch, they would have seen some mistakes made by others or fantastic ideas that were employed.

In creating and using an interactive hypermedia learning environment, the program staff and participants could be exposed to deal with and save for future use, discussions, theories and concepts used by other educators, students, student analyzers, and reviewers of the previously produced digital video and its various styles, forms and/or embraced other genres in the making of their digital film.

The in-class, or rather in-person program sessions, combined with the online hypermedia video learning environment could extend review time, discussions, analysis and teaching methodologies of digital film and tools. Reflections and comments could also cover contexts, political settings, institutions, systems and social situations. It will emphasize the comparative and divergent angles from which digital film production can be approached and offered within various learning environments from elementary through to high school.

Figure 8. Reflection of filmmaking experience

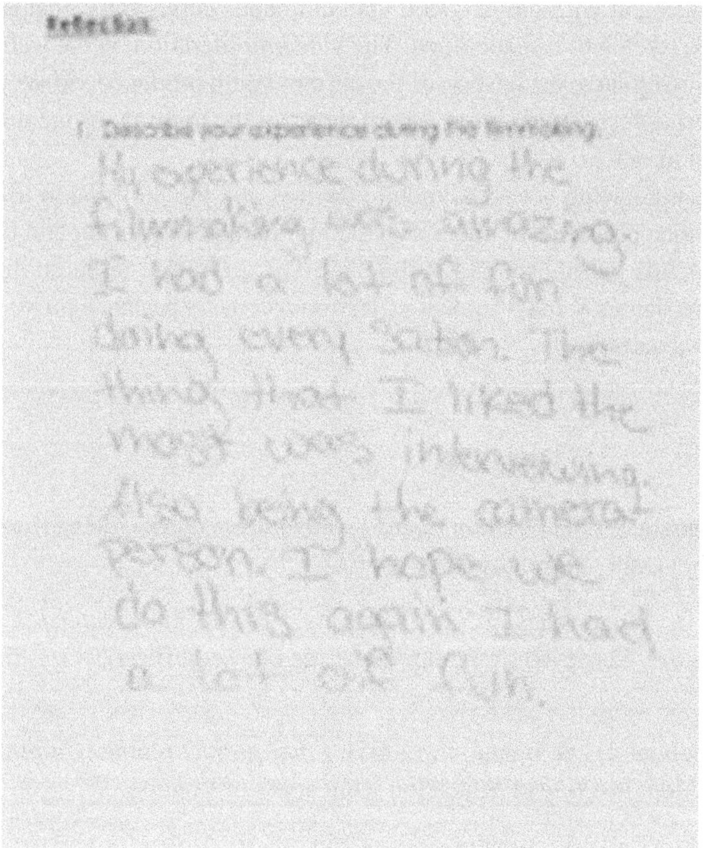
Reflection

1. Describe your experience during the filmmaking.

My experience during the filmmaking was amazing. I had a lot of fun doing every station. The thing that I liked the most was interviewing. Also being the camera person. I hope we do this again I had a lot of fun.

Figure 9. Reflection of filmmaking experience

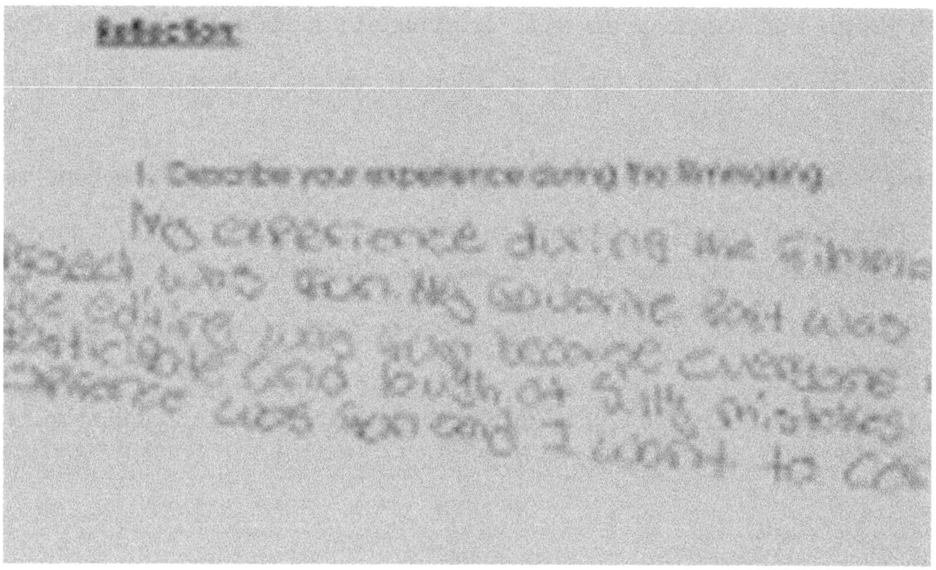
Reflection

1. Describe your experience during the filmmaking.

My experience during the filmm[illegible] project was fun. My favorite part was the editing [illegible] because everyone [illegible] laugh at silly mistakes [illegible] experience was fun and I want to [illegible]

For the program, taking movement within the context of our current digital lived lives of online video repositories and engagement with online video via comments, community engagement and impact of different scales can be taken into consideration. The video proliferation on the web via YouTube, plus a host of other video hosting sites, the hosting of the current youth produced videos of the program could be turned into interactive engagement tools, as either part of the program or included as anchors within other educational opportunities.

The increase in youth having access to mobile devices and the permission to use these devices in class give equity to video programs to underserved communities. Capturing and editing mobile videos to further promote mobile journalism and mobile activism, as future research directions will also be explored. As shown in figures 8 and 9 student reflections on the experience prove to pursue the path of youth generated digital narratives.

REFERENCES

Boykin, W. A., & Noguera, P. (2011). *Creating the opportunity to learn moving from research to practice to close the achievement gap*. Alexandria, VA: ASCD.

Caro-López, H. (2010). Ecuadorians in New York City 1990—2008. *Center for Latin American, Caribbean & Latino Studies*. Retrieved from http://clacls.gc.cuny.edu/files/2013/10/Ecuadorians-in-New-York-1990-2008.pdf

de Block, L., Buckingham, D., & Banaji, S. (2005). *Final project report: Children in communication about migration*. London: Institute of Education, University of London.

Delgado, R. (1989). Storytelling for oppositionists and others: A plea for narrative. *Michigan Law Review*, *87*(8), 2411–2441. doi:10.2307/1289308

Goodman, S. (2003). *Teaching youth media*. New York, NY: Teachers College Columbia University.

Goodman, S. (2010). Educating for democracy: An uncommon standard. *The New York Times,* pp. 1-6.

Ladson-Billings, G. (1995). Toward a theory of culturally relevant pedagogy. *Educational Research*, *32*(3), 465–491.

Lewis, O. (1959). *Five families: Mexican case studies in the culture of poverty*. New York: Basic Books.

Niesyto, B., Buckingham, D., & Fisherkeller, J. E. (2003). Video culture: Crossing borders with young people's video productions. *Television & New Media*, *4*(4), 461–482. doi:10.1177/1527476403255813

Nieto, S. (2002). *Language, culture and teaching critical perspectives for a new century*. Mahwah, NJ: Lawrence Erlbaum Associates, Inc.

Solorzano, D., & Yosso, T. (2002). Critical race methodology: Counter-storytelling as an analytical framework for education research. *Qualitative Inquiry*, *8*(1), 23–44. doi:10.1177/107780040200800103

Suarez-Orozco, C. S.-O., & Todorova, I. (2008). Learning a new land. Cambridge, MA: The Belknap Press of Harvard University Press.

Vasudevan, L. (2004). *Telling different stories differently: The possibilities of (counter)storytelling with African -American adolescent boys* (Doctoral Dissertation). Retrieved from http://repository.upenn.edu/dissertations

Yosso & Solorzano. (2001). Critical Race Theory in Chicana/o Education. Paper presented at the National Association for Chicana and Chicano Studies Annual Conference, San Jose, CA.

ENDNOTE

1 Student shared after film workshop addressing women's rights May 2015.

Chapter 8
Investigation Into the Selection of Online Learning Platforms and Tools in Higher Education

Paul Evan Acquaro
Teachers College, Columbia University, USA

ABSTRACT

Selecting and implementing the platforms and tools to support online learning effectively in higher education is currently a challenge for which there is not a one-size-fits-all solution. Finding the right balance between supporting strong pedagogy, offering training and support, providing data security and privacy, ensuring ease of use, among other factors, shape the decisions that leaders in higher education make as they develop and implement online learning environments. This chapter explores the results of a study conducted during the Fall of 2016 to better understand the efforts higher education experts undergo to develop online learning environments. The study considers the sometimes-competing pedagogical and pragmatic needs such efforts entail and seeks to identify trends and best practices.

INTRODUCTION

The process of selecting and implementing the platforms and tools that support online learning in higher education offers no clear and well-defined approach. Faced with an expansive selection of tools, competing interests, limited time and budget, stakeholders involved with the creation of an online environment make many important decisions with uncertain outcomes. In an effort to better understand the opportunities and challenges that decision makers face in their efforts to develop their institution's online learning environments, a study was conducted during the 2015/16 academic year probing the influence of the myriad technical, pedagogical, and experiential factors on the choices that decision makers face when evaluating and purchasing learning platforms and tools.

Thirty people in decision-making roles at higher education institutions were surveyed about their experiences in teaching and/or managing online learning programs, exploring their backgrounds, teaching and learning preferences, and satisfaction with their current online environments. Then, going beyond the

DOI: 10.4018/978-1-5225-4111-0.ch008

survey results, one-on-one interviews with five of the subjects were conducted to provide additional insight and explanations. While their stories are but a small sample of the many unique and diverse experiences occurring every day, they provide some awareness of both the challenges and success that institutions have had in their approach to addressing the needs of online learners. The findings ultimately point to the pivotal role that institutional governance plays in shaping the decision making and implementation process suggesting nascent best practices that can guide others in their work.

BACKGROUND

According to a study by Picciano & Seamen (2009) it was estimated that over a million K-12 students took an online course during the 2007 -2008 academic year. In higher education, it is estimated that in the United States, that at the same time over 3.9 million higher education students were taking a course online (Allen & Seamen, 2008). By 2012, according to federal data, the percentage of American higher education institutions offering some level of online offerings had reached 86.5% (Allen & Seaman, 2014). Considering the growth of online education, it is important think of the impact that online learning will have on the structure of higher education in the 21st century. A report by the Chronicle Research Service (2009) made the case that the traditional four year college experience is no longer the norm, "as demonstrated by the proliferation of colleges (particularly for-profit institutions), hybrid class schedules with night and weekend meetings, and, most significantly, online learning" (p.1) This change in higher education norms is outlines by Ortagus' (2017) examination of the characteristics of students enrolling in online courses and found that "postsecondary online learners are not necessarily confined by niche characteristics that require the convenient and flexible benefits of online instruction to gain access to higher education, particularly for students who supplement their residential education with some online courses." (p. 52).

Businesses and governments have also experienced the transformative changes made possible by online collaborative tools (O'Reilly, 2005; Tredinnick, 2006). Grant (2006) in an article about using the knowledge building Wiki noted that "changes in the developed world are leading to a move away from economies based solely on 'traditional' industries toward a new 'knowledge economy'" (p.1). It stands to reason that if one of the goals of higher education is to help students become competent, collaborative, problem solving employees of tomorrow (Scardamalia, 2002), a logical extension that higher education should be preparing students to be critical producers and consumers of information (Buckingham, 2003; Buraphadeja & Dawson, 2008; Marks, 2009). To this end, the development of critical thinking and communication skills is an important component of higher education (Burbach & Matkin, 2004), and collaboration with others is a key element to truly support thinking critically (Choy & Cheah, 2009).

Studies have shown pedagogy that builds upon social learning theories help learners in retaining and applying learned information in other situations. However, researchers and practitioners have called attention to a gap between the learning approaches considered most effective for learners and the actual instruction found in the online classroom (Meier, 2016) and it is noted that both the pedagogy used and the tools available for educators tend to favor more traditional methods (McLoughlin & Lee, 2007).

Before delving into the role of the online tools in closing this gap, it is helpful first to recognize that different philosophies toward learning will suggest different tools, and different tools can be used to reach students in different ways (Kanuka, 2008). While the unique learning styles of students should be taken into consideration (Ally, 2008), one constant is that college students today are not expecting traditional pre-digested, sequenced learning (Tapscott, 1998), instead they require a curriculum that teaches them to be active interpreters and competent participants in a "new paradigm emerging from the networked information economy in contrast to the production models of the industrial age" (Bruns, 2009, p. 227), where they are producing knowledge, interacting with each other, and collaborating across distance and time.

With the vast array of tools available and for most educators a paucity of experience with online teaching and learning, some "online learning theory needs to help educators decide which of the numerous technological options is best suited for their application" (Anderson, 2008, p. 54). Heather Kanuka's chapter in Anderson's *Theory and Practice of Online Learning* (2008) is important in framing this line of thought, as it outlines the need for educators to be aware of their own teaching and learning preferences, since they impact the tools selected for teaching. She writes, "the debate over whether or not we need to prepare our learners for a pervasively networked world revolves around what types of persons we expect our education systems to produce." (p. 93). Essentially, if we ignore the changing nature of the workplace and stick to the script that treats education along the mid-20th century industrial approach, then collaborative, knowledge-building tools are much less a necessity.

However, the assumption is that decision makers in higher education, and specifically for online learning, do want to embrace the actual learning needs dictated by the forces of industry and society. Shifting the perspective to the instructional designer whose pedagogical choices are impacted by the tools, the deliberate and thoughtful selection of the platforms and tools then comes into focus. The importance of tool selection on the interactions within an online learning environment was captured effectively by Oubenaissa, Giardina & Bhattacharya (2002) when they wrote, "the dynamics allowed by the environment became the main element that orients the design approach" (p. 42). The researchers further underscored this when they explain that "with learning spaces becoming more complex and dynamic by the potential of sophisticated and intelligent technologies, these spaces are now defined by their tools, type of activity and interactivity, and work sharing allowed to learners" (p 43). Chu, Reynolds, Tavares, Notari, and Lee (2017) reinforce this when discussing how using the proper collaborative tool encourages the exchange of ideas and support of activities to foster learning.

At this point, considering the growth in the numbers of students taking online courses, the question here is how can schools be sure that the tools they are selecting can support the type of instruction that meets the needs of tomorrow's workforce.

Creating the Online Learning Environment

Acknowledging the depth of research concerning social learning approaches, Brown and Adler (2008) explain that educators should build learning communities around the social learning activities, aided by collaborative online toolsets and informed by their communities. On their own, online resources do not have the important reflective qualities that learning requires and online tools do not have value without participants (Oreilly, 2005; Shirky, 2008). However, by utilizing the resources available in terms of open education resources, collaborative tools, and LMS systems, while borrowing heavily from social learning approaches and adult group learning, formal courses of study can be brought to life for the online student.

It is, of course, not quite as simple as the above paragraph would suggest. A ubiquitous component of the online learning environment is the Learning Management System (LMS), onto which criticism has been leveled, often surrounding their rigidity and lack of social networking (Meier, 2016; Watters, 2014; McLoughlin & Lee, 2007). For the most part, the LMS presents courses in the typical semester long format and in contrast to most students highly connected lives, approaches social media as a 'walled garden' (McLoughlin & Lee, 2007). On the flip side, there is acknowledgement of some of affordances that LMS offer, including, being interfaced with library holdings, student information systems (among other digital resources), and log in via a secure server to the course content (Burke & Tumbleson, 2016; Davis, Little and Stewart, 2008).

LMS aside, it is also important to consider what practical factors like training, data privacy, policy, support, and financial resources may impose on the learning environment. Decision makers responsible for the development of a learning environment in a higher education institution must take all of these concerns into consideration as they select tools, train faculty, and ensure that students information is private and secure. Practically speaking, harnessing the strength of social learning theories and collaborative knowledge building with web based tools, may also present conflicting security, cost, and usability issues.

Tool Selection

In *Theory and Practice of Online Education*, Alan Davis, Paul Little, and Brian Stewart (2008) wrote about the complexity of setting up an online learning environment:

Building the infrastructure for online learning has many interconnected components and many factors must be considered, so it is hard to provide a straightforward checklist or recipe to follow. Distance education has provided an understanding of how the entire system of course design, development, and delivery occurs, and how these link to related learner services and other components, all of which are vital aspects of ensuring effectiveness and quality. (p. 122)

This is still a concern today, Walker, Lindner, and Murphrey (2016) see the issue of finding the right tools and technology in relation to competition in the online learning space:

As technological advances continue in the higher education field, it will become vital for universities to take advantage of them to enhance teaching and learning. The higher education market is becoming highly competitive, and having an LMS that is highly adaptable and flexible to integrate with future technologies will prove to be a wise investment for any university. (p. 42)

In addition, the results from this study underscore that while decisions makers in the online learning field believe the general approach to good online learning should embrace social and collaborative teaching and learning methods, there is still very little agreement on how to achieve and support such an environment. In terms of current tool selection, the findings even suggested a mismatch between the desired learning environment and the most popular tools. While the results are drawn from only a small fraction of the many institutions involved in online education, there still seemed to be suggestion of trends and best practices in setting up the environment.

One important trend at the institutional level, which emerged from the data analysis, is that some success appears to have been found by employing different governing approaches, a theme that emerged in the findings but had not been a consideration. In some cases, institutions have begun approaching tool adoption and implementation via setting up institutional processes, taking into consideration instructional approaches and school-wide resources. Such undertakings typically came from self-assessment processes spanning technical and academic areas. It seems to have placed the institutions that have conducted such reviews into more secure positions supporting their online teaching efforts.

Though the original intent of this study was focused on what influences the selection of the online tools and platforms being implemented in online programs, in seeking these answers, a larger picture formed in which the tools themselves are only part of the story.

METHODOLOGY

The study employed a mixed-methods approach, using an online survey that was administered to a targeted group of participants, mostly via reference as well as follow-up interviews. The goal of the survey was to develop a broader understanding of the factors that impact decision makers in the selection and implementation of the LMS and other tools. This was followed by a series of one-on-one interviews to dig deeper into some of the findings. A variety of methods were used to identify participants, including attending professional conferences, requesting participants recommend peers and contacts, and appealing to members of interest groups via the social media tool LinkedIn. After a preliminary analysis of the 30 surveys, 5 people were interviewed, selected on their background experience in teaching and information technology and reported satisfaction with the tools.

The final pool of participants represented undergraduate, graduate, private and public colleges - including two institutions that offer their degrees almost exclusively online. The subjects had professional titles that ranged from senior instructional designer to Provost for online learning; all of the respondents had a primary role in making decisions regarding online tool selection.

Survey Results

Ranking the Influencing Factors

One of the goals of the survey was intended to determine the way tools were chosen, or at the very least, narrowing in on what phenomena influences the decision-making process. To this end, one of the survey questions asked participants were asked to rank the top three factors out of a list of factors which could possibly influence their platform and tool choices. Out of the total 30 respondents, *recommendation from faculty* came in at the top spot of 47% (n=14), with 9 participants ranking it in first place and 4 in second. *Research studies* came in second at 40% (n=12), with 8 participants giving it the top choice, and 2 each for second and third place. *Ease of use for faculty* came in third overall (33%, n=10), split evenly between the first and second spots (See Figure 1).

Preferred Approaches to Teaching and Learning

Participants were asked to rank their preferred learning approaches along a five point Likert scale ranging from *very effective* to *not effective*. The list contained several social learning approaches, along with more direct teaching methods: case-based, client-based, discussion-based, problem-based, project-based, self-paced, lecture based, and other.

Figure 1. Top three ranked factors that impact tool selection as ranked by participants

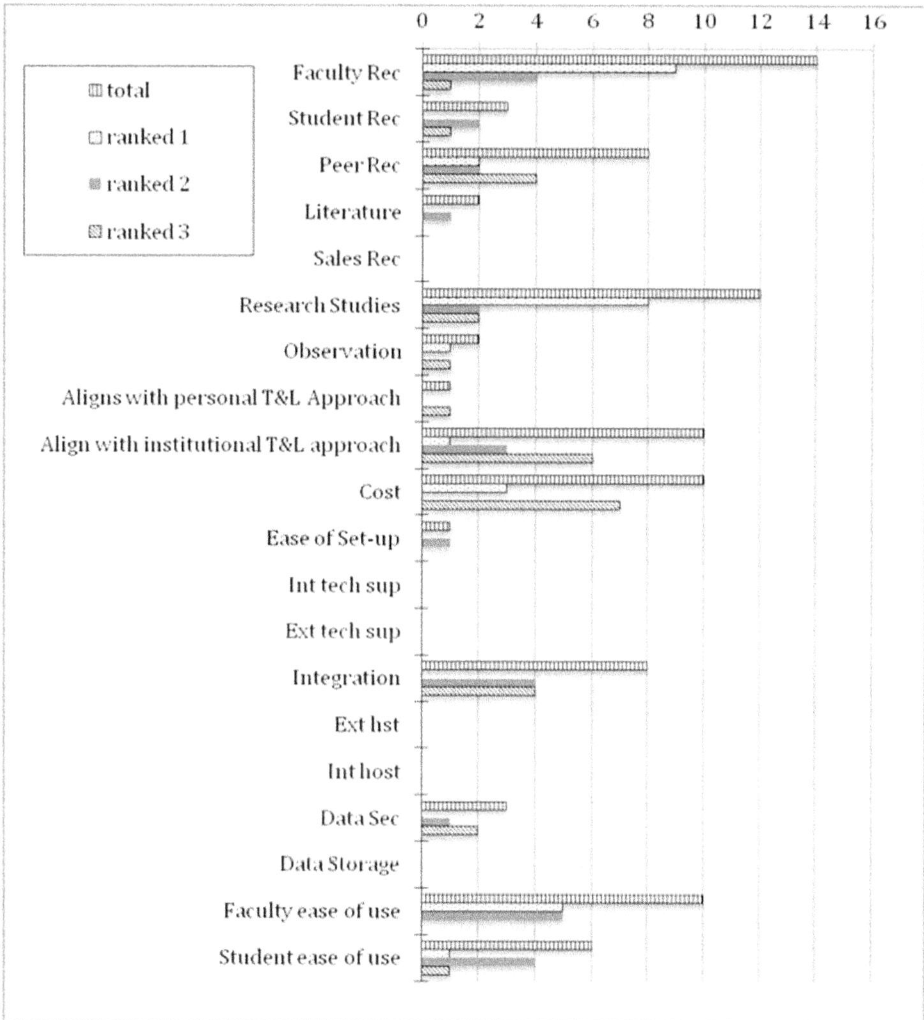

The term *effective*, as used here, is drawn from the definition proposed by Bradford Bell and Jessica Federman in their 2013 meta-analysis of e-learning in post-secondary education. Bell and Federman wrote that the term is often applied in e-learning studies as comparisons between in-person and online. However, the researchers suggest a more expansive definition of effectiveness and instead take a cue from Richard Clark's (1994, as cited in Bell & Federman 2013) agnostic point of view that technology is a tool to be employed by instructional designers and its pedagogical use is paramount in its proper use. Bell and Federman cite the work of Steven Ross, Gary Morrison, and Deborah Lowther (2010) in their definition of education technology "as a broad variety of modalities, tools and strategies for learning. Its effectiveness, therefore, depends on how well it helps teachers and students achieve the desired instructional goals" (p. 19). In his UNESCO working paper, Lubin, I. A. (2016) also found Ross, Morrison, and Lowther (2010) and Clark (1994) to be important references in emphasizing pedagogical ends over tools, but being mindful of the technologies affordances. Thus, the term *effectiveness*, in this study, is seen as the degree to which the tool or learning approach in question meets the participant's pedagogical goals.

Among the 30 respondents, it was generally agreed that *lecture-based* learning was the least effective approach to online learning, with all 30 indicated that it was either *not effective* or only *somewhat effective* (total: 70%, n=21), and all believed that *problem-based (87%, n=26), project-based* (87%, n=26), and *case-based* learning (90%, n=27) were either *very effective* or *moderately effective. Discussion-based* learning was considered *very effective* by 53% (n=16) or *moderately effective* by 33% (n=10) (see Figure 2).

Which Types of Tools Are Considered Effective?

Because the names of tools are often in flux, a question was presented that listed tools by category. Participants were asked to rank what they believed were the most effective tools in teaching and learning. The scale offered the following options: *very effective, moderately effective, somewhat effective, not effective,* and *unknown.*

Across all respondents, the top three tools that were considered to be very effective were videos 73% (n=22), simulations 72% (n=21), and discussion boards 53% (n=16). Blogs and chats tied for fourth place at 50% (n=15), followed by social networking 43% (n=13) and wikis 40% (12). At the bottom of the effectiveness ratings were Podcasting, Drawing, RSS Feeds, and Social Bookmarking, all coming in at 10% (n=3) (see Figure 3).

How Does a College Respond to Support Calls and Training on New Tools, Which Sometimes They May Not Even Know Are in Use?

To explore this question, the survey asked what level of institutional circumscription should exist for faculty tool use. The survey asked:

Figure 2. Preferred approaches to teaching and learning among all participants

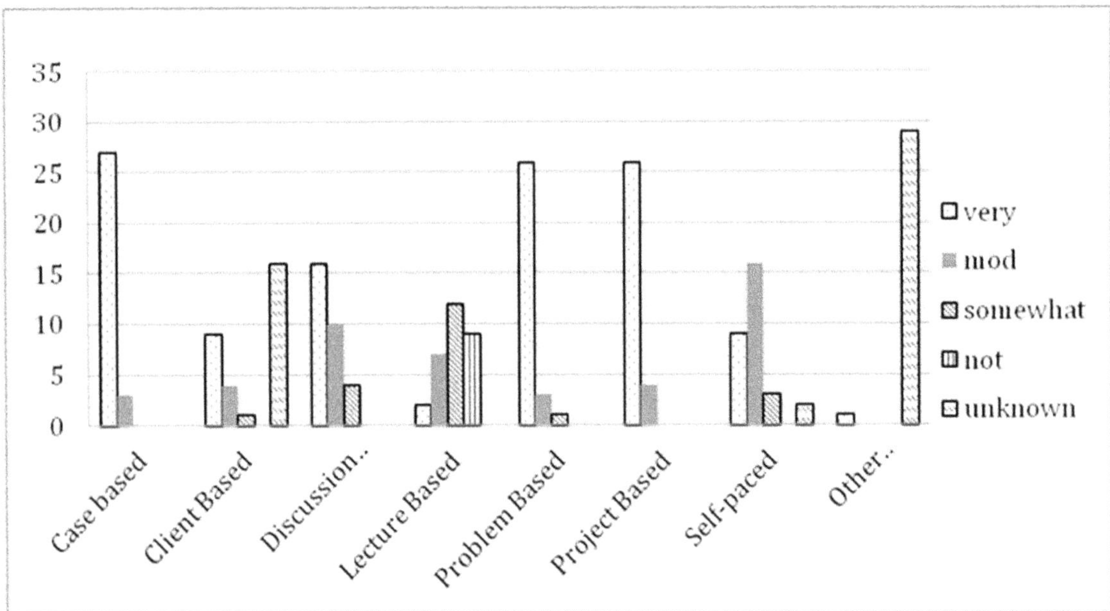

Figure 3. Effectiveness of online tools by type

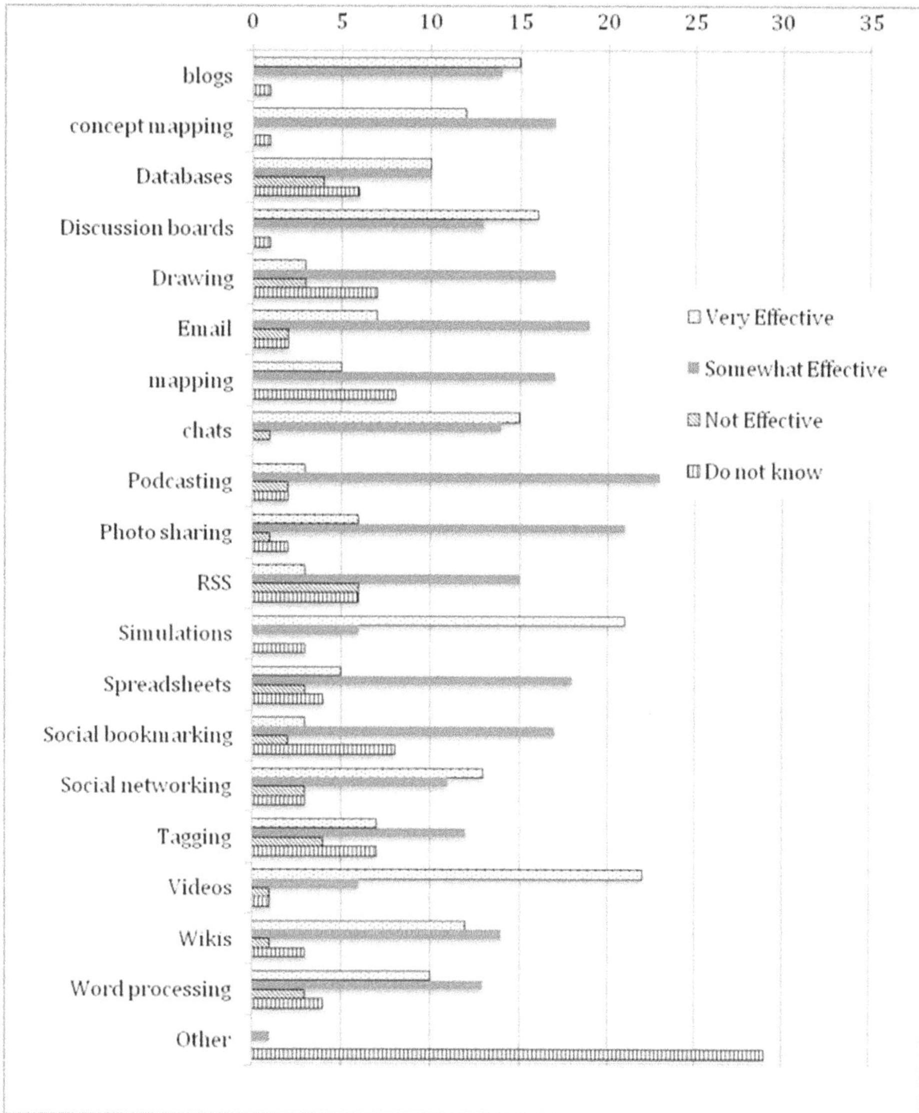

Do you feel that faculty should be able to:

- *Use any learning tools and platforms tool they wish*
- *Use any learning tools that they wish, but must use the institution's learning platform*
- *Use only the learning tools and platforms that your institution provides*
- *Undecided*

In the results, the majority (56%, n=17) indicated that faculty should be able to use any tool that they wish, as long as they use the school's LMS as the basis for the class. Following this, 20% (n=6) took a harder stance, indicating that faculty should only use school-provided tools, while 7% (n=2) were open to any tools. The remaining 13% (n=4) were undecided.

While there were many other data points in the survey, these preceding findings offer a small window in what drives tool selection, and which types of learning approaches are valued.

Interview Subjects

Turning to the follow up interviews, the following five subjects took part in one on one interviews to help give more background to the survey findings. For some context, here are some short descriptions (and pseudonyms) of the interviewees:

Brad works at a large, private, multi-campus university with an international reach. His school offers both undergraduate and graduate degrees. He has an extensive background in administration, some background in IT and teaching, and has been working with the administration of online programs for 5 years.

Jake works at a university extension. The university itself is a public, multi-campus institution that offers both undergraduate and graduate degrees. His background is exclusively in IT, with little formal teaching experience, and he has been working with the administration of online programs for 5 years.

Kate works at a large public college that is part of a state education system. Her school offers a great deal of online courses at both the graduate and undergraduate level. She has extensive background in administration, some background in IT and teaching, and has been working in online learning for 14 years.

Pat works at a small private college that offers graduate degrees. Her background is primarily in teaching, and she has little formal background in IT, and has been working with in online learning for 18 years.

Scott works at a medium-size private graduate college. He has an extensive background in both teaching and IT, and has been working with the administration of online programs for 15 years.

Interview Results

The interview protocol followed a general set of questions, with each one taking its own shape as the interviewees expanded upon their answers. In considering the role of tools first and foremost, the idea that tools themselves are less important than the learning environment and approach came across in several of the interviews. Another important idea that came up was the idea that selecting an LMS – hastily – can actual be a 'trap' for a school.

First however, Scott speaks specifically to the point of his efforts to ensure that faculty feel supported and are happy with the tools. In general, he sees tools more as a set of options:

I think you need to have access to open tools for the faculty where they can go and experiment and not feel like that they have run off the range. That's what I love about Google Apps—it really gives you a way to say—instead when people like I just found this new tool let's buy it—you can say, "Oh, sorry, well you didn't like WebEx, would Google Hangouts help you? Maybe you would just like to try a Skype with this? Maybe that would be all you need."

Pat's observation about ignoring tool specificity serves well to underscore this point:

I can't tell you how many times I've hammered a nail into the wall, using a shoe—that's the analogy that I use all the time. I will use any tool if it works. It may not be the tool that was created for that purpose, but I'll use it for something.

Jake indicted that he was satisfied with the learning environment at his school, but acknowledged that platforms have their limitations and a perfect online environment is probably not a reasonable expectation:

I actually think that there are some big problems with the LMS and with the way some things are currently handled. So in that sense I'm a little dissatisfied, but our tools, in comparison to what is available in the world, I'm pretty happy with what we've got.

However, it was Kate who went further in suggesting that tools, specifically LMS, can be a "trap" for an institution:

I think as an institution we fell into the trap that almost every institution falls into when they are going online—and maybe it's not fair to call it a trap but you sort of go "OK, I'm going online, I need an LMS, and there is a thing called a CMS that someone that some said once, I think I must need that too." And you have vendors lining up, saying, "Oh you know Blackboard can do that for you, or whatever ... Canvass ... etc."

Outside Tools

Over half of the survey participants indicated that they felt faculty could bring in any tool they wish, as long as they use the college's LMS. Twenty percent of the participants indicated that faculty must use the tools that are supported and available, and the remaining percentage was comprised of participants either more lenient who thought faculty could use any tool they wish, or they were undecided. It seems that there is a bit of division at this point between creating an environment that can be easily supported, data-secured, and quality-controlled versus one that is entirely open for exploration, trial, and error.

For example, Kate's college offers a set curriculum and a suite of supported tools. Kate is also in the small percentage of respondents who feel that faculty should *only* be using provided tools. Depending on the scale and reach of the school, this is a pragmatic approach. With this scalable, replicable, but restrictive model come options that other schools may not have:

When you have a traditional university, you have tenured faculty, and each faculty sort of designing their own course. We can't do that. If we did that we'd be teaching statistics 500 different ways, and with 100 different textbooks, so we actually have a standardized curriculum. So, that in turn allows us to make deliberate choices of instructional design at scale and technology at scale.

It would seem that speaking with Kate, that having the curriculum and teaching approach established, the challenge moves onto selecting the right technology:

I think I spend most of my time thinking about the technology and how we can get it to do what we need it to do over anything else. The curriculum, we have subject matter experts and we know good instructional design, and we have the content. It's really the technology that's often the challenge.

Scott and Pat discussed the importance of having a space to experiment. Pat spoke about a center for technology that was a place for faculty to experiment and try out tools, and pilot courses using new technologies. Scott spoke about how it is important for faculty to drive the course and, in a sense, choose the tools that work best for them:

Even if it's not faculty-driven, it should be faculty-guided, there should be somebody there who is engaged with the institution and understands the institutional mission, and not just institutional mission, but departmental and programmatic mission that at the end of the day they are not just cookie cuttering programs out.

What seems promising is a middle ground that Jake and Brad have separately staked out. Brad discussed how his institution established a yearlong project during which they canvassed the university and came up with a representative set of tools in use. They then tested and learned them, so that they could be supported and integrated properly, from an IT standpoint, into a larger, flexible environment. In Brad's words:

So we ran a support pilot for a full year that just kind of said, "Hey, are our school based and central teams able to provide support if faculty are interested in innovating?" Because there was one big discussion around "Well, I don't want to do anything because the university cannot support me so I'm not going to get started and have my dreams dashed, when they can't be achieved."

Jake discussed the need for the support team to know the tools that are being used, as they are on the front-line supporting users. He also wants to ensure that software is adhering to best data security and privacy standards, and explained how a systemized vetting helps him:

Sometimes people have an out of pocket solution that needs to be evaluated in relation to a particular class of software. For example, telecommunications, we already know what we need to know about WebEx and Adobe and Skype, there's the whole lot. We don't have to think too hard about those things, but sometimes unique challenges come forward and we have to spend a little bit more time evaluating those, and depending if it's within our domain, evaluating it in-house here. If it's institutional, then we participate in the committee through the IT department.

The least structured environment seems to be the one that Scott described. With multiple LMS, faculty are encouraged to bring in any tool they wish. The risks and rewards may balance out, as the college encourages exploration and discovery on one's own, however the technology may be being implemented less systematically and/or without proper support.

As factors like *security*, *cost*, and *usability* come into play, some sage advice comes from Kate explaining how her college is now rethinking – and reevaluating its LMS strategy from a needs based perspective:

We're really starting off with functional requirements. We looked in the past and we said we need an LMS, OK let's go look at LMSs, OK we saw there three, I like that one better, let's go with that. Now what we're trying to do is determine what are they key things that we need to do.

Role of Pedagogy

Moving beyond the tools, the role that pedagogy plays in tool selection was also made by several interviewees. For example, Scott discussed his experiences in online education; he explained that students want strong pedagogy, not cumbersome tools:

My concern isn't the overall online experience in terms of is it the most beautiful, does it have all this branding to it, is it the best experience online. Because at the end of the day what the students want is the experience, and they really want strong transformative pedagogy.

And also what faculty should ultimately be concerned with:

In our meeting yesterday we started talking about "how do I do this," and the first focus is that they want to talk about the technology but then it starts to have to do more with the pedagogy.

Pat underscored these statements with a discussion how the tool is a vehicle for getting something done, for driving pedagogy:

Faculty who come in may need anything ... for me technology has always been the tool that we slap on top of what we're trying to get done. So I've set up the environment that I'm working in to be that way.

Avoiding the Trap and Look Ahead

Scott, who reported to be *satisfied* with the tools, also said he liked his LMS situation because he had several to choose from. His satisfaction seemed to be rooted in the fact that there was a suite of options, and that he also takes an agnostic view of learning tools in general. "I like my current LMS situation. Which is just that I can just cherry pick from different LMS, that's something I've never had before." However, he also tempered this with the following observation about the transient nature of the technology: "I've never been anywhere that we haven't changed LMS within three years."

Pat (who is not satisfied with the tools), echoed Scott's sentiments:

The LMS usually is given to you at the institution you are at. I've been through two or three new LMS integrations, but I wasn't a part of the selection committees.

This question may be best served by Kate's view of the LMS and higher education. As mentioned, her institution has an LMS in place, but there is now a very deliberate process in place to find a replacement that is better suited to the needs:

What I think has ended up happening in this space is that everyone searches for an off-the-shelf solution and then once they really get into it they find out that it doesn't really work all that well, because it was not ever geared to what their functional requirements were.

Brad, whose institution has gone through a self-assessment and has settled on a particular LMS, echoed this type of deliberate and thoughtful needs assessment approach:

We worked this year with Gartner on an external review and also some interests groups both in [city] and [city] to try to understand some of the LMS of note, this year we put together a rubric of evaluation that Gartner helped validate and review and what their results suggest are that there are minimal differences between some of the core LMS administrative systems, the systems that connect to your student system and help you do grade books and rosters and what we call add/drops and stops, those elements that extend to grade books, basic quizzing functionality, lessons, tools.

The goal for Brad is a solid but flexible architecture:

That is what we are all looking for—balance in how to sustain a set of services that are flexible from an architectural stand point, secure and reliable from a data standpoint, and still allow for some innovation. For us, it's building that architecture that you can add things in and take things out as they are needed.

This is a goal shared by all of the interviewees. Everyone is seeking a tool set that is flexible, can be mixed and matched, and supported in an integrated manner. Like, for example, Jake:

I think that a whole lot of things like LMSs and much of the software is just a front end on a database anyway. So if we can keep the database part but maybe make the interface more in line with synchronous content, while also gaining the advantages of asynchrony and distributed content, I'd be excited for that to happen.

While many people in the survey said that they were are satisfied with the toolset they had (60%, n=18), it seems that true satisfaction may be tempered by the search for the right balance between innovation and integration. The dream of the perfect toolset was said well by Scott, who would like to see the tools to be so transparent that students only see the learning:

My dream environment would be an institution that would be flexible enough to bring in technologies as they become relevant and also focused enough to understand that the online program needs to have a certain experience to it. That experience can be pedagogical, but I think any student coming into it—I want to remove the technology burden on the students so they notice how strong the pedagogy is.

DISCUSSION

If as the survey results suggest, that the pedagogical goal is to support *project-based, problem-based*, and *case-based* approaches, as they are thought to be the most effective ways to teach and learn online, then selecting the platform and tools that can support these approaches should be the guiding criteria. Following the survey results, the main factors to consider then is ease of use by faculty and students, published research on the tools, and faculty recommendation. Of course, these top factors are tempered by finding the right balance between cost, security and support.

In terms of current platform and tool selection, the survey findings suggest a possible mismatch between the reported most effective teaching and learning approach vis-à-vis the tool type rankings. What stands out about the results seen in Figure 4 is that when considering how social teaching and learning approaches were highly regarded, *video*, which does not typically require much social interaction, is considered a more effective tool than collaborative knowledge building ones like *wikis* or *social bookmarking*. However, this survey did not ask how lessons and activities were constructed and how videos may be situated in another tool. It seems though that the top ranked tools should be ones that support rich social interaction like *wikis* and *social bookmarking*.

Stepping back a bit from the tools themselves, the most important step in selecting and implementing an online learning environment is assessing the type, scale, and program model that makes the most sense for the institute. However, it was beyond the scope of the study to go much further into the types of models, but just from the interviews, it is apparent that there exists a contrast between a faculty-driven model and a top-down professional teaching model, with variations in between.

There are many ways to undertake this type of assessment, and it will differ from school to school. One approach to identifying the correct online learning tools is through an assessment of the institution's needs, from which a criteria matrix could be used to help select LMS and other tools (Mentor, 2016). In the interviews conducted in the study, the two large university systems, one private and the other public, that did an assessment across all of its schools, agreed upon a single LMS and came up with a process to support new tools as they are brought into the environment. Another large public school already had a strong idea of how they should operate and built their online program to fit this approach. Two smaller schools, both private, seemed to be either about to enter this process or will soon be thinking about it.

Following the assessment phase, a governance structure should be developed around a set of policies and procedures which can then help in the adoption of technologies in a supportable and sustainable manner, and help bridge any gaps between the IT and academic divisions. In the cases of the schools that already have a governance model, the assessment helped to guide the creation of learning environment.

Support and training is also an requisite component; as one interviewee noted, faculty are the last mile in delivering online education, and another interviewee explained how group support of learning technologies is an integral component in creating the overall online learning experience. Questions that emerge are how to best offer assistance to faculty and how to open up a process where tools can be selected and vetted so that meet certain support and training criteria.

Experimenting is important, too, and the creation of centers that offer faculty the opportunities to test and explore tools can be a key development in this space. Again, according to the survey, the most influential factor in tool selection is faculty opinion, first in terms of ease of use of the tools, but also their overall input and recommendations, thus a process that bridges experimentation and implementation, space to explore, and proper pedagogical and technical support can possibly help bolster a program's success.

RECOMMENDATIONS FOR FUTURE STUDIES

This study very much raised as many questions as it began with, but with the intent of honing the line of questioning from information gathering to more actionable information. However, the first suggestion would be to expand the amount of subjects. Leaders of online learning are busy people with multiple, competing demands on their time, participating in a survey and a 45-minute conversation can be difficult to arrange. That being said, it would help to check, and even expand on the studies conclusions by running this study with a larger sample.

Additionally, the focus so far has been solely on the administrator, yet at many times it seemed that getting the opinion of faculty and students would add another voice to the mix, which could help place findings into a larger context. Adapting this study to be done with faculty and students would provide two other important perspectives into the adoption of preferred tools and learning approaches. Within these three population groups, one would likely find a wealth of information that could help better inform and improve the setup of the teaching and learning environment for online programs.

Finally, it would be of interest to catalog and evaluate the different models of online learning to see if there is a suite of online platforms and tools best suited to a particular approach. While this study has begun exploring these models from an administrative perspective, this is an aspect that emerged during the course of the study.

CONCLUSION

While the size of the sample is small, the findings do seem to suggest that the decision-making process in online tool and platform selection is heading into a new phase. As online learning programs evolve, moving away from the exploration stage to a more stable and institutionalized one, the road to tool adoption must become easier to navigate. As there is no 'right way' for going forward, institutions must look at embracing compatible teaching and learning frameworks, developing strong technological and pedagogical support systems, and ensuring that faculty voices are heard in the selection process when developing a learning environment that supports meaningful online education. The notion of the LMS 'trap' that one of the participant's coined, must be a goal of the institution to avoid, rather the adoption of tools, or creating an environment, needs to be a deliberate and thoughtful process with an established policy for new tool selection.

In the case where there may not yet be such a comprehensive policy for tool, a path to developing one begins with an evaluation of what online education should look like at the institution. Again, there is no doubt that there are many ways to make this type of assessment, and it will most likely differ from school to school; however, as seen from examples in this study, out of an assessment, a governance structure and set of policies can be developed around a set of processes and procedures which can help to regulate the adoption of technologies in a supportable and sustainable manner.

REFERENCES

Allen, E.; Seaman, J. (2008). *Staying the Course: Online Education in the United States, Survey report.* Sloan Consortium (Sloan-C).

Allen, E., & Seaman, J. (2014). *Grade change: Tracking online education in the United States*. The Sloan Consortium.

Ally, M. (2008). Foundations of Educational Theory for Online Learning. In T. Anderson & F. Elloumni (Eds.), Theory and Practice of Online Learning (pp. 15-44). Athabasca, Canada: Athabasca University Press.

Anderson, T. (2008). Towards a theory of online learning. In T. Anderson & F. Elloumni (Eds.), *Theory and practice of online learning* (pp. 45–74). Athabasca, Canada: Athabasca University Press.

Bell, B., & Federman, J. (2013). E-learning in postsecondary education. *The Future of Children*, *23*(1), 165–185. doi:10.1353/foc.2013.0007 PMID:25522650

Brown, S., & Adler, R. P. (2008). Minds on fire. Open education, the long trail, and learning 2.0. *Educause*, *43*, 17–32.

Bruns, A. (2009). *Blogs, Wikipedia, second life and beyond: From production to produsage*. New York, NY: Peter Lang.

Buckingham, D., (2003) Media Education and the End of the Critical Consumer. *Harvard Educational Review, 73*(3).

Buraphadeja, V., & Dawson, K. (2008, July). Content Analysis in Computer-Mediated Communication: Analyzing Models for Assessing Critical Thinking Through the Lens of Social Constructivism. *American Journal of Distance Education*, *22*(3), 130–145. doi:10.1080/08923640802224568

Burbach, M., Matkin, G., & Fritz, S. (2004). Teaching critical thinking in an introductory leadership course utilizing active learning strategies: A confirmatory study. *College Student Journal*, *38*(3), 482–493.

Burke, J., & Tumbleson, E. (2016). Choosing Content-Creation Products. *Library Technology Reports*, *52*(2), 23-27.

Choy, S. C., & Cheah, P. K. (2009). Teacher Perceptions of Critical Thinking Among Students and its Influence on Higher Education. *International Journal on Teaching and Learning in Higher Education*, *20*(2), 198–206.

Chronicle Research Services (2009). *The College Of 2020: Students Executive Summary*. Author.

Chu, S. K. W., Reynolds, R. B., Tavares, N. J., Notari, M., & Lee, C. W. Y. (2017). *Twenty-First Century Skills Education in Switzerland: An Example of Project-Based Learning Using Wiki in Science Education. In 21st Century Skills Development Through Inquiry-Based Learning*. Springer.

Davis, A., Little, P., & Stewart, B. (2008). Developing an Infrastructure for Online Learning. In T. Anderson & F. Elloumni (Eds.), *Theory and Practice of Online Learning*. Athabasca, Canada: Athabasca University Press.

Grant, L. (2006). Using wikis in schools: A case study. *FutureLab*. Retrieved from https://www.nfer.ac.uk/publications/FUTL98

Henriksen, D., Mishra, P., & Fisser, P. (2016). Infusing Creativity and Technology in 21st Century Education: A Systemic View for Change. *Journal of Educational Technology & Society*, *19*(3), 27–37.

Kanuka, H. (2008). Understanding e-Learning Technologies-in-Practice through Philosophies-in-Practice. In T. Anderson & F. Elloumni (Eds.), *Theory and Practice of Online Learning*. Athabasca, Canada: Athabasca University Press.

Lubin, I. A. (2016). *Intentional ICT: Curriculum, education and development* (IBE Working Papers on Curriculum Issues No. 17). Geneva, Switzerland: IBE-UNESCO.

Marks, D. (2009). Literacy, instruction, and technology: Meeting Millennials on their own turf. *AACE Journal, 17*(4), 363–377.

McLoughlin, C., & Lee, M. (2007). Social software and participatory learning: Pedagogical choices with technology affordances in the Web 2.0 era. ICT: Providing choices for learners and learning. Proceedings ASCILITE Singapore 2007.

Meier, E. (2016). Beyond a Digital Status Quo: Re-conceptualizing Online Learning Opportunities. *Occasional Paper Series, 2015*(34), 2.

Mentor, D. (2016). EMxC3 = e&mLearning Cultivating Connected Communities: Sustainable Workforce Talent Development. In D. Mentor (Ed.), *Handbook of Research on Mobile Learning in Contemporary Classrooms* (pp. 240–259). Hershey, PA: IGI Global. doi:10.4018/978-1-5225-0251-7.ch012

O'Reilly, T. (n.d.). *What is Web 2.0: Design Patterns and Business Models for the Next Generation of Software*. Retrieved from: http://www.oreillynet.com/lpt/a/6228

Online Courses Have Three Critical Components (and Learning Management Systems Are Not One of Them). (2017). *Distance Learning, 14*(2), 64-63.

Ortagus, J. C. (2017). From the periphery to prominence: An examination of the changing profile of online students in American higher education. *The Internet and Higher Education, 32*(Supplement C), 47–57. doi:10.1016/j.iheduc.2016.09.002

Oubenaissa, L., Giardina, M., & Battacharya, M. (2002). Designing a framework for the implementation of situated online, collaborative, problem-based activity: Operating within a local and multi-cultural learning context. *International Journal on E-Learning, 1*(3), 41–46.

Picciano, A. G., & Seaman, J. (2009). *K-12 online learning*. The Sloan Consortium.

Ross, S. M., Morrison, G. R., & Lowther, D. L. (2010). Educational technology research past and present: Balancing rigor and relevance to impact school learning. *Contemporary Educational Technology, 1*, 19.

Scardamalia, M. (2002). Collective cognitive responsibility for the advancement of knowledge. In B. Smith (Ed.), *Liberal education in a knowledge society* (pp. 67–98). Chicago: Open Court.

Shirky, C. (2008). *Here Comes Everybody: The Power of Organizing Without Organizations*. Penguin Press.

Tapscott, D. (1998). *Growing Up Digital: The Rise of the Net Generation*. New York: McGraw-Hill.

Tredinnick, L. (2006). Web 2.0 and Business: A pointer to the intranets of the future? *Business Information Review, 23*(4), 228–234. doi:10.1177/0266382106072239

Walker, D. d., Lindner, J. R., Murphrey, T. P., & Dooley, K. (2016). Learning management system usage. *Quarterly Review Of Distance Education, 17*(2), 41–50.

Watters, A. (2014). *The monsters of education technology. Creative Commons Attribution-ShareAlike 4.0 License*. Retrieved October 15,2016 from http://monsters.hackeducation.com/

KEY TERMS AND DEFINITIONS

Governance: Refers to the definition and implementation of policies and procedures to assist in the management and development of institutional initiatives.

Learning Management System: Software applications for the administration, documentation, tracking, reporting and delivery of educational courses or training programs. LMS facilitate the delivery of instructional material to students, assists in the administration of assignments, quizzes, tests, and other activities, can track student progress, and manage data.

Online Learning: Courses offered by institutions that are conducted in a virtual space. Learners can engage with an academic institution in the traditional method of in person classes or through the virtual method computer mediated instruction. A blended-learning approach mixes in-person and online sessions.

Online Learning Platform: See learning management system.

Online Learning Tool: Software that student can use to work through ideas, concepts or processes to create, perform, or respond to assignments and problems. LMS will often offer a collection of integrated online learning tools.

Social Learning Theory: The theory that people learn by observing others and explains how people learn new behaviors, values, and attitudes. The Internet offers ways to collaborate in learning despite being physically separated.

Chapter 9
Virtual, Augmented and Mobile Learning in Health Education

Ozan Karaca
Ege University, Turkey

Kadir Demir
Dokuz Eylul University, Turkey

S. Ayhan Caliskan
Ege Unversity, Turkey

ABSTRACT

Currently, technology is used for various purposes in every aspect of life. One of these areas is the field of education, which has an important role in advancing the technologies. Educators often desire to concentrate on their own courses or fields. However, the changing perception of the students and the technologies they are interested in are necessitating educators to follow technological innovations and developments in instructional technologies. The objective of this chapter is to attract attention to some of the global virtual, augmented, and mobile applications that are currently utilized within the medical learning arena. The authors also address technologies that aim to train human resources in health and other fields. The chapter illuminates the use of virtual, augmented, and mobile learning environments via their use in health education.

INTRODUCTION

Today, technology is used for multiple purposes in various fields, while smartphones in particular, enhanced opportunities for formal and informal learning. Rapid technological advances created a paradigm shift towards the idea of autonomous learning in virtual environments by removing the physical boundaries in education. During recent years, this shift also saw an increase in the number of tasks assigned to virtual environments in required training of the private and public-sector workforce. Although the classification descriptions of the related learning environments seem heterogeneous, they change over time (Alessi, Giangreco, Mainetti, Mighali, Patrono, Pinnella & Storelli, 2017; Boulos, Lu, Guerrero,

DOI: 10.4018/978-1-5225-4111-0.ch009

Jennett, & Steed, 2017; Müller, 2012; Noll, von Jan, Raap, & Albrecht, 2017) influenced by new and old theories, old and new technology, and the repurposing of technologies.

It is a common practice in social sciences to focus on, and analyze the reflections of learning theories, methods, styles or multi-generational people and their environments when defining these concepts. Thus, the definitions of virtual and mobile learning environments should be discussed with respect to the traits and requirements of all generations and the ever-growing populace of people becoming digitally proficient to quickly adapt and utilize current technologies in different areas of life.

BACKGROUND

Digital Impact and Learning Styles

Digital natives are considered to be 21st century children and young adults, who started life with current technologies and carry out much of their work via these technologies (Prensky, 2001). On the other hand, digital immigrants are adapting to these technologies. Occasionally refered to as Net Gen (Muhammad, Mitova, & Wooldridge, 2016) or D-Gen (Jenkins, 2007), those considered as digital immigrants are supposedly following in-step or asked to be re-thought as a concept (Jenkins, 2007) or target audience. They can easily share photos, personal and location information with people of their choosing. However, while this distinction made by Prensky had a slight resemblance of what felt like a reality at the time, the rapid innovations and growth of technology has blurred this reality as all generations are introduced. And there are varying numbers of early adopters, intermediates and laggards (Sahin, 2006) across the generational spectrum of adopting new technologies. In ever-increasing fashion, individuals use online tools to learn or train themselves, dating apps to seek out intimacy, relationships, or marriage after meeting and spending time online with one another (Mazur, 2017; Rochadiat, Tong, & Novak, 2017; Rosenfeld, 2017). The Pew Research Center's FactTank, reports an increase in online dating for adults under 25 as well as those in their late 50's and early 60's (Smith, & Anderson,2016), and a significant drop was observed about the previous stigma of online dating as conveyed by the Pew Research Center in 2005 (Smith & Anderson, 2016). While Forbes and others report that a third of recently married couples met online, and claim that those couples are more satisfied and less likely to split-up (Magid, 2013; Ortega, & Hergovich, 2017; Paul, 2017). Additionally, Rosewarne (2016) writes about the power of people seeking intimacy on the internet is more a sign of people seeking connectedness. This social connectedness is also prevalent in other social media (Mentor, 2018), such as Facebook, Weibo, Ren Ren, Instagram and LinkedIn, nonetheless, the latter predominantly with a workforce networking motivation. Digital natives, and all those who adopted online practices in their daily lives, possess qualities and competencies that are sometimes unlikely to be attributed to older generations and their use of technology (Kleiman, 2004). Yet, there are many senior citizens utitilizing social media (Bonsón, Royo, & Ratkai, 2017; Li, 2017; Teng, & Joo, 2017). According to Bilgiç (2011) digital natives, and others who engage online more and more via their smart phones for informal learning (Şahin & Mentor, 2016):

- Want to access information quickly,
- Prefer video and graphic instead of text,
- Prefer to read smaller chunks in hypermedia fashion instead of reading a piece linearly,
- Prefer games and informal learning,

- Multi-task through simultaneous multimedia,
- Feral learning by searching and discovering.

To attract the attention of learners, whose lives are integrated with physical, augmented and virtual realities, it is necessary to engage them via digital media. The latter should also now include mobile learning from smart phones which is now more than just a communication tool (Mentor, 2014). Learners and adults from all industries prioritize communication, and constantly monitor the flow of online information, while feeling the need to be constantly connected to social networks (Kuzu, 2014; Pedró, 2007). With the expansion of digital engagement in all spheres of life, mixing reality, including the workplace, there is a need to also look at the adoption of other emerging technologies like virtual and augmented reality and how it could impact work and workplace learning.

In this part of the chapter, concepts such as virtual reality (VR), augmented reality (AR), mixed reality, AR and VR mobile applications, as well as wearable technologies were examined in the context of the learning environment for health education. All of these have the potential to reinforce learning, enhance motivations and provide concrete experiences. Innovative examples of applications, both local and global are shared demonstrating theoretical infrastructure and technologies. The concluding part of the chapter include several claims on the future of mixed reality as well as wearable technologies, and their relationship with health education in accordance with the existing literature on virtual learning environments.

VIRTUAL LEARNING ENVIRONMENTS (VLEs)

VLEs, as the name suggests, emphasize learning that takes place in digital environments, also known as e-learning systems. Several learning approaches such as flipped learning, mobile learning, and distant learning are conducted with VLEs. VLEs provide the ability to overcome geographical limitations by facilitating face-to-face learning (Kartsonaki, Lawrie, Sharpe, & Bailey, 2017; Liebowitz, & Frank, 2016; van Raaij & Schepers; 2008). Additionally, the flipped classroom method (King, Crawford, San Miguel, Barrie, Spencer, Tyransky, & Kman, 2017), where learning is commenced before real time engagement, was claimed as an effective method by several researchers due to its capacity to be utilized in medical and health education (ibid; McGrath, Mesiwala, Barrie, Boulger, Thompson, Mitzman, & King, 2017). While the method demonstrates capability to enhance pre-engagement and to develop skills such as critical thinking and evidence-based reasoning both in science education and professional development (Şahiner, 2016), other research claims no significant difference between flipped and traditional approaches to educational engagement (Arnott, & Planey, 2017; DeLozier & Rhodes, 2017; Njie-Carr, Ludeman, Lee, Dordunoo, Trocky, & Jenkins, 2017).

VLEs occcasionally allow students to adapt to individual differences and learning speeds. VLEs, which are frequently preferred in both formal and informal instruction, are widely used in vocational education due to their advantages (Mcann, 2016; Mentor, 2016). In the context of vocational education, VLEs are used in formal secondary, undergraduate, and graduate education, in informal, lifelong learning or professional development.

In this section, VLEs are explained through e-learning systems, which offer advantages of independence from time and place. In addition, it will be examined from the perspective of Virtual Reality (VR) technology, which provides stimulating visualization and immersive experiences. In addition, this information will be offered with examples by providing theoretical information about virtual learning environments in particular for health education.

E-Learning Systems

Ifenthaler (2012) defined learning management systems (LMS) as web-based software that enables users to manage and utilize learning content in interactive learning environments. Owing to these systems, educators, users, and courses could be managed with comprehensive tools for assessment, grading, feedback on assignments, communications, and other educational engagement purposes that are now more and more, a part of learning management systems or that could be integrated into these systems. On the other hand, Paulsen (2002) defined these as systems that enable students, teachers, and administrators to access and manage the systems through access control, hosting learning content and organizing communications tools and user groups.

LMS users include students, teachers, administrators, education administrators, content developers, and educational administrative inspectors, and superindendents. Several institutions provide online courses, classroom event registration, live virtual courses and distance learning on these central learning portals. LMS incorporates features such as:

- Secure access and entry
- E-learning record keeping,
- Student questionnaires and assessments,
- Test creation and application,
- Classroom management,
- Grading, collaborative learning,
- Certificate management,
- Multiple language support,
- Skill gap analysis, and
- Advanced learning analysis (Chapman, 2008).

Previosuly, mobile web sites running learning management systems, which are compatible with mobile web browsers, were developed in order to utilize LMSs efficiently on mobile platforms. Currently, web responsiveness of sites occured offering a strong impetus towards fundamental mobile first e-learning designs (Curinga & Svaranos, 2016). Although several mobile device features could be accessed by these mobile websites, LMSes also released native and hybrid applications (apps). This was due to the ability of local and hybrid applications to interact with the advanced user interfaces, to access the device hardware and software features. Furthermore, the demand for continuous internet connection to utilize LMSs via mobile web browsers revealed several drawbacks whereas some native LMS apps allowed offline use (Müller & Dikke 2012).

With an LMS, it is possible to conduct vocational education with both holistic and conventional methods thanks to educational content, tests that could provide automated feedback, cultivate discussion forums, offer wikis, dictionary activities, and animations or simulations integrated with the LMS creating VLEs. There are also specialized VLEs to train the field-specific workforce in certain occupational groups. Medical Education field is one such example. Simulated and standardized patient applications, which are widely used to train physicians in conventional medical education, are used as virtual patient applications in web-based VLEs (King, Crawford, San Miguel, Barrie, Spencer, Tyransky, & Kman, 2017, McGrath, Mesiwala, Barrie, Boulger, Thompson, Mitzman, & King, 2017; Şahiner, 2016, Arnott, & Planey, 2017; DeLozier & Rhodes, 2017; Njie-Carr, Ludeman, Lee, Dordunoo, Trocky, & Jenkins, 2017).

Global VLEs: Massive Open Online Courses (MOOC)

From constant access to LMSes to VLEs, learners who thrived in, or developed autonomous learning skills, could individually conduct professional development processes through Massive Open Online Courses (MOOCs). These mass educational platforms such as EdX, Coursera, Udemy, Khan-Academy, Skillshare, Udacity, to name but a few, emerged during the last decade. Skillshare provides an experience-oriented virtual learning opportunity to small groups to develop their psychomotor skills. Of these, Udacity claim "job-guaranteed" virtual learning opportunities for future professions through "Nanodegree" and "Nanodegree Plus" programs, which offer courses in the field of information technologies and software development (Knox, 2017; Mirrlees & Alvi, 2016; Pathak, 2016). Some of the computer programming MOOCs, are supported by corporations and companies like AT & T and Google, while several institutions (Harvard, Columbia University) opened their own in the form of Edx. Some argued that MOOCS could evolve into an alternative to formal education (Littlejohn, Hood, Milligan, & Mustain, 2016), but still has much to prove, given the high attrition rate.

Among the key features of MOOCs are the opportunities to attract highly motivated individuals. Students who are obliged to take "compulsory" courses in formal education have the opportunity to take the courses they are interested in; from the schools or educators they want, through MOOCs. 155,000 students enrolled in the "Circuits and Electronics" course, which was held from March to June 2012 and delivered via EdX. 89.7% of the students explained the reasons for signing up for this course as a chance to acquire knowledge and skills, personal development and advancement in professional life (Breslow et al., 2013).

Due to the flexibility that these open environments offer in terms of pace, MOOCs could be especially useful for continuing medical education (Swinnerton, Morris, Hotchkiss, & Pickering, 2017; Pickering, Henningsohn, DeRuiter, de Jong, & Reinders, 2017). Since 2010, many respected higher education institutions such as Stanford, Harvard, Pennsylvania, and Toronto begun their own MOOC offerings. While many organized courses, in collaboration with common platforms also came to light. For example, in Canada, a MOOC platform was organized by following the training of radiologists working in rural areas in cities (Sezer et al., 2016).

Virtual Reality Learning Environments (VRLE)

Another technology that could be used as VLE is "Virtual Reality" (VR) environments, which is a prominent immersive solution for learning by experience. Virtual Reality, commonly abbreviated as "VR", is an environment that is designed to run gaming, education, promotional, service, or simultaneous applications in different locations with, or without user interaction, utilizing realistic three-dimensional environments or 360-degree multimedia recordings presented through stereoscopic headsets and an all-encompassing screen.

According to a definition by Kayapa and Togan (2011), VR environments offer computer-generated three-dimensional simulations, presented to the user to manage a simulated environment that the individual perceives up close, through enclosed media rich goggles. These environments include a set of technologies that aim to convince the user that s/he is in a different environment, deceiving the senses of the user.

Virtual Reality Working Principles

Today, there are two types of VR media: monoscopic and stereoscopic platforms. Monoscopic platforms display the VR environment via a screen display and provide interaction using conventional input devices like a mouse, keyboard, joystick, or physical cabinets and cockpits as in-flight simulators. Although this technology does not provide the user with a sense of a complete environment, the operation principles are simpler when compared to stereoscopic platforms.

Stereoscopic platforms are mini-screened helmets that could display the image of the two perspectives of the same object, to both eyes of the user like a binocular (Limniou et al., 2008, Galkandage, Calic, Dogan, & Guillemaut, 2017; Shao, Tian, Lin, Jiang, & Dai, 2016). These helmets provide the user with a small image for each eye, which allows the brain to combine the two images to create a three-dimensional experience. These platforms offer the VR environment-user interaction via sensors that can follow movements of hands, arms, eyes, and the head, as well as integrated changes in location (Hannaford, Barral, Rephaeli, Ching, & Bajaj, 2017; Sherman, 2002; Walsh, Daems, & Steckel, 2017). Furthermore, it could also reflect the interaction between the VR environment and the user through physical control objects such as headmounted controlling displays, joysticks, mechanical and sensory objects like gloves, vests, and integrated body kits that could be worn on various parts of the body and involve multiple parties. These VR devices with bi-directional data transfer features are called haptic devices, derived from the term "haptikós", which means "tactile sensation" in ancient Greek.

CAVE (Audio-Visual Experience Automatic Virtual Environment) expensive VR platforms, which were widely used in the 90's in university technology development departments or driver training (pilots, ship captains, race car drivers and others) and were replaced by easy to access Head-Mounted Display (HMD) technologies today. VR headsets, one of the most common HMD-type technologies, include a pair of small stereo screen covering the eyes and various motion tracking sensors that differ, based on the different techniques used in the device (Cruz-Neira et al., 1992; Walsh, Daems, & Steckel, 2017).

Demand for software compatible with VR headsets has also increased in line with the availability and affordability of the VR headsets. SuperData report (2016) predicted that 13 million Americans plan to buy at least one VR headset in 2016 and at least 7.2 million VR headsets will be sold in 2017. Based on this prediction, it could be argued that the demand for VR equipment will increase even further in the coming years and it could be expected that the VR software would affect all areas of life, including education.

How Did VR Become a Learning Environment?

Based on Dewey's (1913) experiential-oriented learning theory, a learning environment should present the students with a powerful sense of being, and the active inclusion in the learning activities should be as comprehensive as possible. This so are able to interact with the environment, and more so a mutlimedia rich environment, rather than just passivley observe, could achieve higher level of learning goals (Moreno & Mayer, 2000). When considered in this context, virtual reality should be explored for large-scale deployment, since it provides a vivid learning environment (Hanson & Shelton, 2008; Murray, 2017; Traxler & Kukulska-Hulme, 2016 Yang, 2017).

In the learning process, in which direct educational engagement of individuals is organized, it is suggested that learners could be faced with tasks that are defined as a problem or a situation from real life (Durak, 2016; Elvezio, Sukan, Oda, Feiner & Tversky, 2017). Virtual reality learning environments (VRLE) offer distinctive features such as tangibility, enhanced visual appeal, and extended interaction opportunities (Ferrington & Loge, 1992; Pfeiffer, Feiner, & Mayol-Cuevas, 2016). Furthermore, these environments allow users or participants to interact with and share the virtual space where learning takes place (Economou et al., 2000). In short, this method could be an effective tool in providing an artificial empirical learning environment for learners (Cheng & Wang, 2011). It is even possible to use this method to provide an environment where surgical field students could experience several procedural knowledge about operations (Sweeney, 2012).

In summary, VRLE offers virtual environments that allows learners to experience reproducible, formal or informal learning, where they can interact with events, objects or individuals in the environment, receive instant explanatory or procedural information in the environment, instead of being a passive recipient. As VRLE becomes more mobile, it should be considered as a significant growth market in achieving the strategic objectives of the just-in-time teaching approach(Novak et al., 1999):

- To structure the out-of-class time for maximum learning benefit.
- To maximize the efficacy of the classroom session, where faculty interacts with students.
- To create and sustain peer interactions. Students continue to work as a team (with instructors as coaches) to help all students achieve the course objectives with the maximum amount of retainable knowledge.

With VRLE, students could discover interesting constructs, disassemble or assemble them, and observe the environment with a 360-degree perspective. VRLE allows a form of highly dynamic simulation that could be constructed via intuitive interactions with the computer. The personalized experience created by this virtual environment, which provides a close to real feeling of interaction, could further motivate students (Lu et al., 2005).

Using VRLE in Health Education

VR owes its popularity to the entertainment industry followed by education and health industries. According to the Virtual Reality Society (2016), the medical health field rapidly adapted this technology; from surgical simulations to treatment of phobias, robotic surgery applications to professional skills training (ibid.). Although the encounter of health education students' with patients at an early stage is generally regarded as a sound educational approach, the hospital environment usually have a different practice. Hence, medical educators and learners could offer realistic audiovisual and interactive virtual reality learning environments due to the learning experiences that could be limitlessly repeated.

It is possible to enhance the training of new employees, or improve the existing workforce using virtual reality and mobile learning experiences (Henderson & Feiner, 2016; Mentor & Ahmad, 2010). Educational or vocational institutions could improve the experiences of learners and their problem-solving skills or test their skills by exposing them to real-world conditions. Providing solutions for health and life sciences, production or manufacturing industries, public employees, and even the service sector, VRLE enables learners to solve their problems more accurately, safely and efficiently, and reduce their mistakes and/or advance their knowledge.

In a study conducted by Limniou et al. (2008) with undergraduate chemistry students, it was determined that virtual reality learning environments helped students to better understand the molecular structure and changes that occurred during a chemical reaction, when compared to two-dimensional simulations. Abulrub et al. (2011) reported that interactive and comprehensive VRLEs were promising technologies in engineering education and a useful tool in training the new generation of engineers with required professional skills. Another study by Saleeb and Dafoulas (2013) on VRLE utilization in workforce education with architecture students found that several professional skills of the students who observed architectural and environmental designs with VRLE and interacted with the environment, developed a higher increase in their level of learning experience and satisfaction. For the health field, VRLEs has the potential to take an active role in training several health care professionals in areas such as medicine, dentistry, pharmaceuticals, and nursing. Learners in these fields can use VR, and mobile technologies to observe and experience the normal structure and functions of, say, the human physiology, pathologies and treatment procedures, and to acquire use of medical equipment or even basic surgical skills.

Innovative VRLE Examples in Health Education

One of the important developments in medical education in the 21st century is the gradual expansion of new technology utilization in education. Simulators, e-learning and web-based learning environments, which are invaluable in terms of competency based education, have broadly increasing application areas in medical education (Sayek, 2016). Innovative virtual and mobile learning initiatives for development of workforce in health industry were investigated and listed below in order to exemplify the topics within this approach. These examples were sorted in the order of local to global initiatives.

1. RCSI virtual reality medical training simulator

RCSI Medical Training Sim, which was developed with the collaboration of Immersive VR Education Ltd. and Royal College of Surgeons in Ireland and could openly be accessed on Oculus virtual reality platform. The module that teaches basic concepts is available for free on the Oculus virtual reality platform and can be customized with in-app purchase options.

The RCSI Virtual Reality Medical Training Simulator provides students, attending faculties or medicine or surgical trainers, the opportunity to observe and intervene in common cases that are risky for the patient outside the clinical setting. This software is designed to offer a complete didactic learning approach and could provide students a much more realistic trauma room experience as compared to plain instruction or videos.

2. **MedicActiV:** virtual platform for medical education

The MedicActiV virtual medical education platform was introduced to the market in 2016 by SimforHealth, a company that designs innovative educational products for educational institutions that train healthcare professionals and healthcare organizations that need continuous professional development training. This platform is an innovative e-learning platform that offers virtual reality clinical cases for all healthcare professionals and students (physicians, surgeons, medical students, interns, nurses, midwives,

Figure 1. The RCSI Virtual Reality Medical Training Simulator screen samples

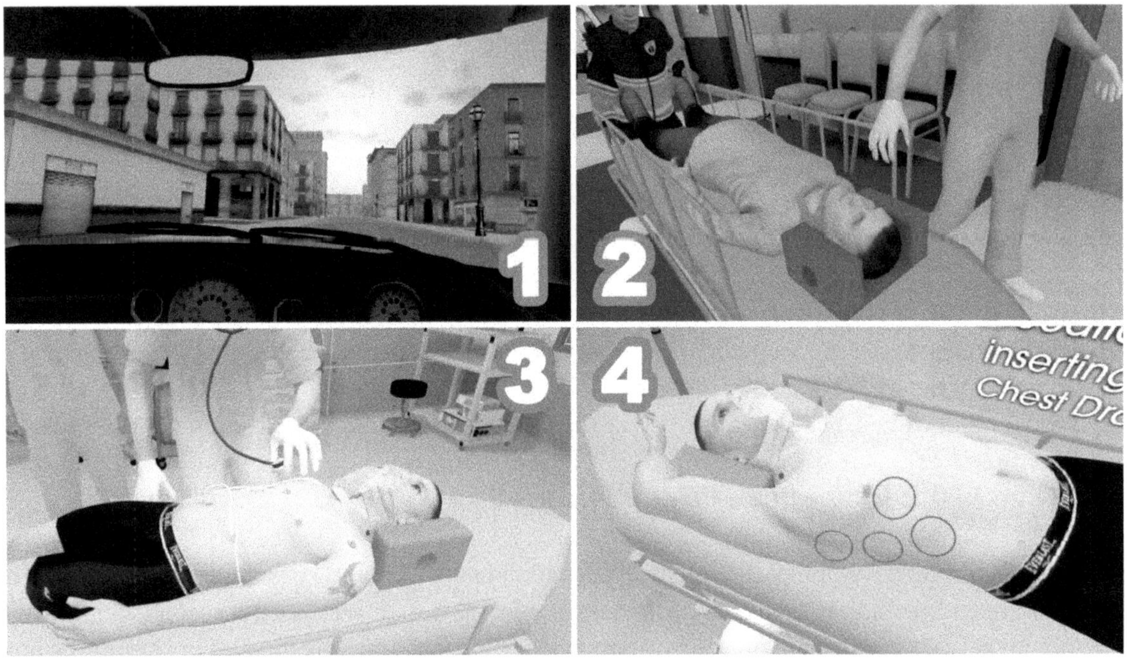

masseurs, and physiotherapists) to acquire the clinical skills they require or to develop their proficiencies. Higher education institutions or institutes that train health professionals can use existing clinical cases developed by SimforHealth for healthcare organizations or learning communities that require ongoing professional development education. In addition, they could create an educational module with Patient-Genesys (Virtual Clinical Case Generation Engine) based on institution-specific cases.

PatientGenesys, a virtual clinical case generation engine, is the product of a joint research program financed by the French government under the FUI 16 R & D support program. After its integration with MedicActiv in 2016, PatientGenesys provides opportunities when creating a clinical case for medical educators to create virtual patients (gender, age, three-dimensional environment, etc.), selection of consultation stages and entry of necessary information (medical records, clinical examinations, treatments, recommendations or supplementary examinations) about a clinical case. Once the instructional design is complete, the system processes all data, creates the three-dimensional environment, translates the input textual data into real-like speech, and creates a three-dimensional virtual clinical case with a virtual specialist that could provide consultation.

This platform allows students or health professionals to work on the same clinical case from multiple locations simultaneously. Multiple participants and different levels could be selected depending on the user knowledge, field expertise or skill levels. The platform could offer spot-on scenarios by suggesting relevant cases based on the participant level. For example, it could be presenting the case about monitoring a patient undergoing cardiac arrhythmia (atrial fibrillation) to a family physician or general practitioner, while proposing the management of prenatal perineal relief to a masseur or physiotherapist.

Figure 2. MedicActiV: Virtual Platform for Medical Education screen samples

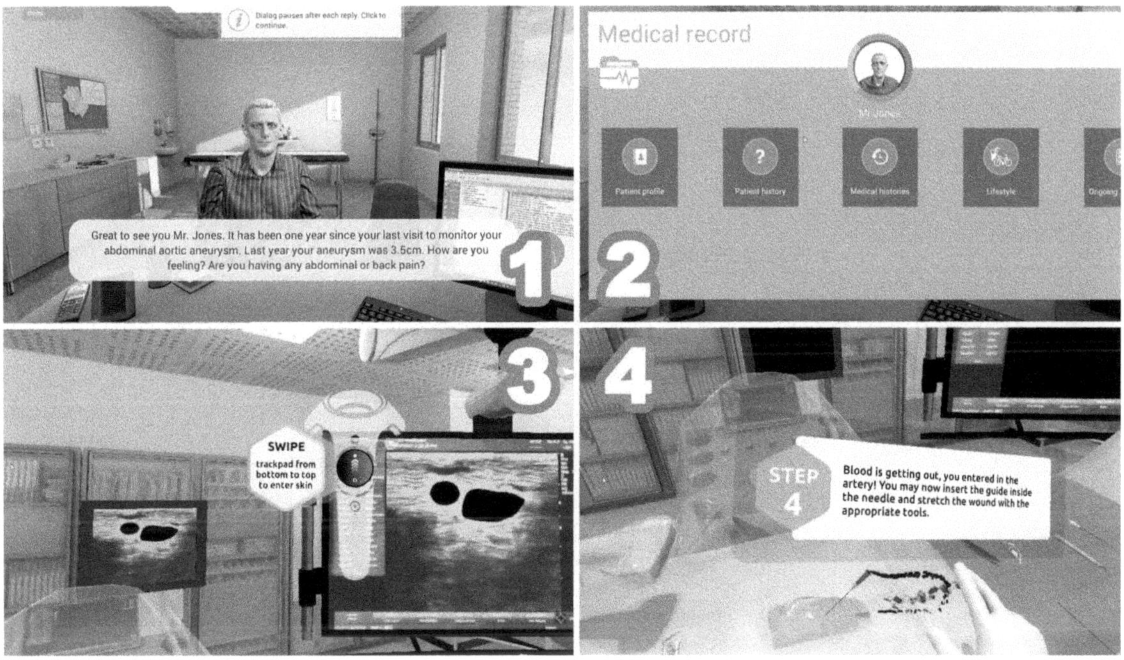

3. HumanSIM: sedation and airway

American company Applied Research Associates (ARA) developed a three-dimensional VRLE application called HumanSIM: Sedation and Airway for use in emergency anesthesia procedures and respiratory management for the American Army Telemedicine and Advanced Technology Research Center (TATRC). This software, which is a part of the medical education provided to military medical personnel before they take the field and which is open to public use, was created using Epic Games' award-winning, real-time physics engine "Unreal Engine", and realistic three-dimensional models with multiple scenarios. HumanSIM: Sedation and Airway can be used in a wide range of subjects, from basic first aid training to continuing professional development education for the physician.

The software allows physicians, nurses, healthcare professionals and other emergency personnel to experience Rapid Sequence Induction (RSI) and Moderate Sedation practices, which are rapid sedation and intubation techniques for injured individuals. Built on a learning by practice strategy, HumanSIM Sedation and Airway empowers health professionals with competencies in rare, complex, or error-prone cases and processes. It also provides health professionals the opportunity to hone evaluation and clinical decision-making skills, to improve realistic, and persistent cases in simulated dire circumstances without taking risks. Thanks to HumanSIM: Sedation and Airway, which is a product of the combination of the most advanced digital gaming technology available with the metaverse physiologic-pharmacological models and Dynamic Virtual Human Technology (DVHT), learners can experience ten different scenarios that include various medical cases with implications such as airway trauma and multidrug interactions.

4. **YOU:** virtual reality medical training

Figure 3. HumanSIM: Sedation and Airway screen samples

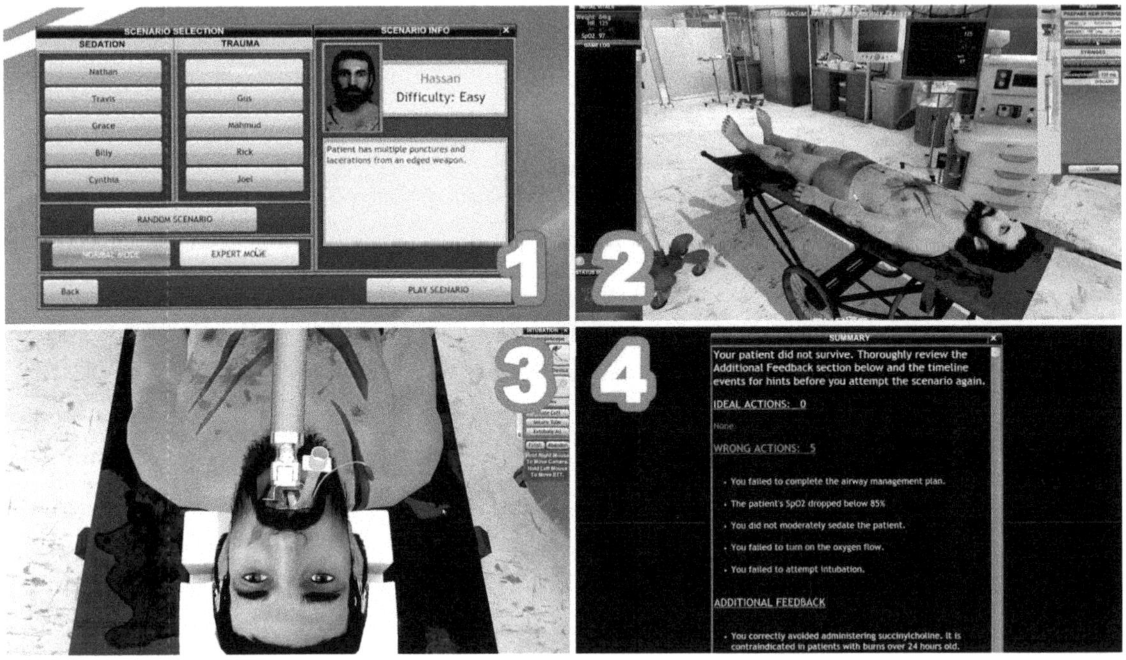

BioLucid, a company that developed healthcare virtual reality products by purchasing the shares of the digital health company Sharecare in 2016, created a high reality audio-visual environment with the software called YOU, developed by the company, for healthcare professionals, educators, students and patients to examine as well as discuss healthy physiology, diseases and treatments. Users could navigate a 'real' three-dimensional human body model and organs freely, in real time, based on anatomical and functional aspects

YOU could visualize the normal structure and functions of human physiology, pathologies and treatment processes in a way that is very close to reality. Through interactive experiences conducted with YOU, it is possible to monitor the entire process from symptoms to treatment, enabling awareness for all stages (Sharecare, 2016). Furthermore, YOU incorporates several features such as physiological functions (e.g. heart rate), personalization of the severity of the illness and treatment procedures for the patient, 360 ° rotation of the organs and systems, video recording of the images, and access through virtual reality headsets, the desktop, and mobile devices.

5. **The Body VR:** Anatomy Viewer

Body VR is a virtual reality platform that is designed to visualize medical diagnoses, observe the effects of treatments and make better treatment decisions, as well as design personalized patient simulations by integrating patient records into the platform. The platform provides solutions to healthcare providers and pharmaceutics and medical equipment manufacturers, as well as an educational institution or patient educational applications. In short, this platform is a virtual reality learning environment that allows users to navigate within the human body and learn all the details using a narrated story.

Figure 4. YOU: Virtual Reality Medical Training screen samples

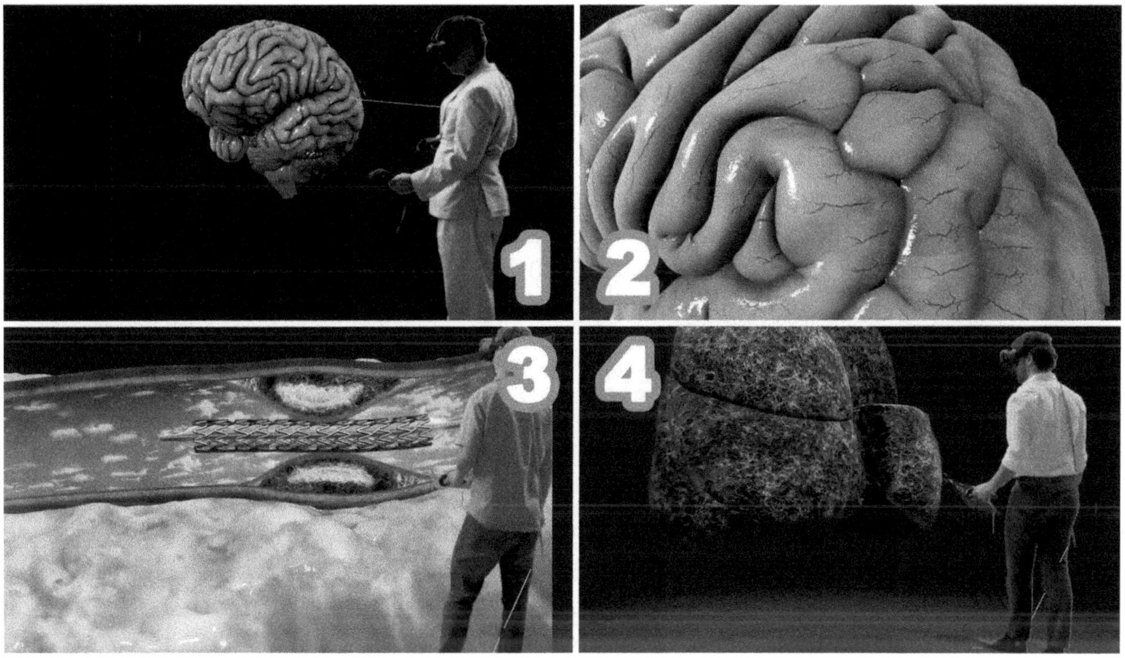

Figure 5. The Body VR: Anatomy Viewer screen samples

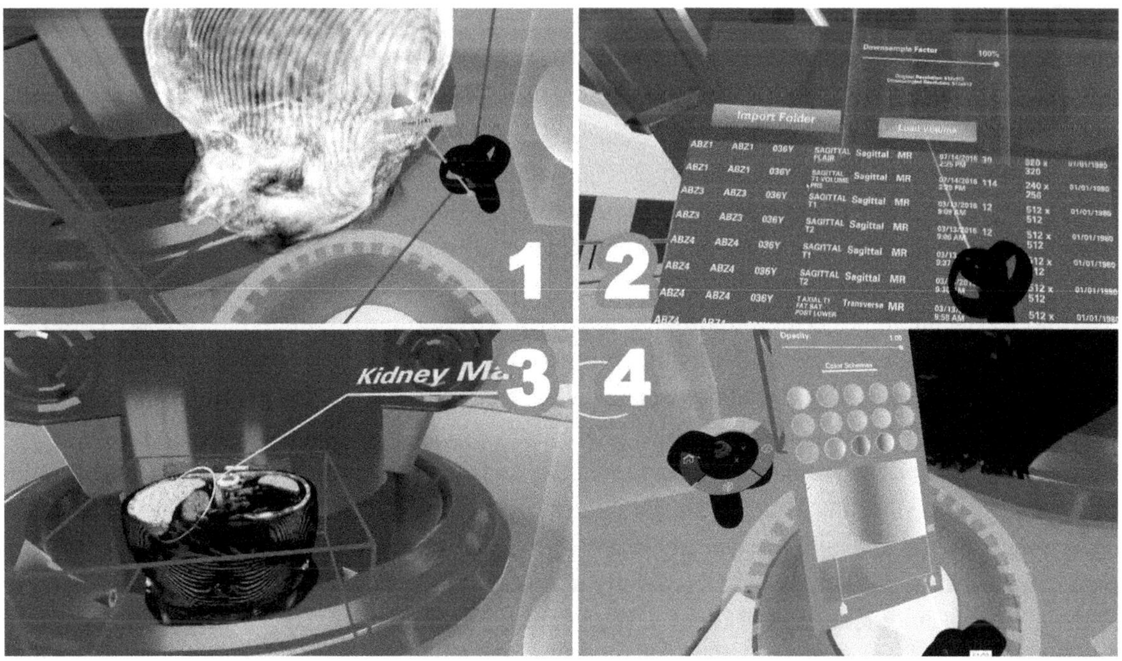

This educational virtual reality experience allows individuals to navigate and interact in an anatomical replica of the human body. Learners can navigate the circulatory system and discover how blood cells work to spread oxygen in the body or enter into one of the billions of living cells and how organelles work together to fight deadly viruses (The Body VR, 2015). The Body VR uses high quality and realistic graphics that fully visualize the human anatomy. The VRLE that presents the impacts of the disease and the treatment on the human anatomy in an audiovisual experience allows the organs and systems to be rotated, zoomed in and out, and be hidden and highlighted. The company that produced the platform works in cooperation with several colleges in the US, and around the world. It also plans to develop a curriculum that covers the entire human body by 2020, which might be available via mobile device.

MOBILE LEARNING ENVIRONMENTS

The concept of mobility was included in the e-learning and virtual learning media that emerged as a result of advances in wireless communication technologies and mobile technologies. Accordingly, since the early 2000s, e-learning studies focused on wireless communication technologies due to increasing access to these technologies (Mentor & Ahmad, 2010; Mentor, 2016, Wu et al., 2013). The rapid improvement in wireless communication technologies also led to an increase in the availability of wireless applications in daily life. A variety of devices capable of functioning in wireless environments, and a variety of software enabling them to perform their functions, were developed and linked to the internet environment (Hwang et al., 2008). Thus, learners could have flexible, faster and efficient access to learning environments using the mobile devices they possess such as smartphones and tablets (Lin, 2013).

The concept of mobile learning is defined as learning and teaching activities conducted with devices such as wearable, tablet and laptop computers and smartphones (Wyne, 2015; Demir & Akpınar, 2016). In general, mobile learning or m-learning could be considered as any form of learning that takes place using a mobile device and utilized by "nomadic" students (Alexander, 2004) or for learning on the go or just-in-time learning (Mentor & Ahmad, 2010, Mentor, 2016).

Educators often desire to concentrate on their courses or fields outside of the physical and time constraints of face to face class sessions. And learners require educators to follow technological innovations and developments in instructional technologies. According to Masters et al. (2016), educators' attempts to use mobile technologies with a traditional approach without leaving their safety zone, could even be detrimental to students beyond the expected outcome. To use mobile technologies effectively, it is necessary to understand the social and pedagogical perspectives and basic principles developed within the context of mobile learning (Mentor, 2014; 2016).

Mobile learning engagement is increasing every day and is consistent with the intensive use of mobile technologies in everyday life. This, in turn, facilitates educational use of mobile technologies and access to information by the students. Masters and Ellavey predicted in 2008 that mobile learning will become a mainstream instructional technology generally accepted by educators and learners in the future and that institutions cannot remain indifferent to the innovations in this field. Over the past decade, it was observed that the general public is still ahead of educational policy makers, educational institutions, and educators (Mentor, 2014; 2016). Students find mobile technologies naturally functional and effective in formal, and informal lifelong learning contexts fueled by the rapid development of mobile technologies, the proliferation and diversity of applications (apps), operating platforms, and the increased bandwidth.

The trends in mobile learning are further fueled by all learners in terms of the "commuters' learning journey" (Mentor, 2016), who can access and interact with the learning material (El-Hussein & Cronje, 2010) while on the go. A common characteristic now of 'nomadic' learners is to want continuous access to services regardless of location and time (Saravani, 2010). Notwithstanding that these devices articulate and connect to the real world and enrich the learning experience through interactive content and applications. Another technology, that is integrating with mobile deives in growing fashion, is Augmented Reality (AR). Examples here that is used in daily life and learning environments with different techniques and methods range from Pokemon Go, to interactive maps or directional guides.

AR Evolution on Mobile Learning

Virtual reality (VR) allows users to interact in a completely artificial environment. Augmented reality provides users the opportunity to observe the real world in which virtual objects or content reside (Azuma, 1997; Greenwald, Kulik, Kunert, Beck, Frohlich, Cobb, & Snyder, 2017) in complimentary fashion. With augmented reality, multimedia content is reflected in real world or in the objects of augmented real-world scenarios where the users could perceive the content (Roo & Hachet, 2017; Yuen et al., 2011). The main difference between augmented reality and virtual reality is the interaction with the environment. In augmented reality, the individual continues to interact with real world, while in virtual reality the individual is engaged in a purely artificial world (Zheng and Waller, 2017). From an educational point of view, advantages of augmented reality are (Yuen et al., 2011):

- Increasing student commitment, interest, and motivation,
- Providing first-hand learning opportunities that the students could not experience otherwise (astronomy, history),
- Improving inter-student and student-teacher collaboration,
- Developing creativity and imagination of students,
- Enabling students to learn at their own pace and the way they like,
- And providing unique learning opportunities that are suitable for various learning conditions.

In augmented reality applications, users observe the real world through smart glasses, smartphones or tablet computers. Controls, joysticks, and stick-like appliances are also used in almost all VR and AR applications to ensure that users could have near-real-life experiences. Augmented reality has a great potential in transforming the theoretical knowledge into practical practice. Augmented reality applications that were increasingly used in all areas of education provide opportunities for conditional and contextual learning. Mobile learning combined with augmented reality applications can enhance those experiences through near-real-life engagements. The above mentioned educational advantages of augmented reality demonstrate that augmented reality could be very beneficial in contextual. Mobile learning applications are frequently preferred to transfer the professional knowledge and skills instructed at the school into the professional realm. Like mobile phones, wearable technologies such as Google Glass, are also converging with augmented reality functionalities (Leue, Jung, & Dieck, 2015) and will continue to become gradually useful in different fields and accepted in educational environments. Wearable technologies are lightweight, easy to transport, smart devices with various sensors to track physical movements (Kuzu Demir & Demir, 2017). Thanks to these sensors, it is expected that students' experiences with augmented reality would be more realistic.

Using Mobile Learning Applications on Health Education

The formal learning experience in workforce training does not provide all the required knowledge to learners and it is expected that learners would develop their occupational knowledge and skills over time. For example, 20% of the knowledge used in professional life is acquired in formal learning environments, while informal learning accounts for the remaining knowledge (Cross, 2005). Thanks to m-learning, learners would have the opportunity to acquire new information everywhere or renew their existing knowledge.

Quinn (2011) reported that an individual can be considered at the beginner level in the occupational sense when the individual starts working for the first time. This can feel as the most intense phase with informal learning. Even when the distribution is equal, the individual becomes the learner and practitioner. When the distribution starts to favor the informal learning, the individual experiences transformative learning (Marsick & Volpe, 1999; Marsick, Watkins, Scully-Russ, & Nicolaides, 2017). Therefore, supporting individuals outside formal educational systems, during or after graduation emerges as an opportunity to extend the educational engagment, and thus m-learning becomes more and more significant.

The number of health educators that utilize mobile learning applications such as mobile applications and educational podcasts to facilitate and improve learning of the students are increasing (Albrecht, & von Jan; 2017; Istepanian, Laxminarayan, & Pattichis, 2014; Kumar, Nilsen, Abernethy, Atienza, Patrick, Pavel, & Hedeker, 2013; Malanga, 2017; Mentor, 2014). The increasing significance of mobile technologies in health education is due to the lack of a range of factors, notwithstanding general medical infrasturcutre, overworked staff, insufficient staff, qualified faculty members and an increasing number of digitally aware freshmen and sophomore students enrolled in health education institutions (Maag, 2006). For example, the use of smartphones as clinical examination instruments in simulations by Taiwanese nursing students proved to lead to higher learning outcomes compared to a control group that utilized pen and paper to record and evaluate patient symptoms (Wu et al., 2011). Alegria et al. (2014) discovered that majority of third-year medical students used tablet computers (iPads) for remote access to patient records, and that certain students had access to point of care reference information. Furthermore, in a systematic review study that compiled the results of several studies, Mi et al. (2016) reported that students, who utilized mobile devices in health education, conducted various activities such as reading the course material, checking e-mails, accessing information remotely and discussing the cases in the classroom at the clinic during their internship.

Innovative Mobile Learning Applications on Health Education

In mobile learning in health education, it is important to offer concrete examples. Thus, a review was conducted and innovative examples were identified and studied from various perspectives. While examples for this section are selected, global applications were selected and described, which embodied mobile learning concepts such as AR or mobile applications.

1. **VIPAR:** Virtual interactive presence in global surgical education

This application provides real-time assistance by an experienced surgeon to advance the training and capacity of surgeons around the world. The virtual interactive presence and augmented reality (VIPAR) is a tablet computer (iPad) based application (Davis, Can, Pindrik, Rocque and Johnston, 2016). VIPAR provides virtual help to far away surgeons. Thanks to VIPAR, surgeons can cooperate remotely in real time during surgery. Composite video feed images can be viewed by local and remote surgeons. VIPAR is used for subjective and objective evaluations through the cooperation between the stations established at Ho Chi Minh City, Vietnam, and Birmingham, Alabama.

VIPAR provides audiovisual communications during surgery between two local or remote surgeons. Since neurosurgery is a critical and extremely important operation, excellent image resolution is provided. These operations, conducted in 15 different occasions between Vietnam and the United States, were completed successfully without any serious complications. This application, which is a good example of on-the-job learning and training, has the potential to improve the capacity of surgeons in several institutions.

2. **MedShr:** Case-based mobile learning platform

Case-based learning, also referred as case study teaching and case method learning, is defined in various forms based on the discipline and type of the related case and used in health education for a long period of time (Sturdy, 2007). Learning activities in health education are generally based on cases. Basic, social and clinical sciences, clinical presentations, small group activities and learning content are generally constructed based on a case or exemplified with a case. This allows the student to relate knowledge to real-life situations (Thistlethwaite et al., 2012).

Healthcare students and professionals commonly use e-mail systems, instant messaging software or social media to discuss cases or share information. This method of discussing cases on electronic media, which is known to be particularly effective since digital natives are proficient in the method, is limited to the individual's own professional networks and has limitations based on file sharing limits, information

Figure 6. MedShr: Case-Based Mobile Learning Platform screen samples

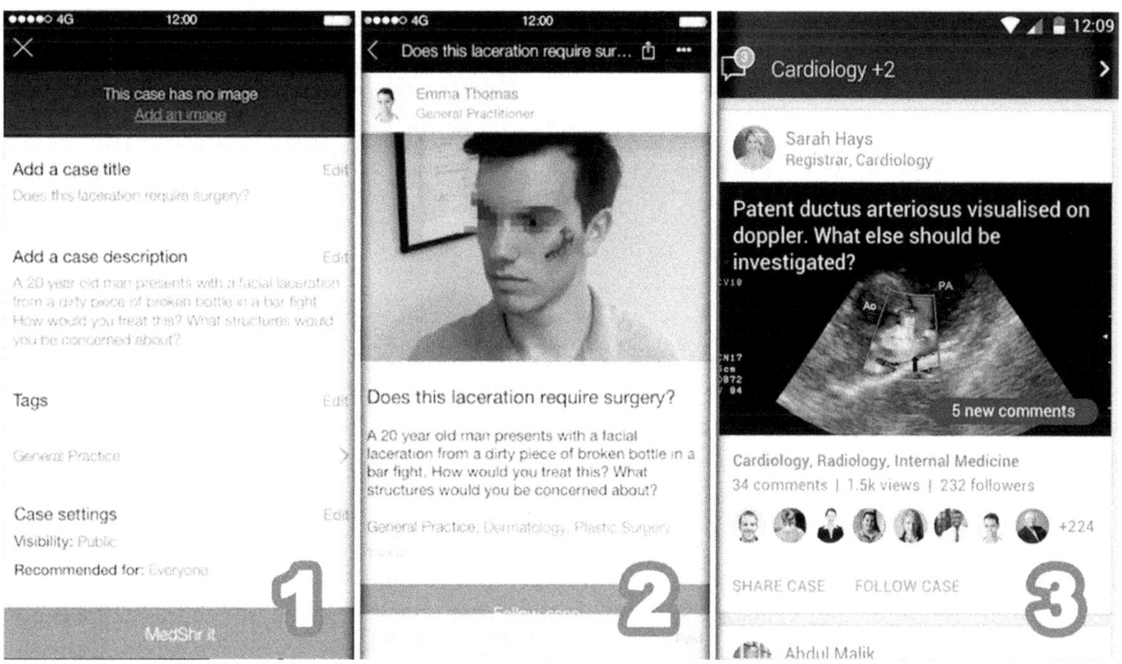

flow continuity problems and the lack of case diversity. Having observed these limitations, the MedShr corporation developed a case based mobile learning platform, which includes specialist physicians in the development team, for medical students and health professionals to discover, discuss and share clinical experiences and medical images. Case Based Mobile Learning (CBML) enables sharing the knowledge and experiences about the cases on private/public discussion groups or individually discussing the above-mentioned items on mobile devices using instant messaging software, social media or specific software.

MedShr could share a wide range of media ranging from ECGs, CT and X-rays to patient photographs and videos and assists colleagues in consulting each other or receiving consultation from specialists and share information in their areas of expertise. MedShr is a completely secure, private network. It emphasizes patient privacy and has a unique in-app system to get patient approval. The user can follow rare or interesting clinical cases, participate in case-based discussions and share experiences.

Medical students, general practitioners, and specialist physicians utilize MedShr groups as a resource to satisfy informal learning needs, before case-based exams and to supplement their formal education. In other words, MedShr is an inclusive case-based mobile learning platform that addresses the entire graduate, post-graduate and continuing medical education fields in the medical education discipline.

This application, based primarily on case discussions, is also a professional network or, in other words, a medical social media platform on a global scale. There are more than 200,000 platform subscribers worldwide that only include physicians, health professionals, and medical students, who share their knowledge and experiences. Furthermore, the software was selected by the Facebook (2016) corporation in 2016 as the mobile application of the year in the social benefit category based on the global contribution of the software.

3. **Medscape:** Mobile application for healthcare professionals

With more than five million users, Medscape is a mobile application, which is an immense source of clinical information for physicians, dentists, specialist physicians, nurses and healthcare professionals. This application, which contains more than over 3200 disease and treatment data, includes more than 7000 drug interactions and dose information.

Users can follow current developments and news related to the health field, learn the interactions or dose information of adult and pediatric drugs, and examine the clinical procedures related to patient care. Thus, Medscape is a powerful source of clinical information for all health professionals.

FUTURE RESEARCH

Revolutionary innovations are introduced regularly to the health field with the advances in virtual reality and augmented reality. Consistent with these developments, new concepts such as Mixed Reality (MR), which was studied in the early 2000s (Liu, Dede, Huang, & Richards, 2017), are becoming popular again with the advances in education technologies. MR combines the virtual world with the real world based on augmented virtuality. In the concept of augmented virtuality, the user interacts with the real object while also interacting with the virtual components located on the real object (Loup-Escande, Frenoy, Poplimont, Thouvenin, Gapenne, & Megalakaki, 2017). When both real and virtual worlds are combined, new environments and visualizations become possible where physical and digital objects could be found together and could interact in real time. It could be argued that when mixed reality technology

Figure 7. Medscape: Mobile Application for Healthcare Professionals screen samples

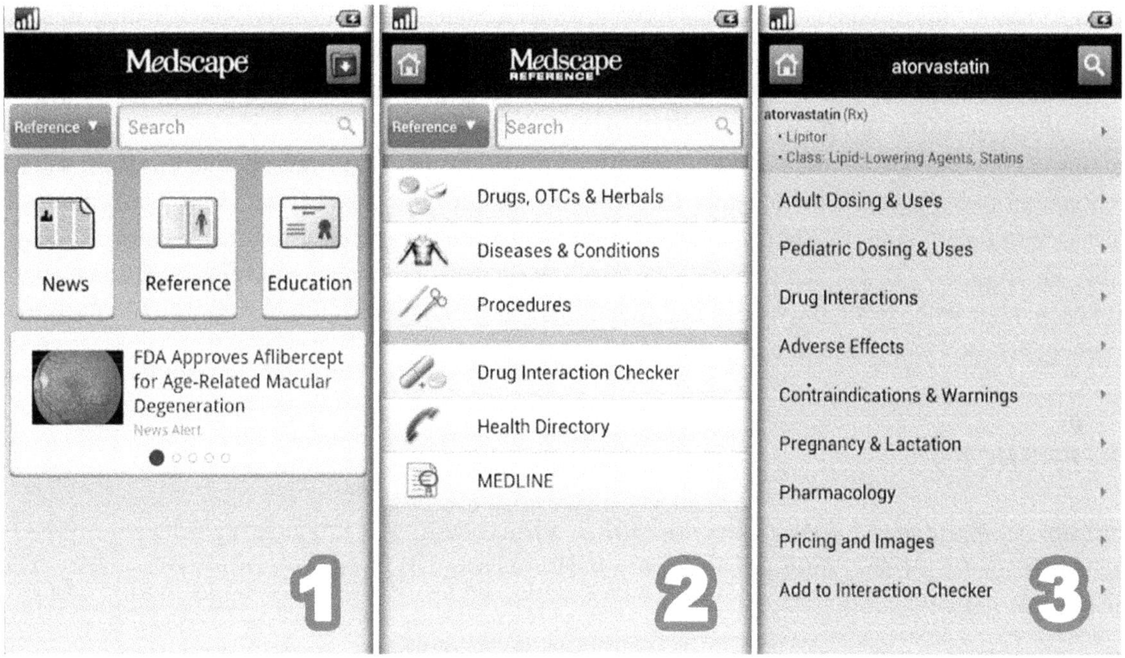

is integrated with haptic devices in health education, and as mass content production is initiated, there would be a shift towards high-reality tactile holograms. A learning age where a medical, dental, or specialty student can gain diagnosis, treatment, or consultation experience on any type of case before graduation, thus acquiring medical knowledge, clinical reasoning, and clinical decision-making skills. In addition to these potential developments, VR, AR and MR technologies will possess a level that can be preferred for real applications for the development of surgical skills in developing hand, arm, etc. movement calibrations and near-reality accuracy in simulations. This will make quality health education to achieve a massive, economic, accessible and multifunctional structure. Thus, with development of content and adoption of VLEs, mobile learning, AR, VR or MR, researchers and institutions would need to create monitoring and evaluation research methods for these specialized curricula and require mixed quantitative and qualitative short and long-term research methods.

CONCLUSION

The studies and applications described so far demonstrated that the health field is rapidly adapting to current advances in teaching technologies. In fact, as demonstrated in this chapter, in certain cases, the health field is the driving force for the future education technologies. Education technologies assist the acquisition and development of qualifications and skills of health professionals. Studies cited in this chapter demonstrated that educational technologies are of great significance even in cases that affect human life (Davis et al., 2016).

Particularly in the field of continuing medical education, physicians or health professionals often attend formal or informal education or incline towards autonomous and individual learning to follow professional trends and expand their knowledge base on diverse cases. Davis et al. (1999) provided evidence that interactive continuous medical training sessions that increase participant activity and enable the acquisition of application skills can positively affect the physician's professional practice and sometimes the health care service productivity. Apart from a few well-conducted trials, didactic sessions fail to be effective in changing the performances of physicians. It could be inferred that it is necessary to focus on mobile learning applications where health professionals can experience virtual reality environments or cooperative learning experiences between peers or professionals and interact with each other when designing active and functional instructional technologies to develop the current or new generation workforce in healthcare.

REFERENCES

Abulrub, A., Attridge, A., & Williams, M. (2011). Virtual Reality in Engineering Education: The Future of Creative Learning. *International Journal Of Emerging Technologies In Learning*, *6*(4), 4–11. doi:10.3991/ijet.v6i4.1766

Albrecht, U. V., & von Jan, U. (2017). Safe, sound and desirable: Development of mHealth apps under the stress of rapid life cycles. *mHealth*, 3. PMID:28828374

Alessi, M., Giangreco, E., Mainetti, L., Mighali, V., Patrono, L., Pinnella, M., ... Storelli, D. (2017). Abstracting IoT Complexity Through an Innovative Virtual Environment. *Recent Advances in Communications and Networking Technology*, *6*(1), 2–14. doi:10.2174/2215081106666170601123935

Alexander, B. (2004). Going nomadic: Mobile learning in higher education. *EDUCAUSE Review*, *39*(5), 29–35.

Archbold Hufty Alegría, D., Boscardin, C., Poncelet, A., Mayfield, C., & Wamsley, M. (2014). Using tablets to support self-regulated learning in a longitudinal integrated clerkship. *Medical Education Online*, *19*(1), 23638. doi:10.3402/meo.v19.23638 PMID:28440172

Arnott, J. A., & Planey, S. L. (2017). Flipped classroom approaches lead to no improvement in learning outcomes or student perceptions. *The FASEB Journal*, *31*(1Supplement), 751–14. PMID:27871064

Azuma, R. T. (1997). A survey of augmented reality. *Presence (Cambridge, Mass.)*, *6*(4), 355–385. doi:10.1162/pres.1997.6.4.355

Bilgiç, H. G., Duman, D., & Seferoğlu, S. S. (2011). Dijital yerlilerin özellikleri ve çevrim içi ortamların tasarlanmasındaki etkileri. *Akademik Bilişim*, 2-4.

Bonsón, E., Royo, S., & Ratkai, M. (2017). Facebook practices in Western European municipalities: An empirical analysis of activity and citizens' engagement. *Administration & Society*, *49*(3), 320–347. doi:10.1177/0095399714544945

Boulos, M. N. K., Lu, Z., Guerrero, P., Jennett, C., & Steed, A. (2017). From urban planning and emergency training to Pokémon Go: Applications of virtual reality GIS (VRGIS) and augmented reality GIS (ARGIS) in personal, public and environmental health. *International Journal of Health Geographics*, *16*(1), 7. doi:10.118612942-017-0081-0 PMID:28219378

Breslow, L., Pritchard, D. E., DeBoer, J., Stump, G. S., Ho, A. D., & Seaton, D. T. (2013). Studying learning in the worldwide classroom: Research into edX's first MOOC. *Research & Practice in Assessment, 8.*

Chapman, B. (2008). Learning Technology Primer. In ASTD Handbook for Workplace Learning Professionals. Baltimore, MD: United Book Press.

Cheng, Y., & Wang, S. H. (2011). Applying a 3D virtual learning environment to facilitate student's application ability–The case of marketing. *Computers in Human Behavior*, *27*(1), 576–584. doi:10.1016/j.chb.2010.10.008

Cross, J. (2005). *Informal learning: Rediscovering the natural pathways that inspire innovation and performance*. San Francisco, CA: Pfeiffer.

Cruz-Neira, C., Sandin, D. J., DeFanti, T. A., Kenyon, R. V., & Hart, J. C. (1992). The CAVE: Audio visual experience automatic virtual environment. *Communications of the ACM*, *35*(6), 64–73. doi:10.1145/129888.129892

Davis, D., O'Brien, M. A. T., Freemantle, N., Wolf, F. M., Mazmanian, P., & Taylor-Vaisey, A. (1999). Impact of formal continuing medical education: Do conferences, workshops, rounds, and other traditional continuing education activities change physician behavior or health care outcomes? *Journal of the American Medical Association*, *282*(9), 867–874. doi:10.1001/jama.282.9.867 PMID:10478694

Davis, M. C., Can, D. D., Pindrik, J., Rocque, B. G., & Johnston, J. M. (2016). Virtual interactive presence in global surgical education: International collaboration through augmented reality. *World Neurosurgery*, *86*, 103–111. doi:10.1016/j.wneu.2015.08.053 PMID:26342783

DeLozier, S. J., & Rhodes, M. G. (2017). Flipped classrooms: A review of key ideas and recommendations for practice. *Educational Psychology Review*, *29*(1), 141–151. doi:10.100710648-015-9356-9

Demir, K., & Akpınar, E. (2016). Mobil öğrenmeye yönelik tutum ölçeği geliştirme çalışması. *Eğitim Teknolojisi Kuram ve Uygulama, 6*(1).

Dewey, J. (1913). *Interest and effort in education*. Houghton Mifflin. doi:10.1037/14633-000

Durak, H. İ. (2016). Eğitimde Temel Kavramlar ve Kuramlar. In Tıp Eğiticinin El Kitabı (pp. 15-22). Ankara: Güneş Tıp Kitabevleri.

Economou, D., Mitchell, W. L., & Boyle, T. (2000). Requirements elicitation for virtual actors in collaborative learning environments. *Computers & Education*, *34*(3), 225–239. doi:10.1016/S0360-1315(99)00047-0

El-Hussein, M. O. M., & Cronje, J. C. (2010). Defining mobile learning in the higher education landscape. *Journal of Educational Technology & Society*, *13*(3), 12–21.

Elvezio, C., Sukan, M., Oda, O., Feiner, S., & Tversky, B. (2017, July). Remote collaboration in AR and VR using virtual replicas. In ACM SIGGRAPH 2017 VR Village (p. 13). ACM. doi:10.1145/3089269.3089281

Facebook. (2016). *FbStart Reveals Apps of the Year Winners and Introduces New Program Updates.* Retrieved May 25, 2017, from https://developers.facebook.com/blog/post/2016/05/31/fbstart-apps-of-the-year-program-updates/

Galkandage, C., Calic, J., Dogan, S., & Guillemaut, J. Y. (2017). Stereoscopic video quality assessment using binocular energy. *IEEE Journal of Selected Topics in Signal Processing*, *11*(1), 102–112. doi:10.1109/JSTSP.2016.2632045

Greenwald, S., Kulik, A., Kunert, A., Beck, S., Frohlich, B., Cobb, S., ... Snyder, A. (2017). *Technology and applications for collaborative learning in virtual reality*. Academic Press.

Hannaford, B., Barral, J. K., Rephaeli, E., Ching, C. D., & Bajaj, V. S. (2017). *U.S. Patent No. 9,645,785*. Washington, DC: U.S. Patent and Trademark Office.

Hanson, K., & Shelton, B. E. (2008). Design and Development of Virtual Reality: Analysis of Challenges Faced by Educators. *Journal of Educational Technology & Society*, *11*(1), 118–131.

Henderson, S., & Feiner, S. (n.d.). *Augmented Reality for Maintenance and Repair (ARMAR)*. Columbia University Computer Graphics & User Interfaces Lab. Retrieved from http://monet.cs.columbia.edu/projects/armar/

Hwang, G. J., Tsai, C. C., & Yang, S. J. H. (2008). Criteria, strategies and research issues of context-aware ubiquitous learning. *Journal of Educational Technology & Society*, *11*(2), 81–91.

Ifenthaler, D. (2012). Learning Management System. In Encyclopedia of the Sciences of Learning. Springer.

Istepanian, R., Laxminarayan, S., & Pattichis, C. S. (2014). *M-health*. John Wiley & Sons.

Jenkins, H. (2007, December 7). *Reconsidering Digital Immigrants...* Retrieved December 09, 2017, from http://henryjenkins.org/blog/2007/12/reconsidering_digital_immigran.html

Kartsonaki, E., Lawrie, G., Sharpe, P., & Bailey, C. (2017, August). Online, virtual, and adaptive learning environments: improving the journey through large first year chemistry courses. In *Proceedings of The Australian Conference on Science and Mathematics Education (formerly UniServe Science Conference)* (p. 74). Academic Press.

Kayapa, N., & Tong, T. (2011). Sanal gerçeklik ortamında algı. *Mühendislik ve Fen Bilimleri Dergisi*, *3*, 348–354.

King, A., Crawford, P., San Miguel, C., Barrie, M., Spencer, S., Tyransky, A., & Kman, N. (2017). Novel Emergency Medicine Curriculum Utilizing Self-Directed Learning and the Flipped Classroom Method: Head, Eyes, Ears, Nose and Throat Emergencies Small Group Module. *Journal of Education and Teaching in Emergency Medicine*, *2*(4).

Kleiman, G. M. (2004). Myths and realities about technology in K-12 schools: Five years later. *Contemporary Issues in Technology & Teacher Education*, *4*(2), 248–253.

Knox, J. (2017). How goes the revolution? Three themes in the shifting MOOC landscape. *Educação e Tecnologia: Abordagens Criticas, 1*, 386–408.

Kumar, S., Nilsen, W. J., Abernethy, A., Atienza, A., Patrick, K., Pavel, M., ... Hedeker, D. (2013). Mobile health technology evaluation: The mHealth evidence workshop. *American Journal of Preventive Medicine, 45*(2), 228–236. doi:10.1016/j.amepre.2013.03.017 PMID:23867031

Kuzu, E. B. (2014). *Hizmet öncesi BT öğretmenleri arasında çevrimiçi sosyal ağların eğitim amacıyla kullanılması. Yayımlanmamış doktora tezi*. Eğitim Bilimleri Enstitüsü, Anadolu Üniversitesi, Türkiye.

Kuzu Demir, E. B., & Demir, K. (2016). Enhancing Learning with Wearable Technologies In and Out of Educational Settings. *Digital Tools for Seamless Learning, 119.*

Leue, M. C., Jung, T., & Dieck, D. (2015). Google glass augmented reality: Generic learning outcomes for art galleries. In *Information and Communication Technologies in Tourism 2015* (pp. 463–476). Cham: Springer. doi:10.1007/978-3-319-14343-9_34

Li, Q. (2017). Characteristics and social impact of the use of social media by Chinese Dama. *Telematics and Informatics, 34*(3), 797–810. doi:10.1016/j.tele.2016.05.020

Liebowitz, J., & Frank, M. (Eds.). (2016). *Knowledge management and e-learning*. CRC press.

Limniou, M., Roberts, D., & Papadopoulos, N. (2008). Full immersive virtual environment CAVE TM in chemistry education. *Computers & Education, 51*(2), 584–593. doi:10.1016/j.compedu.2007.06.014

Lin, H. F. (2013). The effect of absorptive capacity perceptions on the context-aware ubiquitous learning acceptance. *Campus-Wide Information Systems, 30*(4), 249–265. doi:10.1108/CWIS-09-2012-0031

Littlejohn, A., Hood, N., Milligan, C., & Mustain, P. (2016). Learning in MOOCs: Motivations and self-regulated learning in MOOCs. *The Internet and Higher Education, 29*, 40–48. doi:10.1016/j.iheduc.2015.12.003

Liu, D., Dede, C., Huang, R., & Richards, J. (2017). *Virtual*. Augmented, and Mixed Realities in Education.

Loup-Escande, E., Frenoy, R., Poplimont, G., Thouvenin, I., Gapenne, O., & Megalakaki, O. (2017). Contributions of mixed reality in a calligraphy learning task: Effects of supplementary visual feedback and expertise on cognitive load, user experience, and gestural performance. *Computers in Human Behavior, 75*, 42–49. doi:10.1016/j.chb.2017.05.006

Lu, J., Pan, Z., Lin, H., Zhang, M., & Shi, J. (2005). Virtual learning environment for medical education based on VRML and VTK. *Computers & Graphics, 29*(2), 283–288. doi:10.1016/j.cag.2004.12.015

Maag, M. (2006). *iPod, uPod? An emerging mobile learning tool in nursing education and students' satisfaction*. Australasian Society for Computers in Learning in Tertiary Education (ASCILITE).

Magid, L. (2013, June 4). *A Third of Recently Married Couples Met Online and They're More Satisfied and Less Likely To Split-Up*. Retrieved December 09, 2017, from https://www.forbes.com/sites/larrymagid/2013/06/03/a-third-of-recently-married-couples-met-online-and-theyre-more-satisfied-and-less-likely-to-split-up/#2c62c61f595e

Malanga, D. F. (2017). Implementation of Mobile Health Initiatives in Malawi: Current Status, Issues, and Challenges. In Health Information Systems and the Advancement of Medical Practice in Developing Countries (pp. 115-128). IGI Global.

Marsick, V. J., Volpe, M., & Watkins, K. E. (1999). Theory and practice of informal learning in the knowledge era. *Advances in Developing Human Resources*, *1*(3), 80–95. doi:10.1177/152342239900100309

Marsick, V. J., Watkins, K. E., Scully-Russ, E., & Nicolaides, A. (2017). Rethinking informal and incidental learning in terms of complexity and the social context. *Journal of Adult Learning, Knowledge and Innovation*, (0), 1-8.

Masters, K., & Ellaway, R. (2008). e-Learning in medical education Guide 32 Part 2: Technology, management, and design. *Medical Teacher*, *30*(5), 474–489. doi:10.1080/01421590802108349 PMID:18576186

Masters, K., Ellaway, R. H., Topps, D., Archibald, D., & Hogue, R. J. (2016). Mobile technologies in medical education: AMEE Guide No. 105. *Medical Teacher*, *38*(6), 537–549. doi:10.3109/0142159X.2016.1141190 PMID:27010681

Mazur, E. (2017). Diverse Disabilities and Dating Online. In *Identity, Sexuality, and Relationships among Emerging Adults in the Digital Age* (pp. 150–167). IGI Global. doi:10.4018/978-1-5225-1856-3.ch010

McCann. (2016). Mobile Learning in Workforce Development: Cultivating Creativity on Action Learning Teams through Higher-Order mLearning. In Handbook of Research on Mobile Learning in Contemporary Classrooms. Ed. Dominic Mentor. IGI Global. doi:10.4018/978-1-5225-0251-7.ch007

McGrath, J., Mesiwala, A., Barrie, M., Boulger, C., Thompson, L., Mitzman, J., ... King, A. (2017). Novel Emergency Medicine Curriculum Utilizing Self-Directed Learning and the Flipped Classroom Method: Obstetric and Gynecologic Emergencies Small Group Module. *Journal of Education and Teaching in Emergency Medicine*, *2*(4).

Mentor, D. (2014). *mClass: How to adapt to ever changing mlearning environments*. Presentation delivered at the 21st Century Academic Forum Conference at Harvard University The 21st century academic forum conference at Harvard, Cambridge, MA. doi: http://www.21caf.org/uploads/1/3/5/2/13527682/21st_century_academic_forum_conference_book.pdf

Mentor, D. (2015). From App Attack to Goal-Oriented Tablet Use. In H. An, S. Alon, & D. Fuentes (Eds.), *Tablets in K-12 Education: Integrated Experiences and Implications* (pp. 1–21). Hershey, PA: IGI Global; doi:10.4018/978-1-4666-6300-8.ch001

Mentor, D. (2016a). The Commuter's Learning Journey: Field Observations Informing Mobile Learning Initiatives. In Handbook of Research on Mobile Learning in Contemporary Classrooms (pp. 315-335). IGI Global.

Mentor, D. (2016b). EMxC3 = e&mLearning Cultivating Connected Communities: Sustainable Workforce Talent Development. In D. Mentor (Ed.), *Handbook of Research on Mobile Learning in Contemporary Classrooms* (pp. 240–259). Hershey, PA: IGI Global. doi:10.4018/978-1-5225-0251-7.ch012

Mentor, D. (2018). Micro to Macro Social Connectedness Through Mobile Phone Engagement. In Encyclopedia of Information Science and Technology, Fourth Edition (pp. 6184-6194). IGI Global. doi:10.4018/978-1-5225-2255-3.ch537

Mentor, D., & Ahmad, N. (2010, September 22). *Teaching America's First Course on Mobile Phone Learning*. Retrieved December 09, 2017, from https://er.educause.edu/articles/2010/9/teaching-americas-first-course-on-mobile-phone-learning

Mi, M., Wu, W., Qiu, M., Zhang, Y., Wu, L., & Li, J. (2016). Use of Mobile Devices to Access Resources Among Health Professions Students: A Systematic Review. *Medical Reference Services Quarterly*, *35*(1), 64–82. doi:10.1080/02763869.2016.1117290 PMID:26794197

Mirrlees, T., & Alvi, S. (2016). *The MOOC: Rhetoric*. Political Economy and the Value of Technological Citizenship.

MIT Technology Review. (2017, October 10). *The way strangers meet via dating websites is changing society in unexpected ways, say researchers*. Retrieved December 09, 2017, from https://www.technologyreview.com/s/609091/first-evidence-that-online-dating-is-changing-the-nature-of-society/

Moreno, R., & Mayer, R. E. (2000). A coherence effect in multimedia learning: The case for minimizing irrelevant sounds in the design of multimedia instructional messages. *Journal of Educational Psychology*, *92*(1), 117–125. doi:10.1037/0022-0663.92.1.117

Muhammad, A. J., Mitova, M. A., & Wooldridge, D. G. (2016). Utilizing Technology To Enhance Learning Environments: The Net Gen Student. *Journal of Family and Consumer Sciences*, *108*(2), 61–63. doi:10.14307/JFCS108.2.61

Müller, D. (2012). *Design Characteristics of Virtual Learning Environments: A Theoretical Integration and Empirical Test of Technology Acceptance and IS Success Research*. Springer Science & Business Media.

Müller, N., & Dikke, D. (2012). Towards Mobile Personalized Learning Management Systems. In *IADIS International Conference Mobile Learning* (pp. 115-122). Academic Press.

Murray, J. H. (2017). *Hamlet on the holodeck: The future of narrative in cyberspace*. MIT press.

Njie-Carr, V. P., Ludeman, E., Lee, M. C., Dordunoo, D., Trocky, N. M., & Jenkins, L. S. (2017). An integrative review of flipped classroom teaching models in nursing education. *Journal of Professional Nursing*, *33*(2), 133–144. doi:10.1016/j.profnurs.2016.07.001 PMID:28363388

Noll, C., von Jan, U., Raap, U., & Albrecht, U. V. (2017). Mobile Augmented Reality as a feature for self-oriented, blended learning in medicine: Randomized controlled trial. *JMIR mHealth and uHealth*, *5*(9), e139. doi:10.2196/mhealth.7943 PMID:28912113

Novak, G. M., Patterson, E. T., Gavrin, A. D., Christian, W., & Forinash, K. (1999). Just-in-time teaching. *American Journal of Physics*, *67*(10), 937–938.

Ortega, J., & Hergovich, P. (2017). *The Strength of Absent Ties: Social Integration via Online Dating*. arXiv preprint arXiv:1709.10478

Pathak, B. (2016). Improving the Educational Effectiveness of MOOCs by Using Web-Based Information and Communication Technology. In *Revolutionizing Education through Web-Based Instruction* (pp. 186–201). IGI Global. doi:10.4018/978-1-4666-9932-8.ch011

Paul, K. (2017, October 21). *How online dating affects divorce rates*. Retrieved December 09, 2017, from https://www.marketwatch.com/story/what-online-dating-could-do-to-divorce-rates-2017-10-16

Paulsen, M. F. (2002). Online Education Systems: Discussion and definition of terms. *NKI Distance Education, 202*.

Pedró, F. (2006). *The new millennium learners: Challenging our views on ICT and learning (No. 9228)*. Inter-American Development Bank.

Pfeiffer, T., Feiner, S. K., & Mayol-Cuevas, W. W. (2016). 6.4 Group 4: Eyewear Computing for Skill Augmentation and Task Guidance. *Eyewear Computing–Augmenting the Human with Head-Mounted Wearable Assistants, 23*(3), 199.

Pickering, J. D., Henningsohn, L., DeRuiter, M. C., de Jong, P. G., & Reinders, M. E. (2017). Twelve tips for developing and delivering a massive open online course in medical education. *Medical Teacher*, 1–6. PMID:28485629

Ponce, L. B., Méndez, J. A. J., & Peñalvo, F. J. G. (2014, November). A systematic review of using mobile devices in medical education. In *Computers in Education (SIIE), 2014 International Symposium on* (pp. 205-210). IEEE. 10.1109/SIIE.2014.7017731

Prensky, M. (2001). Digital natives, digital immigrants part 1. *On the Horizon*, *9*(5), 1–6. doi:10.1108/10748120110424816

Quinn, C. N. (2011). *Designing mLearning: tapping into the mobile revolution for organizational performance*. Wiley.

Rochadiat, A. M., Tong, S. T., & Novak, J. M. (2017). Online dating and courtship among Muslim American women: Negotiating technology, religious identity, and culture. *New Media & Society.*

Roo, J. S., & Hachet, M. (2017, October). One Reality: Augmenting How the Physical World is Experienced by combining Multiple Mixed Reality Modalities. In *Proceedings of the 30th Annual ACM Symposium on User Interface Software and Technology* (pp. 787-795). ACM. 10.1145/3126594.3126638

Rosenfeld, M. J. (2017). Marriage, Choice, and Couplehood in the Age of the Internet. Academic Press.

Rosewarne, L. (2016). *Intimacy on the internet: Media representations of online connections* (Vol. 86). Routledge.

Şahin, F., & Mentor, D. (2016). Using Mobile Phones for Assessment in Contemporary Classrooms. Handbook of Research on Mobile Learning in Contemporary Classrooms, 116-138.

Sahin, I. (2006). Detailed review of Rogers' diffusion of innovations theory and educational technology-related studies based on Rogers' theory. *TOJET: The Turkish Online Journal of Educational Technology, 5*(2).

Şahiner, M. (2016). Öğrenci Merkezli Eğitimde Yeni Bir Yöntem: Ters Yüz Edilmiş Sınıflar. *Turkiye Klinikleri Journal of Medical Education-Special Topics*, *1*(1), 24–32.

Saleeb, N., & Dafoulas, G. (2013). *Effects of Virtual World Environments in Student Satisfaction: An Examination of the Role of. Governance, Communication, and Innovation in a Knowledge Intensive Society*. Academic Press.

Saravani, S. J. (2010). Standards informing design of library service delivery to mobile devices and nomadic learners. In *Vala2010 abstracts* (pp. 1345–1415). Melbourne: VALA2010 Conference.

Sayek, İ. (2016). Tıp Eğitiminin Gelişimi ve Değişim: 21. Yüzyılda Tıp Eğitimi. In Tıp Eğiticinin El Kitabı (pp. 3-11). Ankara: Güneş Tıp Kitabevleri.

Sezer, B., Onan, A., & Elçin, M. (2016). Sürekli Tıp Eğitiminde Bilişim Teknolojileri. *Turkiye Klinikleri Journal of Medical Education-Special Topics*, *1*(3), 1–6.

Shao, F., Tian, W., Lin, W., Jiang, G., & Dai, Q. (2016). Toward a blind deep quality evaluator for stereoscopic images based on monocular and binocular interactions. *IEEE Transactions on Image Processing*, *25*(5), 2059–2074. doi:10.1109/TIP.2016.2538462 PMID:26960225

Sharecare. (2016). *Sharecare Acquires Virtual Reality Company to Transform Patient Engagement and Medical Education - Sharecare*. Retrieved May 17, 2017, from https://about.sharecare.com/press-releases/sharecare-acquires-virtual-reality-company-transform-patient-engagement-medical-education

Sherman, W. R., & Craig, A. B. (2002). *Understanding virtual reality: Interface, application, and design*. Elsevier.

Smith, A., & Anderson, M. (2016). *5 facts about online dating*. Retrieved December 09, 2017, from http://www.pewresearch.org/fact-tank/2016/02/29/5-facts-about-online-dating/

Smith, A., & Anderson, M. (2016, February 29). *5 facts about online dating*. Retrieved December 09, 2017, from http://www.pewresearch.org/fact-tank/2016/02/29/5-facts-about-online-dating/

Sturdy, S. (2007). Scientific method for medical practitioners: The case method of teaching pathology in early twentieth-century Edinburgh. *Bulletin of the History of Medicine*, *81*(4), 760–792. doi:10.1353/bhm.2007.0093 PMID:18084106

SuperData. (Ed.). (2016). *Virtual reality's hardware hurdles hurt forecast*. Retrieved May 16, 2017, from https://www.superdataresearch.com/virtual-realitys-hardware-hurdles-hurt-forecast/

Sweeney, W. B. (2012). Teaching surgery to medical students. *Clinics in Colon and Rectal Surgery*, *25*(03), 127–133. doi:10.1055-0032-1322525 PMID:23997667

Swinnerton, B. J., Morris, N. P., Hotchkiss, S., & Pickering, J. D. (2017). The integration of an anatomy massive open online course (MOOC) into a medical anatomy curriculum. *Anatomical Sciences Education*, *10*(1), 53–67. doi:10.1002/ase.1625 PMID:27315160

Teng, C. E., & Joo, T. M. (2017). Analyzing the Usage of Social Media: A Study on Elderly in Malaysia. *World Academy of Science, Engineering and Technology, International Journal of Social, Behavioral, Educational, Economic. Business and Industrial Engineering*, *11*(3), 713–719.

The Body VR. (2015). *Want to stay up to date with our announcements? Sign up below*. Retrieved May 23, 2017, from http://thebodyvr.com/

Thistlethwaite, J. E., Davies, D., Ekeocha, S., Kidd, J. M., MacDougall, C., Matthews, P., & Clay, D. (2012). The effectiveness of case-based learning in health professional education. A BEME systematic review: BEME Guide No. 23. *Medical Teacher*, *34*(6), e421–e444. doi:10.3109/0142159X.2012.680939 PMID:22578051

Traxler, J., & Kukulska-Hulme, A. (2016). Introduction to the next generation of mobile learning. *Mobile learning: The next generation*, 1-10.

Van Raaij, E. M., & Schepers, J. J. (2008). The acceptance and use of a virtual learning environment in China. *Computers & Education*, *50*(3), 838–852. doi:10.1016/j.compedu.2006.09.001

Virtual Reality Society. (2016). *Virtual Reality in Healthcare*. Retrieved May 16, 2017, from https://www.vrs.org.uk/virtual-reality-healthcare/

Walsh, E., Daems, W., & Steckel, J. (2017). Assistive Pointing Device Based on a Head-Mounted Camera. *IEEE Transactions on Human-Machine Systems*, *47*(4), 590–597. doi:10.1109/THMS.2017.2649884

Wu, P. H., Hwang, G. J., Tsai, C. C., Chen, Y. C., & Huang, Y. M. (2011). A pilot study on conducting mobile learning activities for clinical nursing courses based on the repertory grid approach. *Nurse Education Today*, *31*(8), e8–e15. doi:10.1016/j.nedt.2010.12.001 PMID:21196068

Wu, P. H., Hwang, G. J., & Tsai, W. H. (2013). An expert system-based context-aware ubiquitous learning approach for conducting science learning activities. *Journal of Educational Technology & Society*, *16*(4), 217–230.

Wyne, M. F. (2015). Merging mobile learning into traditional education, 2013–2016. *The International Conference on E-Learning in the Workplace*.

Yang, H. (2017). *Exploring a Semi-Virtual Reality System Impacting Learning Curves of College Students*. Academic Press.

Yuen, S. C. Y., Yaoyuneyong, G., & Johnson, E. (2011). Augmented reality: An overview and five directions for AR in education. *Journal of Educational Technology Development and Exchange*, *4*(1), 11. doi:10.18785/jetde.0401.10

Zheng, M., & Waller, M. P. (2017). ChemPreview: An augmented reality-based molecular interface. *Journal of Molecular Graphics & Modelling*, *73*, 18–23. doi:10.1016/j.jmgm.2017.01.019 PMID:28214437

KEY TERMS AND DEFINITIONS

Augmented Reality: Augmented reality provides users the opportunity to observe the real world in which virtual objects or content reside.

Case Based Mobile Learning (CBML): It enable to sharing the knowledge and experiences about the cases on private/public discussion groups or individually discussing the abovementioned items on mobile devices using instant messaging software, social media or specific software.

Learning Management System: Web-based software that enables users to manage and utilize learning content in interactive learning environments.

Mobile Learning: Any form of learning that takes place using a mobile device.

Virtual Reality (VR) Environments: Environments that have attracted attention as an immersive solution for learning by experience.

Chapter 10
mClass Planet of the Apps: The Rise of Mobile Learning

Dominic Mentor
Teachers College, Columbia University, USA

ABSTRACT

This chapter introduces how mobile devices can be used on a personal and macro level for professional development. The chapter also covers theories posited for practical applications from pedagogy and andragogy perspectives. There are multiple layers of considerations in terms of context, content, and collaboration factors to optimize mobile learning. There are more mobile devices in the world than people, and more than three quarters of the world's population already has some type of mobile phone, making it the most wide-spread technology and most common electronic device in people's hands. Untapping this ubiquitous technology creates a wide array of educational possibilities. Hence, a mobile first learning design is crucial in organizational leadership and professional development, to help bridge the gap between personal lives, schools, colleges, and the workplace. Furthering the concept of learning is everywhere as a natural segue for ownership of learning and engagement.

INTRODUCTION

Mobile technologies are challenging the role of teachers, students, workers and instructional designers. Mobile learning is offering micro-moments of learning in a multitude of multimodal stimulating means cultivating opportunities for transformation. People now have access to a wealth of multimodal information and stimuli, which can either be forces of distraction, tools of engagement or convert waiting times into productive learning opportunities. Mobile phones and tablet devices are devices of choice for consuming and producing content. Mainly because of ready-in-hand access, easy interfaces, quick start-up times as well as the availability of a plethora of applications (apps). However, policy and professional development offerings are still slow in aiding mobile learning to become fully adopted. The same is true for K – 12 teachers, who are already suffering from major time poverty, which makes it difficult to seek more professional development. Other industries and arenas can also benefit from mobile learning (m-learning). However, mobile learning can help to save time, energy and can help to put their

DOI: 10.4018/978-1-5225-4111-0.ch010

students' learning interest first while making them active participants rather than passive receivers. The same learning and productivity advantages are true in the workplace, but with additional adult learning theories needed. Of course, work has gone beyond the office or site of work, but mobile learning is still off to a slow start in terms of generating mlearning opportunities as well as adoption in formal space. Mobile devices are changing the way we work and offers a multitude of formal and informal mlearning opportunities.

Mobile technologies can be incorporated into formal and informal learning spaces as they extend the formal educational time, and bring to life a basic tenet that learning is everywhere. Furthermore, mobile technologies extend and blend formal and informal educational spaces in a manner that can help to engender authentic and autonomous learning habits. Through the multi-screen world that many people live, mobile learning can help for seamless movement between desktop and mobile device. The mobile phone can either be the main or supportive, just-in-time device. Still, learning designs need to be updated to accommodate this new extension of learning possibilities. Updates to or new learning designs can be adjusted or accommodated in innumerable ways from updating to mobile friendly Learning Management Systems or leveraging various mobile applications including social media or mobile learning in training programs. Additionally, mobile learning offers to overcome challenges or extend opportunities related to workplace training. And if not affected by the app-attack hype (Mentor, 2014), then delivering content on devices with different operating systems, catering to different worker's schedules, or learning styles, as well as real time and asynchronous learning analytics. If the mobile learning is designed well, with a smooth user experience, workplace or educational organizations can benefit from individuals' mobile engagement. Through push and pull, an adult or any learner can be offered a personal learning journey or agency. Some examples of push learning are reminders for bio-behavioral change or needs like quitting smoking, adherence to medication regimen, sending words of affirmation, or pushing one new word per day to learn. Push examples range from the receipt of MCAT, TOEFL or GRE offering one question

Figure 1. Mobile learning

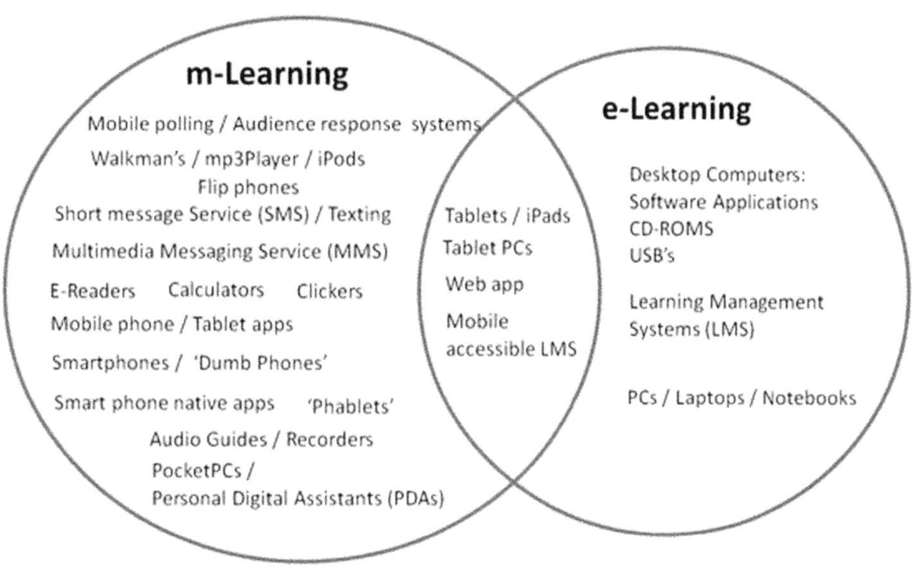

at a time, or notifications of news, updates, or information on a topic of interest. There are also native and web-based apps that can pull users back during inactivity with reminders, or flash alerts or alluring offers of app-based rewards. While behaviorist in form, both the push and the pull learning, offers its own strengths and cons. The push and pull also aids breaking the dichotomous view of education, taking learning beyond the confines of formal settings. And with informal learning probably occurring more frequently, naturally occurring, and aided by the ubiquity of the mobile phone with internet access, the opportunities of push and pull mobile learning is that the motivation differs as assigned or self-sought.

Currently, very few dedicated teacher or instructional design preparation programs exist for exploring the relationship between mobile technologies and learning. By offering a course or program of mobile learning technologies with solid theoretical foundations, teachers-in-training, adult trainers, curriculum designers, strategic learners, and instructional designers can learn which theories and practical pedagogical considerations should inform their mobile and multi-screen teaching and learning initiatives. One graduate course in Organizational Leadership and Adult Learning (ORLD) hosts the possibilities, and promises to overcome the perils, of incorporating mobile learning how it can offer transformative learning opportunities. Interlocking, hybridizing or taking into account all of the theories present in a context, or which theories is being ignored and/or can be added with a new initiative, is a valuable and useful approach. It allows for multi-lens processing and provides one with multiple perspectives to view mobile learning projects, research, readings, field observations, classroom contexts, online learning and many other formal, and informal learning contexts. Hosting a hybrid of theories, especially given different contexts, a plethora of devices, goals, objectives, target audiences and levels; will aid the mobile mediated learning. For example, incorporating behaviorist theory is still a factor, especially when it comes to adding gamification elements, or incremental goal attainment during mobile and computer mediated learning is one to note. Or considering from a mobile learning perspective when to employ situated cognition, perhaps through Near Field Communication (NFC) or Bluetooth technology while on the move.

But why offer a course on mobile learning? With the exponential use of mobile phones, it is imperative to understand how mobile technology can be used for learning and how to design for, and/or incorporate micro-learning. Mobile learning is still slow out of the gate and not widely adopted. By offering a course on mobile learning technologies, students will have the opportunity to keep abreast of current mobile technology that people are already using as part of their daily lives and activities. And to provide a clearer framework for analyzing the growth of mobile and micro-learning, as laid out in the background below.

BACKGROUND

Naismith, Sharples, Vavoula, and Lonsdale's literature review published in 2004 is still considered to be the seminal piece of research that proposed mobile learning as a field of study, different from electronic learning (eLearning). Their approach "move[d] away from the dominant view of mobile learning as an isolated activity, to explore mobile learning as a rich, collaborative and conversational experience, whether in classrooms, homes or the streets of a city. In this way, they hosted the term mobile learning for the mobile learning opportunities available to individuals, a bigger target audience, its scope and the increasing scalability of mobile learning opportunities. Through the sheer ubiquity of mobile devices, mobile learning out rivals the size, breadth, and depth of a computer and access to the internet, influencing how people work, and, even though still slow to catch on, how people learn.

Mobile technology has changed how people work, and how people learn. While slow to catch on in corporate, higher education and K-12 settings, mobile learning, even in micro learning format, in informal settings, has become the norm. The individual to mass-mobile online culture has proliferated to be adopted as common practice in the form of comments to media accessed via mobile means, sparking conversations and debate, influencing beliefs and policy. With mobile devices becoming the computer technology of choice, it is well perched and exploited for mass distribution of digitized content. Albeit, with a strong mobile first design approach aiding the mass distribution of digital content. Yet many of their problems persisted during the initial uptake of mobile devices – from delivering content to various mobile device sizes, to content push and pull aggregation, to content development for diverse operating systems. Somewhat thought to be addressed by native apps, but not necessarily a sustainable method with high costs and native app design talent not as plentiful as one would expect. The native app approach is still popular, but the agnostic web app approach is gaining ground, and web responsive designs are aiding the conversion of content to being more mobile accessible and mobile friendly, with mobile learning initiative following suite.

This chapter describes the characteristics of mlearning, the value of mobile first design approaches, micro moments of learning, and showcase examples of impact and untapped formal learning opportunities. One of which is learning from mobile learning, especially with regards to digital data tracking and analytics. However, mobile learning plans can be undone by not considering the plethora of diverse devices. mLearning still struggles to be adopted as part of traditional formal training or learning; even though, mobile learning ideas are abounding and have been created to serve communities. In some cases, breaking out of defined restrictions according to geography, intellectual discipline, or specific end users. mlearning has been fueled by rapid technological growth trajectories, but still struggles to live up to its promise as people still think of mobile learning designs with traditional lenses and expect it to transcend such limits as output and input. Small screen real estate and input options on mobile devices require a cognitive load consideration of element interactivity (Pass, Renkl & Sweller, 2003).

A graduate course designed and offered at Teachers College, Columbia University, explores how one can utilize the mobile technologies for adult, K-12, and organizational learning as well as the factors to consider for designing m-learning (Mentor & Ahmad, 2010). The course highlights theoretically informed processes that can stand up to the ever-changing mobile technology landscape and support practical pedagogy and andragogy. Adding practical pedagogy and andragogy, the project based mlearning designs takes a progressive structure to elucidate lessons learned and offer directions for growth to adapt to changing mLearning environments. The mClass offers in-service and pre-service teachers, as well as adult learning designers and practitioners from many different disciplines, examples, and opportunities to directly engage learners virtually, in real-time and asynchronously. The mobile learning course also highlights the multimodal stimulation of digital curation through mobile technologies and their ability to prepare and engage any learner, child or adult, better autonomous learning opportunities within different learning contexts. Apart from the theories, practical pedagogical guides could be served well if informed by the m-Learning guild's three C's of context, content, and collaboration (2011), which further frames the course. Context informs which theories and practical processes to consider as well as the how, when, and the what of educational devices and/or apps to use, to meet the planned goals and objectives of interactive educational engagements. While context is crucial, core curriculum and content on a mobile device is dynamic. Taking note of the goals, objectives, context and content can promote learners as active agents and collaborative participants, rather than passive receivers. Many other factors to consider range from teacher preparation, teacher engagement, student personalities, students' prior experience and more. All of which could influence the class dynamic, and how that plays out in a F2F or online space. At times, even in blended or mobile learning contexts, over-talkative personalities could even influence the virtual engagement space. In short, drawing on several theories and practices, a hybrid approach to teaching and learning with tablets is necessary for successful and engaging activities that address the needs of mobile technological integrated teaching and learning. With the ubiquity of mobile devices, there is a general need for a mobile first approach (Wroblewski, 2011), as well as a mobile first e-learning design need (Curinga & Svaranos, 2016). To note, as with many new approaches and technologies, there is still resistance and many questions holding back the adoption and integration of m-learning into the learning and development processes. Some questions that are asked as people try to differentiate m-learning from e-learning are:

- What does the phrase "mobile first learning design" mean?
- What are some of the educational possibilities offered by mobile devices?
- What are concrete examples of mobile learning?
- How does e-learning differ from mobile learning?

The next section of this chapter will offer a response to those questions. The chapter will also offer some diagrams to help with the understanding and differentiation.

MOBILE FIRST LEARNING DESIGN

A "mobile-first learning design" responds to the critique that mobile learning is a weaker alternative to traditional eLearning accessed via computers with bigger screens. Instead, this approach assumes that people will more likely access a website or query via a mobile device.

Given the rise in mobile device use, in particular mobile phones, that growth has prompted software and learning designers to take a "mobile first" approach—which demands an alternative, high-quality mobile experience to maintain learner interest. Mobile-first learning design recognizes the limited element interactivity afforded by a device with a small screen without compromising a quality user experience. A mobile first design approach can aid instructional designers to adapt the way they design for larger screens to a more realistic multi-screen learning experience.

Luke Wroblewski (2011) introduced the concept of a mobile first design approach. He posits that—by designing with mobile devices foremost in mind—learning designers are better prepared for the current explosive growth and new user opportunities via the mobile internet.

An approach of making everything mobile is similarly fraught with issues. The conversion of everything to PDF approach was, and still is a danger to e-Learning, so to is a make everything mobile approach dangerous without considering the form factors from an element interactivity perspective. Make everything mobile makes an assumption of universality, but becomes a disservice to all as it does not take into consideration the multitude of diverse devices, and it does not address limited input and output realities of mobile devices. From an element interactivity perspective, input and output considerations tie in with physiological conditions, ranging from eye-strain on small screen sizes, thumb tendinitis on small keyboard inputs. Why is this important to consider from a mlearning as well as cognition and handheld device perspective; because it will help you the progenitor in many contexts, or rather facilitator of mobile learning to consider where and how the receiver or dynamic active learner agent will interact with the learning and assessment pieces. Endeavoring to 'mobilize' or convert everything, is not so much an approach to open learning to everyone, but rather a flawed approach giving in to more perils and pitfalls than living up to the promise of mobile learning.

EXAMPLES OF EDUCATIONAL POSSIBILITIES OFFERED BY MOBILE DEVICES

While teaching English as a Second Language in a rural area 60 km outside of Cape Town in South Africa, there were many creative efforts needed to convince the majority of students to learn English as a second language. Speaking in English on the farms and in that rural area was frowned upon, ridiculed, but moreover envied. Computers were a scarce resource, but calculators were required and abundant. The simple idea of making words with the numbers, turned into a competition of generating English words and building vocabulary. This was the start of an idea that would develop into wanting to leverage the power of mobile devices for learning. And as calculators became more powerful and PDA's hit the scene, the idea grew exponentially. Prior to the launch of the iPhone, a mobile learning course was conceived and co-designed to offer graduate students a theoretical and practical foundation to leverage alphanumeric mobile phones for learning. Even with technology advancements of PDAs and early smartphones from Palm, the idea of mobile learning was not necessarily understood or warmly welcomed. However, with the advent of the iPhone the same year that the graduate mobile learning course was piloted, the interest shot up well beyond the peak of Gartner's hype cycle (Fenn, & Linden, 2005), and maintained its interest beyond the trough of disillusionment (ibid).

By 2009, a graduate seminar and project-based course was piloted (Mentor & Ahmad, 2010). The course was designed for students who were not programmers, but who were interested in exploring theories and practical pedagogies that inform app development. While working on the design of a mobile phone app or mobile phone learning activity, students gained valuable design insights that would empower them to speak to a programmer on what and how they wanted their app designed. In this intensive course, students learned how to design and develop mobile learning activities and could design and build native or web apps for mobile devices. If a native app, they could decide whether they wanted to design for iOS or Android, and could do so from multiple perspectives of UI\UX designer, develop and run user testing, and also as a developer. Throughout this course on mobile phone apps the instructors melded theory, design, and the development with real-world experiences.

Students who have taken a graduate mobile learning course have been from the USA as well as from around the world. The adult learners came from many different professional backgrounds ranging from learning & development consultants, professional developers, medical educators, coaches, teachers, principals, as well as law firms, policy and government institutions. Some examples of student projects range from informing mobile app development and strategy at the Metropolitan Museum of Art in New York City, as well as the highly successful Explain Everything app. Other examples that have been discussed in the past depending on students' interests were:

- Designing virtualized or augmented reality tours via mobile devices to help prepare a learner for a new context or orient them to learning objectives/content, as in museum education.
- Designing a math ninja game with interaction between parents and young children.
- Exploring film noir snippets and concepts via mobile device engagement.

- Invigorating the teaching of credit bearing Tai Chi in a college setting with videos and using the Learning Management System (like a social media platform).
- An alphabet learning activity teaching and offering learning of identifying and writing alphabet letters.

One area that has seen useful in mobile development is audience, or participant response systems. Classtalk mentioned in the seminal mobile literature review (Naismith et. al., 2005), and iClickers were different types of classroom response systems (Mentor, 2011). Classtalk as cited in the Naismith *Literature Review in Mobile Technologies and Learning* was used with first year physics students at the University of Massachusetts. Classtalk offered students opportunities to participate in class discussions via mobile devices, and exploited the near synchronous aggregating and display of collated responses. While the first iterations of iClickers came with its own devices that could be numbered as a means of tracking students' in class responses, that system was soon abused by students when they would be absent from class, but their friends would have their clickers to answer a professor's question. iClickers could also integrate with learning management systems, and be used for both formative and summative assessments (iClickers conference 2011). In addition to the mobile clickers, referred to as student remotes, the new versions of iClickers include Android and iOS native apps accessible via mobile phones, as well as mobile friendly browser access. In addition to these live class audience response systems, Poll Everywhere which launched in 2007 and others like Quizbox, Merida and Nearpod have grown in popularity. While Quizbox hosts their questions online, Merida also offers native app incorporation. Additionally, Nearpod is a newer option. When it was first introduced it was only available for iOS' iPads, but it has since become hardware agnostic. Nearpod mimics are considered older technologies where teachers could share their screen on their students' devices, guiding them through a lesson. Students can now answer questions via their mobile phones either through native or web apps when prompted. They can do so either with text, or even drawing an answer. Additionally, there are numerous collaborative tools (i.e. Google Docs, Popplet, Padlet, Mural.ly, Minecraft, Peergrade to name a few that allow for brainstorming and collaboration in real time. While QR codes have also been researched as mobile learning interaction and engagement tools with mixed results (Gao, Liu & Paas, 2016; Jamu, Lowi-Jones, & Mitchell, 2016; Mousa, & El-Salam, 2016; Rikala & Kankaanranta, 2012).

Many of the mobile learning interventions, applications and others are shared with the mobile learning graduate class as examples of approaches with applied theories. The graduate course was the USA's first graduate course on mobile learning (Mentor & Ahmad, 2010) and catered to students from various industries. As part of our teaching and learning philosophy, we surveyed the students at the start of the semester, at the end of each class session, and at the end of the semester.

SURVEY RESULTS

When we started the course in 2009 we introduced the concept of anonymous electronic session reviews after every session. We used that feedback on a weekly basis to tailor the class to the students' areas of interest regarding mobile learning, their course of study and disciplines, and their preferred mix of learning styles and approaches, as well as the technical or non-technical level. In 2015, a self-reflective pre- and post-review was added to the survey. Students were asked to rate themselves about their mobile learning knowledge at the start of the semester. They rated themselves about mobile learning as a whole, as well as on specific topic areas like mobile learning literature, mobile assisted language learning, mobile activism, mobile learning in K- 12 settings, with adults in the work settings, and other areas like mobile health, mobile journalism and mobile gaming for educational benefit. Students were asked to reflect and rate themselves before a specific session on their knowledge of that area as it pertains to mobile learning. Apart from students rating themselves with an overall rating at the beginning of the semester, they completed a post-survey at the end of the semester, and rated their mobile learning knowledge, as well as perceived growth in mobile learning. Apart from demographic questions like their level of study (masters, non-degree, or doctorate), country of origin and length of stay in the USA up until taking the mobile learning course and their program or area of study. The latter questions were included to aid discussion in class around mobile technology, mobile learning and mobile cultural practices.

About 55 percent of the students taking the class were from the USA, while Chinese students made up 35% of the class participants as captured from 2015 data. Students from other countries listed themselves as being from Argentina, Brazil, Canada, Hong Kong, Japan, Korea, Indonesia, Mexico, Philippines, Thailand, Turkey, South Africa, and the Ukraine. Students came from various disciplines over the years: Arts and Humanities, Applied Linguistics, Bilingual/Bicultural Education, Communication, computing and technology in Education, Dental School, Human Development, Instructional Technology and media, Mathematics, Science and Technology, Teaching English as a Second Language, Organizational Leadership and Adult Learning, Transcultural and International Studies, School of International Policy and Administration, and the School of Public Health. Twenty-five percent were doctoral level students, while 75% were at the Masters level.

The following data comes from surveys and graphs from students' collated responses for Summer A 2015, Spring 2016, Spring 2017, Summer A 2017, Summer B 2017, and Fall 2017.

An example of survey results for one Spring semester asking students about their own rating of their prior knowledge regarding mobile learning literature is shown in Figure 2.

Also see Figure 3, self-rated prior knowledge about mlearning in the workplace.

Then knowledge of mobile activism (Figure 5).

Figure 2. Mobile learning literature prior knowledge self rating. Self-rated prior knowledge where zero reflects no knowledge of the area and five indicated very high knowledge of the mobile learning area.

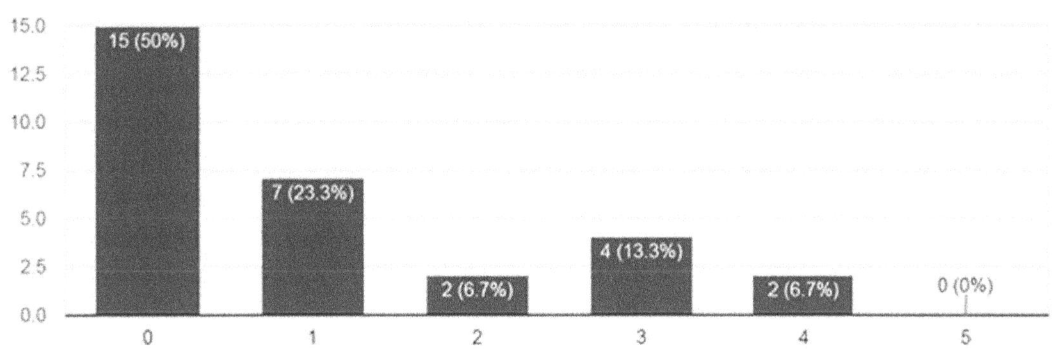

Figure 3. Self-rated prior knowledge about mlearning in the workplace

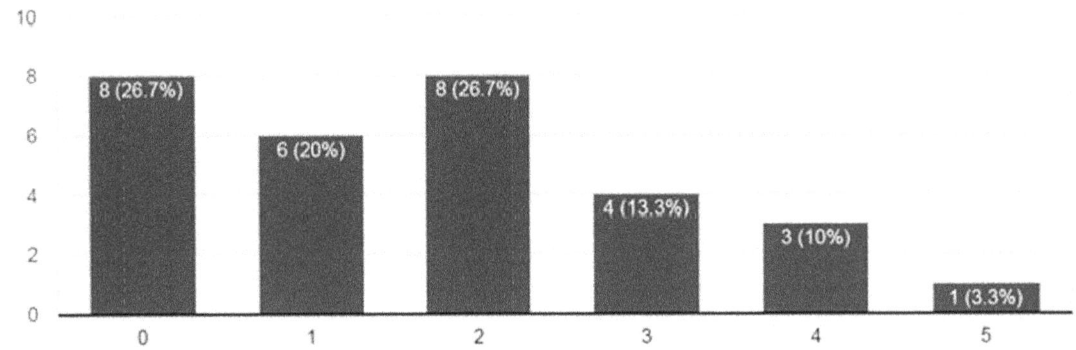

Figure 4. Designing mobile learning educational games

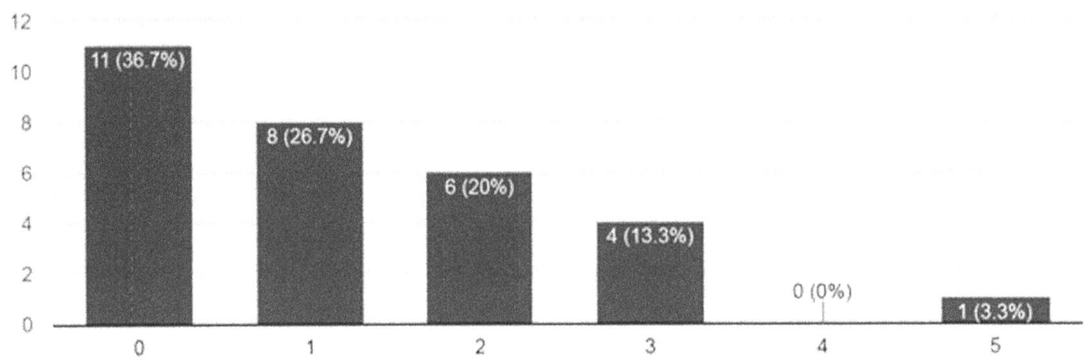

At the end of the semester, students were shown their responses on where they self-rated their knowledge about mobile learning before commencing with the course. They are then asked to complete a post-survey, where they enter open ended qualitative feedback about their experience learning about mobile learning during the semester (Figure 6).

As can be observed from the graph, students rated themselves as having low knowledge and experience of mlearning at the start of the semester, but the overwhelming majority rated themselves very highly at the end of the semester. In the qualitative comments, one point to highlight was that the notion of learning is everywhere took root, as well as the possibility of entertaining micro-learning moments as part of a mobile learning initiative.

These mobile learning activities or applications also had the potential to be used in many different ways within one context or outside of its intended target audience or setting. For example, depending on the museum or subject area (e.g., history, mythology, natural sciences, commerce, culture), a targeted approach can aid follow-up experiences or bring a distally located experience within reach of students and teachers. Similarly, other museums for art, and history embraced mobile learning to engage both their physical and virtual target audiences. Contrary to some beliefs, designs for museums' mobile first engagement approach, has actually increased physical visitors to the museum. Additionally, mobile on-

Figure 5. Knowledge of Mobile Activism

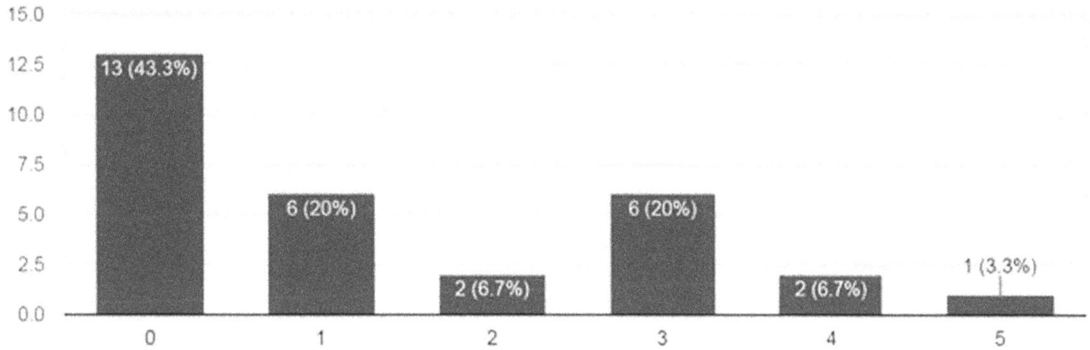

Figure 6. Pre- and Post Self-Rating. Self-rating on their growth with regards to mobile learning at the end of the semester.

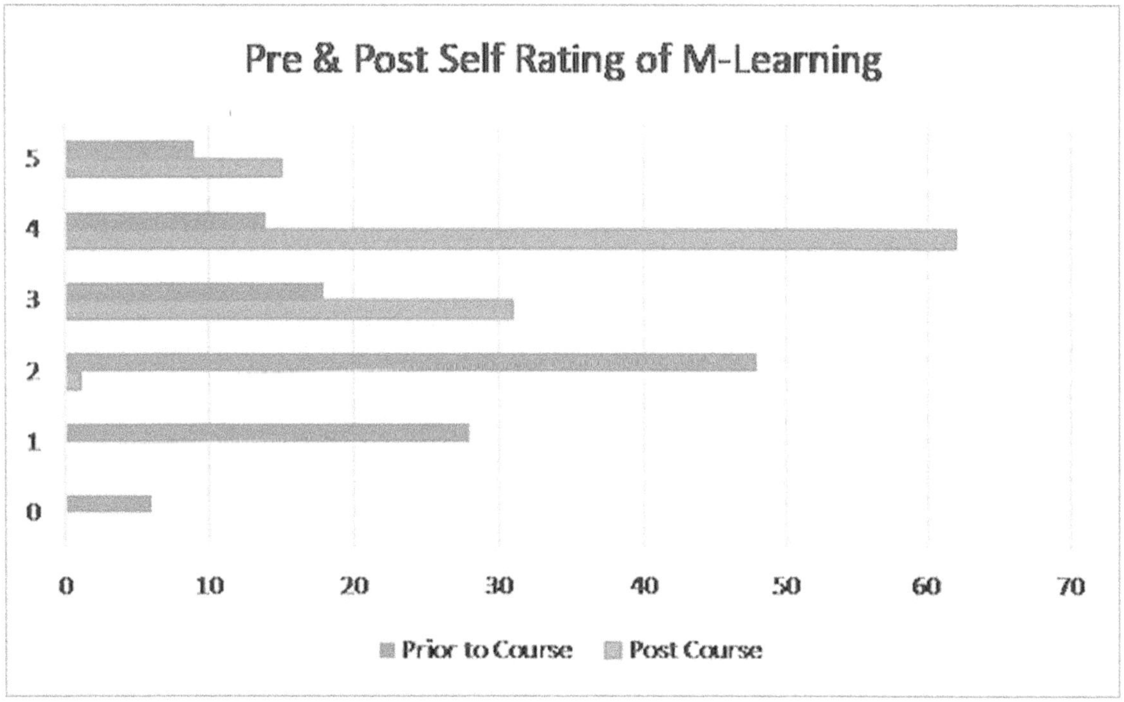

line engagement increases the possibility for fundraising due to its delivery of valuable click and view statistics. Five minutes of mobile access by a virtual or physical visitor can offer valuable moments of micro-learning for both the museum staff interested in the data analytics.

Mobile Learning for Micro-Learning

Apart from mobile devices being one tool on the utility belt of educational engagement that can be used in tandem with others, mobile learning can further support a micro-learning approach. Learning in short bursts can be described as stringing learning moments together in bite size fashion or as ComLabIndia calls it, snack learning (2016). In this way, mobile learning supports micro-learning by accommodating adults suffering from time poverty, and accommodating the commuter's learning journey (Mentor, 2016)

It offers valuable micro-learning of the museum and the artifacts. Grovo (2017) defines micro-learning as possibly consisting in part, of:

- *Five minutes of learning a day*
- *Digestible, point of need learning experiences*
- *Learning, and applying what you have learned, in small, focused steps*
- *Learning done within the limits of working memory*
- *The shortest path between aha moments*
- *Not a weapon of mass instruction*

Of course, micro-learning could be longer or shorter than five minutes a day. Also it could be more than one instance of five or more minutes, but can be moments strung together like a string of pearls and contribute to transformative learning. While some consider micro and mobile learning not as a tool for mass instruction for the most part, there are contexts where mobile learning can be pushed out to a virtual mass audience or delivered in-person to a large audience via something as simple as a mobile digital project like the earlier InnoCube Pico Projector (UNESCO 2014 Mobile Learning Week exhibit) or an add-on click on device such as a Moto Mod (2016). Alex Khurgin does make a strong point in the title of his webinar "*Microlearning for Transformation, Not Information Transfer* (Grovo, 2017)", that micro learning is not just information transfer, but rather micro-learning for transformation (2017). Mobile devices can offer unique opportunities for mobile and micro learning to support transformative learning opportunities.

Looking at mobile learning as additional tool as part of a utility belt, to make available at the right time, or as part of the overall holistic approach, moves away from the mistaken view of mobile learning as an isolated activity to explore. But rather "to embrace mobile learning as a rich, collaborative and conversational experience, whether in classrooms, homes or the streets of a city" (Naismith et al. 2004), or the workplace.

Workplace M-Learning

The increases in mobile phone ownership and people's reliance on their devices, has driven the Bring Your Own Device (BYOD) to work movement. Less than 25% of corporations and other workplaces are making sure that their learning suite of compliance training or data dashboards are available for their workers (2015 Bersin, Agarwal, Pelster, & Schwartz). Mobile learning designs take into consideration various theoretical frameworks to inform the adult learning experience and differentiate instruction. Introverts participate more easily, for example, where mobile learning is a part of the blended learning approach (Costa, Murphy, Pereira, & Taylor, 2018). Both children and adults thrive in personal learning environments.

Enabling Personalized Learning Environments (PLEs)

The hypermedia capabilities of mobile devices correspondingly enable learners to map out and navigate their own learning path and they can do so on-the-go—thus meeting differentiated learning styles, interests, levels, and needs. The mobile learning approach can convert students into active agents as opposed to passive participants. If the mobile learning design is done well, students can take on new roles, for example with mobile journalism, mobile activism, filmmakers or documentarians. Moving learners from passive consumers to active producers of learning content through mobile devices are now simpler and tapping into practices many participate in already as active agents. The latter could be used as part of formative or summative assessments and both of the latter assessment approaches should form part of a PLE design (Patterson, Stephens, Chiang, Price, Work, & Snelgrove-Clarke, 2017; Hricko, 2017; Mentor, 2014; Mentor, 2016).

Catering for mobile friendly PLE's, demand that instructional designers cater for automated learning designs where both linear and non-linear paths are made available to the learner. In a rich hypermedia world, accessible via mobile device, the learner should be able to have the freedom to create his or her own learning path. Allowing the mobile learner to jump to any learning link, just like they would be navigating a hypermedia or wikimedia environment or any online source, jumping from interest to interest or whatever fits their available time. Be it that they decide to watch shorter videos first due to time limitations, or to build up a knowledge or an understanding in bite-size fashion, the freedom to choose, could aid and increase motivation/engagement.

The freedom to choose from various hypermedia artifacts can also aid transformative learning if it encourages and incorporates learner feedback where the learner is navigating its own learning path and journey towards the established objectives of the assembled artifacts working together towards the learning goals. While there might be a fear of losing a clear vision of the learner's growth and proficiency, rather than just reporting on learning metrics such as content watched, quizzes answered, or lesson completions, there is instead opportunity to assess and report on transformative learning that can be observed through data analytics beyond just the learning analytics of clicks and completions.

Taking a hypermedia approach, can open the learning design for learners and their learning experience, and instead of restricting learning to a single event, they can follow up on previous learning, stitch learning pieces together in their own way, or re-watch videos or content, spaced out within a specific required or their own time frame, and do so for review, and reinforcement. Rather than offering grades or badges, or progress bars, the mobile learning design could proactively ask the learner to apply what they've learned and to supply examples of that learning application as part of the mobile learning process.

While providing performance progress and perhaps learning support, learning checklists, and to make transformative learning more visible to other learners in the virtually accessible mobile learning space, it can similarly have the effect of putting current and future learners at ease, gauge and manage their own learning expectations against the previous learners and either match the expectations or set the bar higher for themselves. Building in a social collaborative approach can also make the learning, less of an information transfer, but more of a social engagement, which can be considered more supportive and less intimidating, and further promote the transformative learning to be more than just on an individual plane, but on a larger departmental level or organizational scale. Turning the goals and objectives from individual accomplishments to an organizational benefit, while transforming and establishing a new status quo.

Students or workers can learn how to use mobile polling for garnering organizational direction or for policy setting or consensus building. Students will learn how to use live polling to check prior knowledge, gauge interests, taste or preferences as well as for formative or summative assessments, and for mobile data collection with personal tailoring via conditional questions. Mobile polling and surveys, if designed from a mobile first approach, yield higher participation rates.

There are also opportunities for students will learn how to use mobile games and learning activities. For example, in health applications, they can be used to stimulate smoking cessation, biometric or nutritional tracking. Other self monitoring or recording frequencies of a target behavior to change (e.g. wellness mapping or nail biting) can make optimal use of the personal mobile device. With the rise of mindfulness practices, mobile learning opportunities abound for users to own and track their learning journeys in this regard.

Many multinational and national corporations, universities and other workplaces host multicultural workforces. Mobile devices in these contexts can aid in leveling the playing field, lowering device learning curves, and increasing accessibility beyond a desktop computer. Course participant learn how to extend the learning engagement on mobile devices and use learning analytics to tailor activities for different kinds of adult learners.

Other mobile learning opportunities lie within the organizational training or leadership development. Students will learn how mobile phones in the workplace can offer electronic performance support systems, peer-to-peer availability, and support as well as personal and organizational tracking of learning.

Through mobile social media and communication, mobile devices offer social connectedness at work—defined as a sense of belonging or affiliation. In higher education and other life arenas, social connectedness can be cultivated through mobile devices at a faster rate and far more frequently than a computer. From practical examples and hands-on projects, course participants will see how mobile devices allow for learning to seamlessly move from personal engagement, to virtual collaboration and to the multi-screen reality of modern learners.

Reliance on mobile devices at work—not just younger generation workers—provides opportunities for just-in-time, anywhere, informal learning when confronted with work challenges; as well as making learning more accessible to busy professionals when they have the opportunity or need to learn. Another challenge to overcome, based on students' positions at the start of the course, is that mobile learning does not have to be thought of a replacement to face-to-face or eLearning models, but rather, as another useful method to pull out of the utility belt of educational tools, within appropriate contexts. Similar to using the right tools for particular tasks. Additionally, in many developing countries, mobile learning is accessible in locations where computers and online learning are not, and some still wonder, how is mobile learning different from e-learning.

HOW DOES M-LEARNING DIFFER FROM E-LEARNING?

For one, the User Interface and User Experience (UI/UX) is completely different from the UI/UX of online teaching and learning. Hence, the form factors of input and output with regards to screen real estate and production or interaction on a mobile device, will influence the cognitive load, from an element interactivity perspective. And these are considerations to be aware of for mobile online teaching and learning with additional considerations, limitations and advantages.

Online teaching and learning are different from mobile learning in its approach, form factors, theoretical frameworks, and affordances. With mobile learning for professional growth, context is key, just-in-time content is core, while collaboration and individual engagement are crucial. Understanding the "why" and "how" of mobile learning is the first step in deciding what kind of training should be designed for mobile devices. Mobile learning is not about putting online or e-learning course materials into a smaller package so that they can be played on a mobile device. Rather, it is thinking differently about learners and the new possibilities that being both mobile and connected can offer to employees who want to learn something "on the go". The opportunity to extend or continue their learning engagement from a previous screen to a mobile device offers just-in-time learning and information. Designing mobile learning content demands considerations of a complex mix of different forms of mobility, different technologies, a variety of types of learners who learn in different contexts, with different learning styles, and trainers with many different approaches to instructional design. Moreover, content and experiences are somewhat restricted by the element interactivity of different mobile devices, which also demands different approaches for engaging children and adults on mobile devices.

MOBILE PEDAGOGY AND ANDRAGOGY

With the planet becoming a planet of the apps, and the onslaught of hype on wanting to create an app for everything ranging from conferences to courses, the question becomes, is an app that will only be used once by a user, or take up unnecessary space and battery power. Additionally, investing in designing an app for one course, a conference or other once off events of learning engagement, how is mobile learning changing the role of teachers? In the broader context, mobile access and mobile learning is already changing the role of teachers, as students now have access to a wealth of multimodal information and stimuli. Mini-computers at their fingertips. These can either be forces of distraction or could be utilized as tools of engagement. Thus, how do mobile technologies improve pedagogy and strengthen education quality and one response here is that mobile technologies offer teachers an opportunity to directly engage the students virtually in real time and asynchronously. Furthermore students can integrate their informal and feral learning into their formal mobile learning. However, how should mobile technologies be integrated into formal and informal learning environments?

Mobile technologies can be incorporated into formal and informal learning spaces through creative means. For one, it can extend the formal educational time and honor a basic tenet that learning is everywhere. Secondly, mobile technologies can extend and blend formal and informal educational spaces in a responsible manner that can help to engender authentic and autonomous learning habits.

From an adult learning perspective, mobile learning lends itself to Self-Directed Learning (Candy, 2004; Marsick & Volpe, 1999) and increases the opportunities for incidental and informal learning (Marsick & Watkins, 2001) and could do so in smaller chunks of time Mentor (2016).

M-LEARNING FOR PROFESSIONAL DEVELOPMENT

How do mobile technologies build the capacities of teachers and support their work with students? Gaible and Burns (2005) show the link between well-informed instructors and Mills (2013) highlights the multimodal wonder of digital curation through tablet computers and their ability to prepare and engage students to become better autonomous learners within different K – 12 contexts; be it grade level, subjects or specific projects inside or outside of the classroom. However, many teachers are still resistant to, or ill prepared during teacher training, and be at odds with students' mobile technology savviness and proclivities to mobile technology adeptness.

Questions then remain; how are educators trained to utilize new technologies to improve teaching and learning? How are training institutions ensuring that pre- and in-service teachers receive adequate and ongoing training about mobile learning? Training for pre- and in-service teachers are still sparse or non-existent and not adequately informed by theoretical foundations. The mobile technology landscape changes rapidly and a theoretical grounding would help any mobile technology integration initiative, with in-services for teachers or during the training of pre-service teachers.

POLICY CHANGES TO ACCOMMODATE M-LEARNING

A mobile first approach can help explore how one can utilize the mobile devices for learning (Curinga, & Saravanos, 2016) and the factors to consider for maximizing transformative learning (Marsick, 1998; Nicolaides & Marsick, 2016). Tablet devices, like the iPad, Kindle and many Android tablet options, are rapidly becoming devices of choice and losing favor in the sway of the ephemeral adoption of tablet only devices for consuming and producing content. For the former, partly because of the simple and easy interface, quick start up times as well as the availability of a plethora of applications (apps).

And for the latter, partly because computers in general have become more powerful and manufacturers are answering the need for hybrid computer and tablet devices. However, apart from policy changes needed regarding students and teachers allowed to use mobile devices, very little guidance or professional development offerings are available for K – 12 and other area teachers. Teachers already suffering from poverty lack of time, could benefit greatly on how to use tablets to save time, saving their energy and putting their students first. But, what policy environments – at the national, district or local levels, could help teachers effectively leverage new mobile technologies to improve educational outcomes? What approaches, such as awareness-raising and advocacy, can increase the efficacy of policy implementations? And how have policy implementations been monitored and/or aligned with ICT competency standards for teachers?

M-LEARNING EXAMPLES

A few examples arise in this regard. Many examples from all around the world, and two examples that the author was directly involved in, saw the benefits of policy changes, and pilot efforts that influenced policy changes. Examples from around the world can be found in recent research on mobile learning initiatives ranging from *Science Learning and Engagement in the Digital Age: Understanding the Effect of Mobile Technology on Adult Engagement Experiences at a Natural History Museum* (Kounaves, Archer, King, & Pegram, 2016), and in the realm of new language learning and teaching environments, there is *Mobile Learning Languages, Literacies and Cultures* by Pegrum (2014), and using mobile game technology to enhance student engagement and learning (McGregor & Bartle, 2016; Villanueva & Vaidya, 2016) to name but a small handful. Apart from the growth at the mLearn Conference (2011 – 2016) and UNESCO's Mobile Learning Week Symposium, mobile learning presentations have grown significantly at other conferences in Asia (Woodill, 2013), and further rapid adoption in the Asia-Pacific region, Africa (Nordin, Embi, Norman, & Panah, 2017), the Middle East (Masrom, Nadzari, Mahmood, Zakaria, & Ali, 2016; Sarrab, Al Shibli, & Badursha, 2016) and elsewhere in the world (Khaddage, Müller, & Flintoff, K. (2016), there have also been research handbooks on mobile learning that has either made the case for the need for a wider adoption of mlearning or that offered research initiatives on mobile learning (Briz-Ponce, Juanes-Mendez, & García-Peñalvo, 2016; McQuiggan, McQuiggan, Sabourin, & Kosturko, 2015; Mentor, 2016; Traxler & Kukulska-Hulme, 2016).

One area that has seen useful mobile development is audience, or participant response systems. Classtalk mentioned in the seminal mobile literature review (Naismith et. al., 2005), and iClickers were different types of classroom response systems. Classtalk as cited in the *Literature Review in Mobile Technologies and Learning* was used with first year physics students at the University of Massachusetts. Classtalk offered students opportunities to participate in class discussions via mobile devices, and exploited the near synchronous aggregating and display of collated responses.

While the first iterations of iClickers came with its own devices that could be numbered as a means of tracking students' in class responses that system was soon abused by students when they would be absent from class, but their friends would have their clickers to answer a professor's question. iClickers could also integrate with learning management systems, and be used for both formative and summative assessments (iClickers conference 2011).

In addition to the mobile clickers, referred to as student remotes, the new versions of iClickers include Android and iOS native apps accessible via mobile phones, as well as mobile friendly browser access. In addition to these live class audience response systems, others like Poll Everywhere which launched in 2007, while others like Quizbox, Merida and Nearpod have grown in popularity. While Quizbox hosts their questions online, Merida offers native app incorporation. Additionally, Nearpod is a newer option. When it was first introduced it was only available for iOS' iPads, but it has since become hardware agnostic. Nearpod mimics older technologies where teachers could share their screen on their students' devices, guiding them through a lesson. Students can answer questions when prompted either with text or even drawing and there is a collaborative tool that allows for brainstorming in real time.

Other examples include the great mobile learning projects created by the 2011 New York Mayor's Office of Adult Education Fellows (Mentor, 2011), as well as with a National training organization, where a mobile-first approach expanded the social connectedness and skills training of entry level trainees (Mentor, 2016).

CONCLUSION

A hybrid consideration of theories is definitely a good approach as you might find different scenarios applicable and helpful for design, testing, use and response to mobile learning contexts. Considering a combination or a mix of theories also caters for differentiated instruction and different learning styles which would then allow for catering to the majority and minority as well as the marginalized mobile technology user, be it a student or adult learner. A cautionary note of interest relates to mobile and cyber privacy issues. Is ethical use of user data and data learning analytics assumed? How is responsible use of user data defined and by whom? Should we consider thoughts on 1984 Orwellian implications - especially given privacy issues highlighted by the repeal of net neutrality and other substantiated news stories over the last few years, as well as complaints about the collusion with major technology companies without proper adherence to constitutional and legal frameworks? Is there also a need for better education or future research in this regard so that users can apply critical thinking and hopefully also make informed decisions on their privacy?

FUTURE RESEARCH DIRECTIONS

Future research will focus on expanding and cultivating other m learning projects, transformative learning opportunities with and through mobile devices. Additionally, future research will focus on the improvements and enhancements regarding mobile devices and m-learning integrated with virtual and augmented reality, privacy, work wearable technology and interplay for mobile learning for situated cognition between sensors in the Internet of Things (IoT) and mobile devices. There is also a need for specific research on database managements systems as it relates to mobile learning data.

ACKNOWLEDGMENT

This research received no specific grant from any funding agency in the public, commercial, or not-for-profit sectors. The author wishes to acknowledge Dr. Nabeel Ahmad as co-designer and co-teacher of the mobile phone learning courses when it ran for the first five years. The author also wish to thank Dr. Victoria Marsick for her invaluable input via correspondence.

REFERENCES

Astatke, Y., Ladeji-Osias, J. O., James, P., Moazzami, F., Scott, C., Connor, K., & Saka, A. (2016). Improving and Expanding Engineering Education in the Middle East and Africa Using Mobile Learning Technology and Innovative Pedagogy. In *Advances in Engineering Education in the Middle East and North Africa* (pp. 235–260). Springer International Publishing. doi:10.1007/978-3-319-15323-0_10

Briz-Ponce, L., Juanes-Mendez, J. A., & García-Peñalvo, F. J. (Eds.). (2016). *Handbook of Research on Mobile Devices and Applications in Higher Education Settings*. IGI Global. doi:10.4018/978-1-5225-0256-2

Candy, P. C. (2004). *Linking thinking: Self-directed learning in the digital age*. Canberra, Australia: Department of Education, Science and Training.

CommLabIndia. (2016, February 24). *Top Mobile Learning Trends*. Retrieved December 02, 2017, from http://blog.tristit.com/top-mobile-learning-trends/271

Costa, C., Murphy, M., Pereira, A. L., & Taylor, Y. (2018). Higher education students' experiences of digital learning and (dis)empowerment. *Australasian Journal of Educational Technology*, *34*(3), 140–152.

Curinga, M. X., & Saravanos, A. (2016). Mobile First E-Learning. In D. Mentor (Ed.), *Handbook of Research on Mobile Learning in Contemporary Classrooms* (pp. 23–36). Hershey, PA: IGI Global. doi:10.4018/978-1-5225-0251-7.ch002

Fenn, J., & Linden, A. (2005). *Gartner's Hype Cycle Special Report for 2005.* Academic Press.

Gao, Y., Liu, T. C., & Paas, F. (2016). Effects of mode of target task selection on learning about plants in a mobile learning environment: Effortful manual selection versus effortless QR-code selection. *Journal of Educational Psychology*, *108*(5), 694–704. doi:10.1037/edu0000080

Hricko, M. (2017). Personal Learning Environments. In *Handbook of Research on Instructional Systems and Educational Technology* (pp. 236–248). IGI Global. doi:10.4018/978-1-5225-2399-4.ch021

Jamu, J. T., Lowi-Jones, H., & Mitchell, C. (2016). Just in time? Using QR codes for multi-professional learning in clinical practice. *Nurse Education in Practice*, *19*, 107–112. doi:10.1016/j.nepr.2016.03.007 PMID:27428702

Khaddage, F., Müller, W., & Flintoff, K. (2016). Advancing mobile learning in formal and informal settings via mobile app technology: Where to from here, and how? *Journal of Educational Technology & Society*, *19*(3), 16.

Khurgin, A. (2017). Microlearning for Transformation, Not Information Transfer (Webinar video). *Training Magazine Network*. Retrieved from https://youtu.be/kGff91akTJ0

Kounaves, S., Archer, L., King, H., & Pegram, E. (2016). Science Learning and Engagement in the Digital Age: Understanding the Effect of Mobile Technology on Adult Engagement Experiences at a Natural History Museum. *Mobile Learning Futures–Sustaining Quality Research and Practice in Mobile Learning*, 361.

Lynch, B. (2008, May 1). Startup Poll Everywhere invented the technology to conduct real time polls through cell phones. *Boston Business Journal*.

Marsick, V. J. (1998). Transformative learning from experience in the knowledge era. *Daedalus*, *127*(4), 119–136.

Marsick, V. J., & Volpe, M. (1999). The nature and need for informal learning. *Advances in Developing Human Resources*, *1*(3), 1–9. doi:10.1177/152342239900100302

Marsick, V. J., & Watkins, K. E. (2001). Informal and incidental learning. *New Directions for Adult and Continuing Education*, *2001*(89), 25–34. doi:10.1002/ace.5

Masrom, M., Nadzari, A. S., Mahmood, N. H., Zakaria, W. N., & Ali, N. R. (2016). Mobile learning in Malaysia education institutions. *Issues in Information Systems*, *17*(4), 152–157.

McQuiggan, S., McQuiggan, J., Sabourin, J., & Kosturko, L. (2015). *Mobile learning: A handbook for developers, educators, and learners*. John Wiley & Sons. doi:10.1002/9781118938942

Mentor, D. (2011). We Are New York Social Media for Adult Education. In C. Ho & M. Lin (Eds.), *Proceedings of E-Learn 2011--World Conference on E-Learning in Corporate, Government, Healthcare, and Higher Education* (pp. 1499-1504). Honolulu, HI: Association for the Advancement of Computing in Education (AACE). Retrieved December 24, 2017 from https://www.learntechlib.org/p/38929/

Mentor, D. (2014, February 14). Symposium Presentations. *UNESCO Mobile Learning Week 2014.* Retrieved March 12, 2017, from http://www.unesco.org/new/en/unesco/themes/icts/m4ed/unesco-mobile-learning-week-2014/

Mentor, D. (2016a). *Handbook of Research on Mobile Learning in Contemporary Classrooms.* Hershey, PA: IGI Global. doi:10.4018/978-1-5225-0251-7

Mentor, D. (2016b). The Commuter's Learning Journey: Field Observations Informing Mobile Learning Initiatives. In Handbook of Research on Mobile Learning in Contemporary Classrooms (pp. 315-335). IGI Global.

Mentor, D. (2016c). EMxC3= e&mLearning Cultivating Connected Communities: Sustainable Workforce Talent Development. Handbook of Research on Mobile Learning in Contemporary Classrooms, 240-259.

Mentor, D., & Ahmad, N. (2010). Teaching America's First Course on Mobile Phone Learning. *EDUCAUSE Quarterly*, 33.

Mousa, A. A., & El-Salam, M. A. (2016). Employing QR Code as an Effective Educational Tool for Quick Access to Sources of Kindergarten Concepts. *World Academy of Science, Engineering and Technology, International Journal of Social, Behavioral, Educational, Economic. Business and Industrial Engineering*, *10*(7), 2358–2361.

Naismith, L., Lonsdale, P., Vavoula, G., & Sharples, M. (2004). *Literature Review in Mobile Technologies and Learning (Futurelab Series Report 11).* Bristol: Futurelab.

Nicolaides, A., & Marsick, V. J. (2016). Understanding Adult Learning in the Midst of Complex Social "Liquid Modernity". *New Directions for Adult and Continuing Education*, *2016*(149), 9–20. doi:10.1002/ace.20172

Nordin, N. M., Embi, M. A., Norman, H., & Panah, E. (2017). A historical review of mobile learning research in Malaysia and its implications for Malaysia and the Asia-Pacific region. In *Mobile learning in higher education in the Asia-Pacific region* (pp. 137–150). Singapore: Springer. doi:10.1007/978-981-10-4944-6_7

Pass, F., & Renkl, A. (2003). *Cognitive Load Theory and Instructional Design: Recent Developments.* Academic Press.

Patterson, C., Stephens, M., Chiang, V., Price, A. M., Work, F., & Snelgrove-Clarke, E. (2017). The significance of personal learning environments (PLEs) in nursing education: Extending current conceptualizations. *Nurse Education Today*, *48*, 99–105. doi:10.1016/j.nedt.2016.09.010 PMID:27744138

Pegrum, M. (2014). *Mobile learning: Languages, literacies and cultures.* Springer. doi:10.1057/9781137309815

Rikala, J., & Kankaanranta, M. (2012, October). The Use of Quick Response Codes in the Classroom. In mLearn (pp. 148-155). Academic Press.

Rikala, J., & Kankaanranta, M. (2013). Mobile learning. A Review of Current Research. *Reports of the Department of Mathematical Information Technology Series E. Educational Technology,* 1–65.

Traxler, J., & Kukulska-Hulme, A. (2016). Introduction to the next generation of mobile learning. *Mobile learning: The next generation*, 1-10.

Villanueva, K., & Vaidya, J. (2016). Transforming Learning with Mobile Games: Learning with mGames. In D. Mentor (Ed.), *Handbook of Research on Mobile Learning in Contemporary Classrooms* (pp. 260–278). Hershey, PA: IGI Global. doi:10.4018/978-1-5225-0251-7.ch013

Woodill, G. (2013, September 03). *MobiLearnAsia: Asia's Growing Mobile Learning Conference.* Retrieved December 03, 2017, from https://gowithfloat.com/2013/09/mobilearnasia-asias-growing-mobile-learning-conference/

Wroblewski, L. (2012). *Mobile first: Préface de Jeffrey Zeldmann*. Editions Eyrolles.

Chapter 11
A Roadmap for Developing Computer-Mediated Solutions for Situated Workforce Learning

Edward Bednar
Teachers College, Columbia University, USA

ABSTRACT

With perceptual capabilities, computers can intelligently function as a part of our everyday lives, helping us to make sense of what is happening and joining in with us as we experience and navigate through many different types of situations. In this way, computing systems can be situated when they combine machine perception with background knowledge to observe, explore, and interpret human and environmental activity in a way that supports decision making. Many of the current, prominent situated computing solutions are consumer-focused in nature. But these systems will, in time, change the way we work as well as how we learn. Many disciplines are adapting and changing in the process, and many opportunities remain, including the use of situated computing systems for workforce education. There are many opportunities to innovate and improve workforce education by leveraging the power and affordances of mobile-device technology to turn everyday situations into everyday learning opportunities.

INTRODUCTION

In the mid-1990s, a group of researchers at MIT experimented with wearable technology. This assembly of researchers, eventually named the MIT Media Lab Wearable Computing Group, carried computers and radio transmitters around in backpacks, stored keyboards in their pockets, and wore eyeglasses with digital displays integrated into the frames (Turkle, 2006, p. 220). By lugging around what was still at that time big computing equipment, these innovative researchers were exploring the future of personal-device technology—technology that is wearable, mobile, and convenient for modern use.

Much later, after many years of futuristic ideas and overly ambitious but underwhelming commercial attempts by many companies, Apple changed the landscape—accessibility, application, and use—of personal-device technology by putting extraordinary computational power and the information affor-

DOI: 10.4018/978-1-5225-4111-0.ch011

dances of the Internet literally in our hands to use anywhere we might be, even on the go. There are many mobile-computing devices that pre-date the iPhone. Some, most notably the BlackBerry, designed and marketed by Research in Motion Limited [now BlackBerry Limited], profoundly impacted workforce productivity, especially for remote communication with email. But in 2007, when Apple released the iPhone, it was a seminal moment in, what is still, our brief history of computing.

For the first time, the physical capabilities of a mobile device were integrated, improved with affordances like a multi-touch interface, and paired with a robust operating system. The iPhone, importantly, also became a must have consumer technology. But more importantly, this mobile-computing system was significantly enhanced when Apple introduced the App Store, a software development and distribution platform. This platform fueled the proliferation of applications, and thus the many uses for these devices. Similarly, Google, together with several hardware vendors, created a competing commercial mobile-computing system based on the Android operating system.

Although they are not exactly wearable, the collective proliferation of iPhone and Android devices has resulted in a new norm where computing devices are now rarely very far removed from our attention. Now, in workforce settings, because of this device proliferation, nearly every employee has a smartphone—likely an iPhone or Android device—which is well suited for many purposes. This situation is in contrast to just a subset of employees previously having specialized mobile devices like BlackBerrys.

As the capabilities of wireless telecommunication networks have dramatically improved, access to fast, seamless Internet connectivity for mobile devices has increased dramatically. This improvement in network access and speed has opened up many possibilities for developing mobile computing systems backed by cloud computing infrastructure and, especially, very powerful data analysis technologies based on artificial intelligence. At the time of this writing, Google, Facebook, and many other companies are developing innovative situational and contextually aware solutions using these technologies.

This chapter begins with a primer on the contemporary technologies needed for developing situated computing systems. It then reviews situated learning and informal learning theories to develop an educational context for situated computing. Then, the chapter aims to provide a conceptual roadmap for developing situated computing solutions for workforce education using existing technologies, leveraging examples from commercial product offerings where these technologies are just now beginning to be effectively used.

CONTEXTUAL AND SITUATED COMPUTER-MEDIATED SYSTEMS

Perceptual Computing

Similar in the ways that the personal computer made computing "personal," the affordances of the class of mobile devices the iPhone and Android achieved unleashed the ability to make computing "perceptual." Perceptuality is an aspect of human-centric computing systems—systems that sense, measure, and monitor data about what a particular individual in a specific context is doing, and then those still external systems react to that person's physical, cyber, and social realms to, in many cases, intimately support decisions and actions (Sheth, Avantharam, & Henson, 2016, p. 64). With perceptual capabilities, computers can intelligently function as a part of our everyday lives—part of our everyday experiences—helping us to make sense of what is happening and joining in with us as we experience and navigate through many different types of situations.

Sensory Capabilities

The first iPhone, for example, had three sensors that provided perceptual input and actuation to its primary user. The proximity sensor enabled the device to detect when objects were near. This feature was mainly employed to disable the touch screen and turn off the display while the device was being used as a conventional phone would be, held close to the ear. The ambient light sensor measured the light intensity of the physical environment. By calibrating the brightness of the display, based on the conditions where the device was being used, this sensor enhanced the visual experience, reduced eye strain, and helped preserve the battery. The accelerometer detected movement of the device. It was initially just used to change the display orientation between portrait and landscape modes. These three sensors were designed, built, and integrated into the device to make it more accommodating to the user—to make the device human-centric—affording simple, immediate benefits.

Along with advances in software, innovative people and organizations will continue to develop and further advance mobile-device hardware. As the technology continues to advance, additional perceptual capabilities will emerge, which will allow developers to create additional human-centric features and situational interventions that leverage those capabilities. Many mobile devices now have, for example, gyroscopes, fingerprint detectors, barometers, compasses, magnetometers, and photodetectors, to name a few other commonly found sensors that can be leveraged for situational data capture. Additionally, wireless connectivity technologies built on standards for data exchange like Bluetooth® enable the integration of mobile devices with external devices. This integration creates a seemingly endless array of possibilities for expanding the sensory data that perceptual computing systems can leverage, including physiologically based sensing capabilities. As examples, external, specialized devices with sensors can capture and relay data like blood pressure and blood sugar levels, making a mobile device much more than just an accessory for a person living with heart disease or diabetes, respectively.

Output and Auxiliary Capabilities

The typical output and auxiliary capabilities of mobile devices—speakers, cameras, microphones—may no longer seem unique or impressive, but their use as a channel for interactivity will continue to develop and will play an important role in perceptual computing over time. For instance, Pokémon Go, released by Niantic in July 2016, bridged a virtual world with the real world simply by taking advantage of commonly available mobile-device capabilities—GPS sensor, Internet connectivity, the touch screen display, and the camera—to create a cartoon monster scavenger-hunt game, which was the first widespread, massive implementation of augmented reality (McDermott, 2016, p. 27). As an additional example, Snapchat allows users to augment pictures with artificial addition of bunny ears and mustaches, allowing for creativity and self-expression within a social media environment. Pokémon Go and Snapchat are using mobile devices to provide augmented reality solutions even though these devices were not explicitly designed to provide that type of capability. These companies—and many others—are simply reimagining and repurposing the output capabilities of today's mobile devices to build innovative products. This cycle of innovation in the mobile-device space is just beginning, but the rate of adoption of their technology affordances and the speed at which solutions are arriving is accelerating to the point where yesterday's ideas are quickly becoming today's must-have applications.

Internet Connectivity and Cloud Computing

While modern mobile devices are powerful and their sensors can effectively collect data, they are still resource-constrained from the standpoint of computational capacities. Using a classical view of balancing computational resources, for example, a mobile device's wireless radio can interact with cellular towers to collect data while the CPUs [central processing units] process instructions that implement positioning algorithms, all in a performant manner, but these computational resources can burn through the device's battery capacity very quickly if the load is not properly balanced. This type of computational processing—massive amounts of data about individual interactions—must also leverage institutional information for enriching that data (e.g., entity masters, lexicons, geolocation anchors), which requires substantially more powerful computational resources than a mobile device can handle on its own. At the time of this writing, cloud computing services have proliferated and are becoming somewhat mainstream, offered as for-rent hosted solutions, and when paired up with resource-constrained mobile devices, can provide a seemingly endless supply of data storage, computation, and energy (Vinh, Bouzefrane, Farinone, Attar, & Kennedy, 2015, p. 235).

Summary

Collectively, these technology affordances, amplified by the pervasiveness of mobile-device technology, present an immediate and accessible opportunity to leverage perceptual computing methods and systems to create computer-mediated solutions that facilitate and/or support many everyday interactions—everyday situations that we commonly find ourselves in, like needing a reservation at a restaurant or finding an alternate route home because of traffic situations. And if we draw on ideas from the learning sciences and computer science, there is a rich collection of theoretical frameworks and practiced work to provide a solid basis for developing these computing solutions—these situated computing solutions—including solutions for situated learning.

Situated Computing

Computing systems are situated when they combine machine perception with background knowledge to observe, explore, and interpret human and environmental activity in a way that supports decision-making (Sheth et al., 2016, pp. 64-68). These systems incorporate appropriation by users within the context of their immediate situation (Beaudouin-Lafon & Mackay, 2000, p. 369). In this way, human interaction with the computer can be an integrated experience from the standpoint of environmental factors like time and location, but also of more personalized factors like preferences and activity patterns that apply within that situational context. As situated computing systems perceive and leverage our physical environment to know who we are, where we are, what we are doing, and what we want, they can be used to bridge the gap between our intentions and the actions we can take to achieve them (Gershman, McCarthy, & Fano, 1999, p. 1). They can, in many ways, intelligently intervene.

Computer Mediation

The opportunity to provide relevant and timely situated information is a result of the affordances of modern computer mediation—using, for instance, the perceptual capabilities of mobile devices and the pattern identification capabilities of artificial intelligence to develop probabilistic models of what someone is doing or about to do. While sensory capabilities serve as the primary entry point of situated computing solutions, the data generated by the sensors must be processed to the extent that meaning is derived from those values. This processing must occur immediately for responsiveness purposes, and, additionally, over time, for relevance purposes. For example, the opportunity to provide a situated navigation service that adds value to the user by proposing alternate routes to someone who is driving somewhere ends when the exit for that alternate route has just passed. Over time, in the same scenario, if the interventions by the navigation service are often, or worse, always, making the same recommendation but that alternative is never chosen by the user, an otherwise potentially useful computer-mediated solution becomes, perhaps, an annoyance, perhaps even one that gets uninstalled from the device.

Much of this processing is a function of artificial intelligence, which is the science and engineering of building intelligent machines, specifically for the purposes of situational computing via intelligent computer programs (McCarthy, 2007, p. 2). Intelligence, in this context, is the computational ability to generate the mechanisms of reasoning through computational procedures (McCarthy, 2007, pp. 2-3). These are procedures that focus on a specific domain of knowledge and/or inquiry, like getting somewhere in the most efficient way possible both immediately and, over time, as we continually approach the same situation.

Artificial Intelligence and Machine Learning

The artificial intelligence upon which situated computing systems function is created by converting massive amounts of raw data (e.g., sensor data) into symbolic representations that are evaluated by computational procedures relative to a variety of known and derived relationships and associations (Sheth, 2010, p. 89)—a collection of well-defined rules for manipulating symbolic representations (Dawson, 2013, p. 7). Examples of artificial intelligence capabilities include pattern recognition, image analysis, text processing, and sentiment analysis (Sheth, 2010, p. 89), which can be used to build computing solutions that, for example, recognize speech, understand natural languages (e.g., search engines), can see, provide expertise (McCarthy, 2007, pp. 8-9), and even beat humans when playing games like chess (McCarthy, 2007, p. 6).

These computing systems can, for instance, combine sensory inputs and artificial intelligence to process, represent, and reason over data points, leveraging that contextual understanding to improve human activities and experiences (Sheth, 2010, p. 89) in many ways. For instance, these types of systems can detect that we are driving home from work and recommend an alternate route due to traffic conditions. They can recognize how fast we are driving and provide subtle, or even not so subtle, alerts or recommendations to slow down. They can recognize that we are entering a secure facility in a workplace, where confidential information is handled, and remind us about policies that are central to interacting with and protecting that information.

As these computing systems detect and interpret aspects of the user's environment and experience, using, for example, even just the types of sensors in the earliest generation iPhone, they can understand and respond to events as they occur (Hull, Neaves, & Bedford-Roberts, 1997, p. 146). At any given time, these responses can be tailored to our multitude of intentions, both conscious and unconscious, to which we assign varying levels of both importance and relevance (Gershman et al., 1999, p. 1). The landscape of potential uses for situated computing solutions is vast, including use as tools for public education that could change some aspects of the culture of a country—of its politics—and its global relationships.

Intuitive and Appropriately Applied Functionality and Interventions

The possibilities with situated computing solutions and their unique perceptual affordances may seem endless, which is good. Still, a level of intuitiveness and appropriateness must be integral to their function if they are to be considered useful and intelligent—they must reflect that in whatever we are doing, we are psychologically tuned into the connections that matter (Turkle, 2008, p. 122)—for instance, using commercial examples: (1) Localized information about nearby restaurants could be provided at any time but should reflect our interests, budgets, and, perhaps, what time it is, and do so in a culturally-aware manner. For example, typical meal times can vary considerably in different worldwide regions. Additionally, social, cultural, and religious norms can vary significantly as well, making a seemingly straightforward situational context considerably complex. (2) Contextually rich information about an artist could be presented while we are standing in front of their work in a museum, but the system better be finely tuned enough to know, with certainty, where we are. (3) Alerts from integrated, connected monitors can be provided to report on physiological matters like a person's pulse, activity level, or blood sugar. But the alerts need to be considerate, able to additionally discern where we are and what we are doing, and only be disruptive if it is absolutely necessary (Hull et al., 1997, p. 151), functioning in a manner that ensures privacy and confidentiality. The boundaries between solutions that are helpful versus inappropriately infringing on privacy must be intentionally balanced and respected when developing these types of situated computing systems.

These examples of situated computing solutions may seem straightforward, but they are also not necessarily as simple as they sound from a design, algorithmic, data engineering, and computational standpoint. Yet many innovative enterprises are finding ways to provide direct, personalized services using situated computing systems, so much so that, while still simple and basic, these solutions are becoming quite common and even relied upon. Remember that as much as every car once held a map, now drivers rely on their nearly ubiquitous GPS systems or GPS applications on their phones. But, thinking ahead, when driverless cars take over some day, then even today's GPS-based computing systems will shift. While the foundational technology constructs for developing and providing these situated computing solutions are cutting edge and relatively new, they are also largely understood and reasonably accessible, especially given the availability of cloud computing infrastructure.

Cloud Computing

In simplistic terms, cloud computing is a systems engineering approach where a collection of configurable computing resources (e.g., networks, servers, storage systems, applications, services) can be rapidly provisioned, released, and operated (Lewis, 2017, p. 8). As an approach, cloud computing is new and emerging but it is also mainstream and accessible at the same time, building on decades of prior and evolving systems design methods. Three aspects of cloud computing are especially relevant to building situated computing systems: (1) Computing infrastructure can be provisioned in small increments, providing flexibility regarding infrastructure costs, especially upfront infrastructure costs. (2) Massive amounts of computing infrastructure are available for provisioning, enabling systematic functions like processing large amounts of sensory information accessible. (3) Many services are offered by cloud infrastructure providers that implement complex system capabilities like machine learning, a form of artificial intelligence. Instead of building an entire machine learning system, including the infrastructure, these services make it possible to build solutions on an existing system rather than dealing with the deep system engineering complexities of the infrastructure, including, to some extent, the scalability and security of the infrastructure.

Summary

While, at this point, many of the current, prominent situated computing solutions are consumer-focused in nature, these systems will, in time, change the way we work as well as how we learn. Many disciplines are adapting—most assuredly fundamentally changing—in the process. And many opportunities remain, including the use of situated computing systems for learning.

Situated Learning

Pedagogy

Learning theories—such as the prominent pedagogical theories of behaviorism, constructivism, and connectivism—provide instructional strategies and techniques for facilitating learning and serve as a foundation for making strategy decisions in that regard (Ertmer & Newby, 1993, p. 50). They provide methods for considering how learning will occur, the factors that influence learning, the role of memory in learning, and how transfer occurs. Additionally, they present a mechanism for evaluating different types of learning (Ertmer & Newby, 1993, p. 53). With any process of knowledge construction in a learning strategy, the instruction should be designed to create the conditions for meaningful learning (Wilson, 2017, p. 61). Put into practice, an important way in which learners interact with their environments is through situations that allow them to bump up against the world in order to test their thinking (Bransford & Schwartz, 1999, pp. 82-83).

Situated Learning Theory

As a theory, situated learning—or situated cognition—presumes that knowledge accrues in meaningful actions that have relative meaning to each other within some cultural system (Driscoll, 2017, p. 55). More specifically, situated cognition theory suggests that learning is tied to the activity, context, and culture within which knowledge is developed and used—through which, an authentic setting for learning can be created (Turkay, Hoffman, Kinzer, Chantes, & Vicari, 2014, p. 5). Generally, human cognition is all the ways in which people draw information from the world, combine that information with the knowledge that they already possess, and interpret and make decisions about the information (Baber, 2003, p. 11). Situated cognition is largely based on embodied cognitive science, which is a theoretical framework for understanding cognition in terms of the whole body and its interaction with the world (Dawson, 2013, p. 11). In embodied cognition theory, the mind uses the external world as scaffolding to enhance the cognitive function that occurs between sensory inputs and behavioral outputs (i.e., sensing and responding to what is happening) (Dawson, 2013, p. 11-12).

When, for instance, we use a map—in digital or analog form—while we are attempting to navigate in real-time, we are learning where to go relative to where we are. We are constructing an understanding of our situation and applying that understanding to our goal of getting to where we want to be. This experience is different, from a learning standpoint, from just perusing a map at home; it is situational with regard to our environment and our actions rather than just informational. Since it is situated, it involves thinking that relates to and expands on the information that has been presented, meaning that the thinking is a form of elaboration (Anderson, 2015, p. 141). Because of the effect elaboration has on memory, information that is presented in the context of solving problems is more likely to be subsequently remembered and used than information presented in the form of simple facts (Bransford & Schwartz, 1999, p. 64). Looking at a map before driving somewhere is different cognitively from simply driving somewhere or just thinking about where you would like to go. The former is a rote activity whereas the latter draws on elaborative processing of the situation.

Cognition in Everyday Situations

Situated learning theory is principally focused on how everyday cognition occurs, specifically that the nature of human knowledge is to dynamically construct meaning in real-time as we navigate situations (Clancey, 1995, p. 2) and as we experience whatever the world has to offer at any given moment. From a neuroscience standpoint, it is likely that human information processing is hierarchically organized by the brain, although we don't yet know how, in a way where some information is processed in detail, with great effort, while other information is processed quickly, with less, or, perhaps very little, effort (Baber, 2003, p. 11). According to embodied cognition theory, the latter, shallower type of cognitive function reduces the information that is physically manipulated and stored by framing it within the situation—in a way, out in the world rather than internally—rather than attempting to mentally store and manipulate everything that is relevant to that given situation (Robbins & Aydede, 2009, p. 7). This cognitive efficiency is a form of mental offloading that significantly reduces the load on working memory (Robbins & Aydede, 2009, p. 6) and is an important method of carrying the cognition of everyday things, including workplace activities.

Anchored Instruction

Much of the historical influence of situated learning theory has been in designs for anchored instruction, which is a means of providing a simulated-situational context for problem-solving (Driscoll, 2017, p. 56). For example, for many years, very sophisticated simulators have been developed for anchored instruction purposes, putting situated learning theory into practice for the training of pilots. But these are big, expensive, highly specialized, and stationary systems. Now, with the proliferation of consumer, mobile-device technology, the intersection between human interaction and technology can be mediated in real-time, not for simulation, but, rather, for the purpose of immediate, informal, active learning within a situational context—technology meeting us in everyday occurrences like GPS-based navigation systems to support, influence, and, hopefully, improve our lives. Thus, anchored learning from a situated perspective occurs through the learner's active participation in real or simulated events and activities (Driscoll, 2017, p. 55)—real-world situations harnessed into learning opportunities.

Informal Learning and Experiential Computing

Informal Learning

According to Carliner (2017), citing a 2005 article, informal learning is a process in which learners set their own learning objectives, which contrasts with formal learning where those objectives are set by an instructor or instructional designer. Oftentimes, though, informal learning activities include some elements of formal learning characteristics (p. 142). Thus, informal learning is linked to formal learning. Much of the success with informal learning is based on motivation and accessibility. If we want to learn something and have an opportunity to do so through accessible content, via a mediated solution, or both, we are likely well positioned to succeed in that effort—even if doing so is unintentional. MasterCard, for example, developed a learning platform called Degreed, which enables individuals and subject matter specialists to curate learning "playlists" mixing internal and external learning content—articles, videos, podcasts to name a few—into a personalized learning experience ("Global Human Capital Trends," 2016, p. 62).

Informal learning is manifest through specific learning processes, including: (1) non-formal learning – instruction that occurs outside schools, workplaces, etc.; (2) incidental learning – knowledge unexpectedly gained through interactions; and (3) self-directed learning – formal educational activities undertaken by the learner in a self-directed manner (Carliner, 2017, p. 143).

It is the second informal learning process [knowledge unexpectedly gained through interactions] that offers a significant opportunity for situated learning solutions. This opportunity as a facilitated learning process is interesting because it can be triggered—activated situationally—and subsequently mediated for the purpose of targeted, directed learning activities. If done well, and cleverly, learners may not even realize that they learned something until long after the situation or experience occurred (Carliner, 2017, p. 143) even, again, if unintentional.

Experiential Computing

Computing as part of experiential interactions is a form of digital mediation for everyday activities through embedded computing capabilities (Yoo, 2010, p. 213), meaning, somewhat, that experiential computing is the mediation of human experiences through digital technology (Yoo, 2010, p. 215). Much of the literature about experiential computing focuses on the ability to mediate experiences because of the intersection with computer-enabled devices that takes place during those experiences. The Internet of Things is a more recent descriptor for this type of computing, where everyday items like thermostats, ovens, refrigerators, etc. are Internet enabled. The idea is that computing is taking place in everyday artifacts, so the interactions with these computers are more of a peripheral activity than a dedicated one (Yoo, 2010, p. 217). An intelligent thermostat can, for example, recognize weather patterns, the user's preferences, and know where someone is and if they are on their way home, so as to intervene to apply preferences. This type of interaction can be completely peripheral—it can be automatic and in an authentic manner.

Additionally, one of the purposes of experiential computing solutions is to assist people in the completion of activities. These activities may be actions the computer completes, something the computer guides a person on in a specific subject, or task assistance that the computer provides (Carliner, 2017, p. 143) like reminders. For example, an Internet-enabled oven might ask, by presenting a banner on a mobile device, if it should pre-heat itself because it knows when a person usually gets home on, say, a Monday, and it knows that a frozen pizza is likely the only thing on the menu for dinner given what day it is.

We are in an age where the systems that are coming online offer unprecedented capabilities which can be harnessed into situationally relevant and personally oriented solutions. Now is a good time to dream about possibilities, for chances are good that there are immediate options for creating these types of situated computing solutions. For instance, the potential for increasing the effectiveness and driving down the cost of workforce education—what we will call situated workforce education—can be addressed and differentiated by these types of solutions.

Computing Solutions for Workforce Education

Creating realistic contexts and situations for learning must be emphasized if learners are to build authentic knowledge from their participation in workforce education and training programs (Park, Kim, & Yu, 2011, p. 37). Additionally, as with conventional academic settings, workforce learners bring their experiences—but within the context of their work and the workplace—to the instructional intervention (Tracey & Morrison, 2017, p. 156). This work and workforce context further enhance the need for education and training that is built to serve realistic, relevant purposes. It can necessitate, for example, not subjecting all learners to all parts of a learning intervention by using a form of personalization, which can be very effectively mediated with computing-based educational programs. And there are many options when deciding how to proceed with workforce education programs that use technology to facilitate and support learning within those programs.

Simulation Technologies

Simulation technologies are, perhaps somewhat obviously, at the very highest end of the options for technology-based, situated workforce education when considered in terms of scope, cost, and, likely, also in terms of results. There is, for example, a long history of effectively using simulators for flight training. These simulation technologies have been in use for more than half a century and are used to improve planning, analysis, and training for flight test operations (Vitsas, 2016, p. 1). Some of a pilot's initial instruction on the normal operation of an aircraft can be primed through situated learning methods. Then, continued education on more advanced situations where the operating conditions are more extreme can also be effectively simulated, allowing pilots to safely experience unusual situations so that they are better prepared for them in the event they encounter similar conditions in actual flight.

Simulation technologies are also commonly used in military field training where sensors—called inertial measurement units—are used for evaluating human movement under the many stressful, situated conditions that are simulated in military training exercises (Davidson et al., 2016, p. 28). The simulation serves as a realistic encounter where sensory-based input and computational analysis can help determine who is best suited for roles in this context and to better equip them for success through realistic, but still only simulated, experiences. In a less technical example, IBM developed a learning program, called Basic Blue, which used simulation modules in an e-learning environment to replicate real-life scenarios to convey critical information to new managers (KPMG, 2015, p. 17).

Classroom and Online Training

In a more business setting, formal education programs—in classroom settings—can be very effective. But they are also expensive to develop, purchase, staff, and maintain. There is also a significant cost when considering these training programs in terms of the lost productivity when employees are away from their primary work responsibilities. It is, for example, expensive in the immediate term for a salesperson to be away from selling things in order to learn about how to sell things. The investment in time, expense, lost productivity, etc. should work out when viewed over a long period of time—it, in fact, has to for the training program to be useful—but may not be of apparent benefit in the immediate term.

Online training programs, where there is no physical classroom but still the structure of classroom training programs, can be used as part of formal workforce education programs as well. They are less expensive but still carry many of the same characteristics as classroom training programs in terms of benefits and downsides. In this way, for example, an employee looking to develop a needed skill might browse through content from Coursera, Udemy, Udacity, and many others to instantly access a lecture, course, or workshop ("Global Human Capital Trends," 2016, p. 59). But the opportunities for incorporating situated learning methods in these educational settings—physical and online environments—are very limited.

Computer-Based Training

On a much smaller scale than simulation and classroom environments, Computer-Based Training systems (CBTs) are also widely used for workforce education, especially for regular and ongoing training on company policies, legal and regulatory matters, and many other legal and human-resource practices (Hawkins, 2011, p. 4). CBTs are non-interactive, self-paced forms of instruction delivered by digital media (Hawkins, 2011, p. 3). CBTs are the least expensive option for workforce education. And while they are highly scalable in terms of reach, and as training programs sit at the lowest end in terms of scope and cost, they are also the lowest in terms of results. There is no opportunity for incorporating situated learning methods in CBTs, other than the use of storytelling, narration, and other more informal methods.

Situated Workforce Education With Mobile-Device Technologies

Given the examples of workforce education and training solutions in the previous section, simulation systems, like aircraft simulators, are the only type that directly situate the instruction. Classroom educational programs—including online programs—can leverage situational mechanisms like the case study method first introduced by Harvard University to provide a more realistic educational setting. These case studies are based on the understanding that problem-solving is at the core of professional experience (Breslin & Buchanan, 2008, p. 37) and can be used, for instance, to describe real-world situations so that the instructor can use these scenarios as a teaching method for creating a conceptual environment for the analysis and decomposition of a situational construct. But, even then, the situational aspects of the experience are just narratives—real or fictitious stories that can be used to portray situations for discussion purposes.

And, lastly, there is nothing situational about Computer-Based Training systems (CBTs). The content is presented arbitrarily, individually, and in a controlled setting. It is possible, using artificial intelligence-based computing methods, to create a synthesized exchange between the learner and the computer for a more elaborative form of instruction. But, otherwise, there is no real opportunity for interaction—for discussion of topical expansion—except for quizzes, rote forms of testing that serve little more than to exercise a learner's short-term memory. Using CBTs does, of course, allow companies to add an employee's name to a compliance checklist for legal and audit purposes to represent that specific areas of training are provided to workers on an ongoing basis, which is valuable but not from the standpoint of situated education.

The use of mobile device technologies offers exceptional opportunities for situated workforce education because of their many sensory capabilities and their uniqueness and accessibility. These amazing computational machines are in our hands, or at least nearby, all the time. They are loaded with an impressive collection of sensors that provide the inputs for advanced, perceptual computing solutions. These mobile devices are connected to the Internet and are likely also connected to a company's internal networks as well. These devices can be backed by the affordances of high-end computational infrastructure, cloud or otherwise. This infrastructure affords us the capability of processing massive amounts of information about a given situation and can use artificial intelligence to develop an understanding of a given situation and grow from that point—the more data, the more a machine learning algorithm can expand and shape the landscape of potential responses and choices involved. These devices can also be used to intervene in clever ways as learners—the workforce—go about their roles and responsibilities in the workplace.

Situated Computing for Workforce Education

A large, global company in the information sciences business provides highly confidential, private, and sensitive information that is used, for example, by law enforcement agencies for criminal investigation purposes. In addition to other security measures, the company must secure specific areas within its buildings and must make certain that its workers, who are cleared for access to any of these secure areas, know and comply with the policies for accessing those controlled areas.

During onboarding, new employees attend an orientation program, in a classroom setting, during which there is a briefing on the policies and procedures for accessing secure, controlled areas. All new employees attend this training program, not just those who will have access to secure areas within the company's buildings.

Ongoing, all employees are required to take several Computer-Based Training (CBT) courses annually, one of which includes information and serves as a refresher on the policies for accessing secure facilities. These courses include several quizzes that test each employee's immediate-term knowledge of the subject matter. At the end, there is a procedural step where employees acknowledge that they have completed the course, understand the policies, and agree to them as a matter of continued employment with the company. This acknowledgment is logged for audit and compliance purposes.

A Common Workplace Situation

Authorized employees regularly enter, work in, and later exit secure areas in the company's buildings. For these employees, the situation—entering, working in, and exiting the secure areas—is an everyday part of working for the company. Their workplace just happens to be a secure area. They may or may not think much about the importance of the locked doors, badge-access controls, the presence of security officers, and/or any other heightened security measures. Their only specific reminder about the security policies, which they have acknowledged and agreed to, is during their annual, compulsory CBT course.

Using the Situation as an Instructional Opportunity

Something as simple as an occasional reminder on an employee's mobile device as they are passing through the entrance to a secure area could provide a situationally significant reminder about the policies for accessing secure areas and the importance of those policies. In this situated context, it may be reasonable to ask that they elaborate on the policy and indicate their experiences, preferences, and reflections of the policy. It may be interesting to share anecdotal information about things that have happened—or could happen—if the security of the area is not preserved. If something bad has happened previously—maybe, even, somewhere else, at a different company, for legal concerns—it may be possible to ask if employees are aware of the situation and the consequences. Or, it may be possible, as a form of simulation, to present a situation and ask if employees consider it acceptable under the terms of the policy, even potentially in a way that encourages collaboration through integration with a social media platform or in a gaming type of interactivity that encourages a competitive response. Any of this would be much more memorable than what a disconnected, arbitrary training exercise, like a compulsory CBT, could provide.

Developing a Mobile-Device Learning System

A situational learning solution of this type will require the implementation and integration of many system capabilities. At the time of this writing, existing technologies can be used to do so and, depending on the scope, a solution can be built relatively inexpensively.

At a high-level, the system must gather and process perceptual data, essentially using direct input from mobile devices and derived data to understand the user's situational considerations. In this way, the system will interact with the user's surrounding environment to collect relevant data to understand the outside world, meaning that it will be interpreting and exploring whatever activities are taking place (Sheth et al., 2016, p. 65).

The sensory data must then cycle through an appropriate number of computational processing steps to evaluate what the sensory data indicates, what the user's preferences are in that regard, and what the system knows, ideally using machine learning techniques that learn and adjust over time. From there, the system can intervene with the user for interactive purposes, meaning that the system can prompt, signal, interrupt, etc. and do so in an intelligent, relevant, and resourceful manner. In this way, the training system can equip learners with feedback so that they can experience the best and worst case scenarios relative to the situation.

There are many options for presentation, interaction, and intervention within most situational contexts—like banner notices and alerts—and many options for programming these parts of systems (e.g., open source frameworks, code libraries). But balancing these options, and getting the most of what each offer, is actually the hardest part. If the information is presented that is out of context or incorrect, or if the interactions are presented in an annoying manner, they will be summarily dismissed. Most of this is a matter of design, creativity, and empathy—artistic creativity—versus a technical or engineering matter.

At the time of this writing, Google is experimenting with messaging solutions for situated notifications that are integrated with Google Maps. Based on perceptual computing inputs—information from proximity sensors, geolocation data, pattern analysis, and predictive modeling services to name a few—solutions are being built and released that, for example, recognize that a user is in their car and about to drive home from work. At that time, Google is positioned to make recommendations on the best route based on traffic conditions, construction activity, weather, etc. This solution, although not an educational use of technology, illustrates how situational events can be targeted for informational purposes and in a very simple, concise, and brilliantly situated manner. This Google service can be considered a perceptual computing exemplar, considering the theoretical basis presented and discussed in this chapter. It can be used conceptually to model any type of situated computing solution, including solutions for situated workforce education that are, again, very simple, concise, and brilliantly situated.

Generalization and Disclaimer

This common workforce education scenario is presented as a hypothetical, yet realistic, situation that occurs in many workplaces. It is not an example taken from any specific company or organization and is not meant as a specific recommendation for a solution for any company or organization. It is merely presented for exploring how a system for workforce education could be built on the affordances of mobile-device technologies.

Limitations, Risks, and Other Considerations

The affordances of perceptual and situational computing solutions may seem like a technology opportunity at the surface. But, more so, these solutions are potentially an opportunity to touch lives and inspire people on the upside or to violate privacy and trust on the downside. To think that we now have both a massive amount of computing capacity and a remarkable tool for connectedness on a personal device that for many people is never more than a few feet away is, to say the least, a new frontier. Building a computing system, for commercial use or otherwise, that operates on a device that will spend a lot of time in someone's pocket is a big responsibility.

Building workforce education solutions on mobile-device technologies—and doing it well—will, of course, require integrating perceptual capabilities like the many sensor technologies in mobile devices with data analytics, cloud infrastructure, artificial intelligence, machine learning systems, etc. But, more importantly, the solutions must be designed in a way that emphasizes an understanding of how people think, behave, and interact while they are using these computing systems. The design must incorporate and accommodate the implications of the many important, personal norms and expectations different people have from the standpoint of culture, society, and human cognition, including how those interactions may vary in different worldwide regions. Leveraging personality, for instance, as a situational aspect of what a certain learner needs in a situation—even going so far as considering auditory or visual learning abilities and limits—can be accomplished by integrating perceptual capabilities.

Solutions for workforce education that leverage mobile-device technologies must do so in an appropriate manner, which can be challenging. What may seem like good information may play out as an annoyance, depending on how people view it. Replacing something compulsory, like annual CBT courses, with situated mobile device interactions may not be well-received if they are compulsory interruptions. If someone must regularly stop what they are doing to read an alert, answer a question, and/or acknowledge a message, the interruption might not be viewed as productive or useful. The interactive nature of design must be emphasized in order to create an experience that is useful, non-disruptive, informative, and personalized. Personalization should also include the ability to delay and/or opt out of the service.

Another important consideration in successfully developing a situated computing system for workforce education is balancing the amount of surveillance and tracking that is part of the system. Just as how badge systems keep offices and schools safer than they would be otherwise, people don't typically appreciate when badge access reports are used to see where they are and when. This type of surveillance isn't necessarily bad or offensive. While possibly counterproductive, it must be used appropriately, disclosed, and used in a matter-of-fact manner relative to that disclosure.

In many, if not most instances, employees are using their own personal mobile devices. This situation means that they are likely not obligated to participate in a workforce education program like the situated learning systems discussed in this chapter. Given that reality, there must be a draw or incentive for workers to participate in the program, for instance, in exchange for something of value, or because there is a gaming or social component that serves as a draw, or maybe because the subjects are interesting and compelling in and of themselves. The important thing is to find something that will work, devise a quality solution for it, and implement it in a way where its use is desired.

CONCLUSION

When Apple released the iPhone in 2007 the company changed the landscape of computing in dramatic ways. Those three little sensors in the first model were just a glimpse of what was to come in this area of technology. Many visionary ideas like those from the MIT Media Lab Wearable Computing Group took a large step forward. With the introduction and many advancements of the Android operating system, the proliferation of this new class of mobile-device technology has become everyday in terms of people's expectations of the technology, their ability to access the Internet remotely, their affordability and accessibility, and the intuitiveness of the products and systems built on them.

As computing systems have become perceptual, the intersection between technology and human behavior has become increasingly important as well and will become even more so in the future. The use of perceptual and situated computing will continue to develop and proliferate, including within an educational context. There are many opportunities to innovate and improve workforce education in that regard by leveraging the power and affordances of technology to turn everyday activities into everyday learning opportunities.

REFERENCES

Anderson, J. R. (2015). *Cognitive psychology and its implications*. New York, NY: Worth Publishers.

Baber, C. (2003). *Cognition and tool use*. New York, NY: Taylor & Francis. doi:10.1201/9781420024203

Beaudouin-Lafon, M., & Mackay, M. (2000). Research directions in situated computing. In CHI'00 Extended Abstracts on Human Factors in Computing Systems (pp. 369-369). ACM. doi:10.1145/633292.633516

Bransford, J. D., & Schwartz, D. L. (1999). Rethinking transfer: A simple proposal with multiple implications. *Review of Research in Education*, 61–100. doi:10.3102/0091732X024001061

Breslin, M., & Buchanan, R. (2008). On the case study method of research and teaching in design. *Design Issues*, *24*(1), 36–40. doi:10.1162/desi.2008.24.1.36

Carliner, S. (2017). Informal learning. In R. A. Reiser & J. V. Dempsey (Eds.), *Trends and Issues in Instructional Design and Technology* (pp. 142–151). Boston, MA: Pearson.

Clancey, W. J. (1995). A tutorial on situated learning. In J. Self (Ed.), *Proceedings of the International Conference on Computers and Education* (pp. 49-70). Charlottesville, VA: AACE.

Davidson, S. P., Cain, S. M., McGinnis, R. S., Vitali, R. R., Perkins, N. C., & McLean, S. G. (2016). Quantifying warfighter performance in a target acquisition and aiming task using wireless inertial sensors. *Applied Ergonomics*, *56*, 27–33. doi:10.1016/j.apergo.2016.03.001 PMID:27184308

Dawson, M. R. W. (2013). *Mind, body, world: Foundations of cognitive science*. Edmonton, Canada: AU Press.

Driscoll, M. P. (2017). Psychological foundations of instructional design. In R. A. Reiser & J. V. Dempsey (Eds.), *Trends and Issues in Instructional Design and Technology* (pp. 52–60). Boston, MA: Pearson.

Ertmer, P. A., & Newby, T. J. (1993). Behaviorism, cognitivism, constructivism: Comparing critical features from an instructional design perspective. *Performance Improvement Quarterly*, *6*(4), 50–72. doi:10.1111/j.1937-8327.1993.tb00605.x

Gershman, A. V., McCarthy, J. F., & Frano, A. E. (1999). Situated computing: Bridging the gap between intention and action. In *Wearable Computers, 1999. Digest of Papers. The Third International Symposium on Wearable Computers* (pp. 3-9). IEEE.

Global Human Capital Trends. (2016). *The New Organization: Different by Design*. Westlake, TX: Deloitte University Press.

Hawkins, J. B. (2011). *Bridging the knowledge gap: The effectiveness of compulsory computer-based training in federal employees' professional education* (Doctoral dissertation). Retrieved from https://search-proquest-com.ezproxy.cul.columbia.edu/docview/916240004?accountid=10226 (Order No. 3489818)

Hull, R., Neaves, P., & Bedford-Roberts, J. (1997). Towards situated computing. In *Wearable Computers, 1997. Digest of Papers. The First International Symposium on Wearable Computers* (pp. 146-153). IEEE. 10.1109/ISWC.1997.629931

KPMG. (2015). *Corporate digital learning*. Retrieved from https://assets.kpmg.com/content/dam/kpmg/pdf/2015/09/corporate-digital-learning-2015-KPMG.pdf

Lewis, G. A. (2017). Cloud computing. *Computer*, *50*(5), 8–9. doi:10.1109/MC.2017.141

McCarthy, J. (2007). *What is artificial intelligence?* Retrieved from http://www-formal.stanford.edu/jmc/whatisai/whatisai.html

McDermott, I. E. (2016, November). Fun with virtual and augmented reality. *Online Searcher,* 27-29.

Robbins, P., & Aydede, M. (2009). A short primer on situated cognition. In P. Robbins & M. Aydede (Eds.), *The Cambridge Handbook of Situated Cognition*. Cambridge, MA: Cambridge University Press.

Sheth, A. (2010). Computing for human experience: Semantics-empowered sensors, services, and social computing on the ubiquitous web. *IEEE Internet Computing*, *14*(1), 88–91. doi:10.1109/MIC.2010.4

Sheth, A., Anantharam, P., & Henson, C. (2016). Semantic, cognitive, and perceptual computing: Paradigms that shape human experience. *Computer*, *49*(3), 64–72. doi:10.1109/MC.2016.75

Turkay, S., Hoffman, D., Kinzer, C. K., Chantes, P., & Vicari, C. (2014). Toward understanding the potential of games for learning: Learning theory, game design characteristics, and situating video games in classrooms. *Computers in the Schools*, *31*(1-2), 2–22. doi:10.1080/07380569.2014.890879

Turkle, S. (2006). Tethering. In C. A. Jones, B. Arning, & J. Farver (Eds.), *Sensorium: Embodied Experience, Technology, and Contemporary Art* (pp. 220–226). Cambridge, MA: The MIT Press.

Turkle, S. (2008). Always-on/always-on-you: The tethered self. In J. E. Katz (Ed.), *Handbook of Mobile Communication Studies* (pp. 121–137). Cambridge, MA: MIT Press. doi:10.7551/mitpress/9780262113120.003.0010

Vinh, T. L., Bouzefrane, S., Farinone, J., Attar, A., & Kennedy, B. P. (2015). Middleware to integrate mobile devices, sensors, and cloud computing. *Procedia Computer Science*, *52*, 234–243. doi:10.1016/j.procs.2015.05.061

Vitsas, P. A. (2016). Commercial simulator applications in flight test training. *Journal of Aerospace Engineering*, *29*(4), 1–12. doi:10.1061/(ASCE)AS.1943-5525.0000589

Wilson, B. G. (2017). Constructivism for active, authentic learning. In R. A. Reiser & J. V. Dempsey (Eds.), *Trends and Issues in Instructional Design and Technology* (pp. 61–67). Boston, MA: Pearson.

Yoo, Y. (2010). Computing in everyday life: A call for research on experiential computing. *Management Information Systems Quarterly*, *34*(2), 213–231. doi:10.2307/20721425

KEY TERMS AND DEFINITIONS

Anchored Instruction: An instructional method employed to focus (i.e., anchor) learning within problem solving in a real or simulated situational context.

Experiential Computing: An approach for implementing computing systems in which computational abilities are embedded into the interactions between people and many common, everyday items. This computing approach is central to the rapidly developing internet of things.

Informal Learning: An educational process which takes place outside a conventional learning environment. The process is typically self-directed and can be structured or unstructured.

Machine Learning: A type of computing system in which algorithms learn as they process data within a given context. These systems then use that derived knowledge to deepen their level of understanding of what is happening within the system.

Mobile-Device Technology: A handheld computing device with specialized hardware and software for connecting to the Internet, measuring environmental conditions, and providing human interaction capabilities.

Perceptual Computing: A computing system where sensory inputs measure the physical characteristics of a situation to assess and assign meaning given those inputs.

Situated Computing: A type of computing solution that functions within a specific situational context to intelligently intervene.

Situated Learning: A learning theory basis through which instruction is designed to take place within a real or simulated situational context.

Workforce Education: A form of education and training designed for the needs of learners in a work environment.

Chapter 12
The Impact of Personal Response Systems on Students' Learning Performance:
Research Implications and Future Research Directions

Simon C. H. Chan
The Hong Kong Polytechnic University, China

Stephen Ko
The Hong Kong Polytechnic University, China

ABSTRACT

Personal response systems (PRSs) are prevalent across a range of educational settings, and this increasing importance has prompted many researchers to examine their various aspects. Their effects on students' learning performance have three main categories of antecedents: the learners' characteristics, the instructors' characteristics, and other contextual factors. A theoretical model is developed on the effects of PRSs on students' learning performance. This chapter describes the characteristics of PRSs, reviews their advantages and disadvantages, and proposes a theoretical model of their antecedents on students' learning performance. It concludes by exploring research implications and directions for future PRS research.

INTRODUCTION

Academic institutions in higher education have used advanced technologies and information systems as strategic resources to create better education environments. The field of teaching and learning is changing at a steady pace, and the learning formats in which instructors prepare their work with information technologies are being re-organized to enable changes in teaching. How students learn to meet challenges is also changing, as are the ways that their learning needs are satisfied (Eastman, Iyer, & Eastman, 2011).

DOI: 10.4018/978-1-5225-4111-0.ch012

A more interactive way to learn new knowledge in an academic environment with a variety of teaching and learning tools in the classroom enables students to be better involved in the learning process.

In the past decade, the use of personal response systems (PRSs) in the classroom has become more common in the academic literature (Moss & Crowley, 2011; Chien, Chang, & Chang, 2016). The design of questions by PRSs for students to respond during lessons in higher education has steadily increased (Stowell, 2015). The use of PRSs has become a popular means to demonstrate the importance of teaching effectiveness and learning for individual learners. PRSs are widely used across disciplines and across a range of educational settings, including accounting, business, management, information systems, psychology, mathematics, biology, chemistry, engineering, and computing programs in higher education (Addison, Wright, & Milner, 2009; Carnaghan, Edmonds, Lechner, & Olds, 2011; Keough, 2012). For example, Rana, Dwivedi, and Al-Khowaiter (2016) reviewed the use of PRSs in the business and management discipline. In business schools, an instructor might pose a theoretical model and cases and ask students to indicate how they would act in a given business environment. PRSs have been used in large introductory courses (Bonaiuti, Calvani, & Piazza, 2015; Trees & Jackson, 2007), in smaller tutorial classes, and in laboratory classes (Rana, Dwivedi, & Al-Khowaiter, 2016). Barnett (2006) discussed the implementation of personal response units in large lecture classes.

Empirical studies have found that feedback is important to the effectiveness of PRSs (Lantz & Stawiski, 2014; Rana, Dwivedi, & Al-Khowaiter, 2016). Their use can help boost students' attention, improve attendance, increase participation, and enhance learning performance (Buil, Catalan & Martinex, 2016; Hedgcock & Rouwenhorst, 2014; Roblyer & Wiencke, 2003). Students must respond to the questions, which helps to hold their attention, provide instant feedback for sharing, and encourage class participation (Boyle & Nicol, 2003). PRSs, such as "clickers" allow students to respond to review questions anonymously, and they have helped students become increasingly responsible for their own learning. PRSs can be one of the most useful tools in the new ongoing-learning process.

PRSs have the capacity to enhance learners' engagement and understanding in the classroom, in real-time. However, a survey of the literature devoted to the use of PRSs from the past 10 years seems to point to a lack of agreement on how to best harness its potential (Kay & LeSage, 2009; Moss & Crowley, 2011). While our chapter examines the advantages and disadvantages in effectively implementing PRSs for effective learning and propose a preliminary framework to guide practitioners in the exploration of using the systems in the classrooms. To support teaching and learning effectiveness, it is beneficial to know the format and type of PRS conducted. For the latter point, this study is significant for academics and practitioners to better understand these issues. The purposes of this chapter are to (a) describe the characteristics of PRSs; (b) evaluate their advantages and disadvantages; (c) develop a theoretical model of their antecedents as strategy tools to support students' learning performance; and (d) generate research implications and directions for future research.

Characteristics of Personal Response Systems (PRSs)

PRSs have been identified using a wide range of labels such as audience response systems, classroom communication systems, electronic voting systems, group response systems, wireless keypad response systems, student response systems, mobile phone polling, and "clicker-based technology" (Moss & Crowley, 2011). Our chapter uses the term "personal response systems" (PRSs) to refer to all of these labels, although despite the increasing use of the term, there is no universally accepted definition. PRSs have been recognized as integrated information systems that can support the learning process in a timely manner (Stuart, Brown, & Draper, 2004).

Some PRSs literature has been known in the context of health and medical professionals. It also applied the PRSs or related responses technology to natural sciences. PRSs had evolved over time from hardware, wireless devices, portable receivers, or software applications via mobile phones or tablet devices. The use of PRSs in the classroom has evolved quickly in the business and management discipline (Rana, Dwivedi & Al-Khowaiter, 2016). The use of PRSs to deliver information and receive responses has revolutionized learning for both instructors and students. They are instructional technology tools composed of proprietary software applications in instructors' computers, students' mobile phones or computers, and other response tools.

As PRSs have become widespread in academic institutions, instructors have begun to use information technology and response systems with the Internet (i.e. Wi-Fi) to deliver content to their classes. They can increase students' engagement, performance, learning, satisfaction, and motivation (Sprague & Dahl, 2010), and they offer an interactive platform to support engagement and feedback from individual learners and provide anonymity in responding to questions (Latham & Hill, 2014; Shapiro et al., 2017). During the lessons, students can simply input their responses into the software system. The applications of PRSs provide real-time polling sometimes within a PowerPoint format. Instructors must prepare the review questions in advance, or they can spontaneously create the questions during class. The scores or responses can be captured by PRSs for record. For example, PRSs can be used as a tool for taking attendance, which effectively raised the participation in the classrooms.

In their implementation, PRSs can operate in two different approaches: anonymous mode and known mode. In anonymous mode, the responses to the questions do not identify the students. In known mode, the instructor can determine which students give which responses. The polling results or feedback collected from the students can be recorded, summarized, and displayed immediately onscreen in class. The polling data can be transformed into a histogram or other format to show the statistical results for each review question, which can provide a basis for class discussion. PRSs are effective at increasing commitment, encouraging interactions, and allowing for immediate feedback by individual learners in the learning process (Keough, 2012). Questions can be asked that require students to give numerical, short text, or multiple-choice answers. After classes, the results can be saved in spreadsheets, and logs of the responses can be kept for records. For example, PRSs allow students to rank the reactions by the use of images instead of text-based materials. The image-based evaluation question can speed up the response time and acceptance of individual leaners to the rates of reactions.

Instructors can collect their identification codes for further analysis. Students can receive instant feedback on the course or multiple-choice tests. The questions can be shown spontaneously via PowerPoint presentation during lectures in formats including yes–no, right–wrong, true–false, multiple-choice, rating-scale, fill-in-the-blank, and answer-specific questions. In the classroom settings, students convey their responses to questions by cell phones, smartphones, and laptops. Every student have to engage in the questions, which lead shy students to be more comfortable in the learning environment. PRSs are a simple technology that provides students with interactive experiences (Blood & Neel, 2008; Trees & Jackson, 2007), formative assessment, homework collection, peer instructor (Rana, Dwivedi, & Al-Khowaiter, 2016). In business and management discipline, many large lecture courses have proposed using PRSs to integrate students' interactivity in the classroom. Students press on their mobile phones corresponding to their answers to business decision-making problems, such as yes-no, multiple choice, or true-false question. The results will then be projected on a screen in which teachers can show the correct answer along with the class distribution of answers. Teachers and students can further discuss the impacts of the responses to various stakeholders, such as organizations, managers, employees. Also, students can be linked to class lists and they can be requested to provide unique responses from different perspectives, such as accounting, marketing, legal compliance.

ADVANTAGES AND DISADVANTAGES OF PRSs

PRSs are personal response devices that allow interactions between instructors and students through responses to review questions (Kay & LeSage, 2009). Teachers can adjust very quickly to fill knowledge gaps and students thus experience change at a faster pace and in an interactive way (Stowell & Nelson, 2007).

The advantages and disadvantages of PRSs are summarized in Figure 1. Advantages of PRSs include positive engagement, active collaborative learning, assessment, and instant feedback. Disadvantages of PRSs include technical problem, access and resources problem, and learners' resistance.

Advantages

PRSs are inexpensive and effective teaching and learning tools to promote students' involvement in classes (Trew & Nelsen, 2012). The use of PRSs can increase students' learning performance. Their effectiveness is indicated by their benefits (Kay & LeSage, 2009). Their four main advantages are that they enable positive engagement, active collaborative learning, assessment, and instant feedback.

Figure 1. Advantages and disadvantages of PRSs

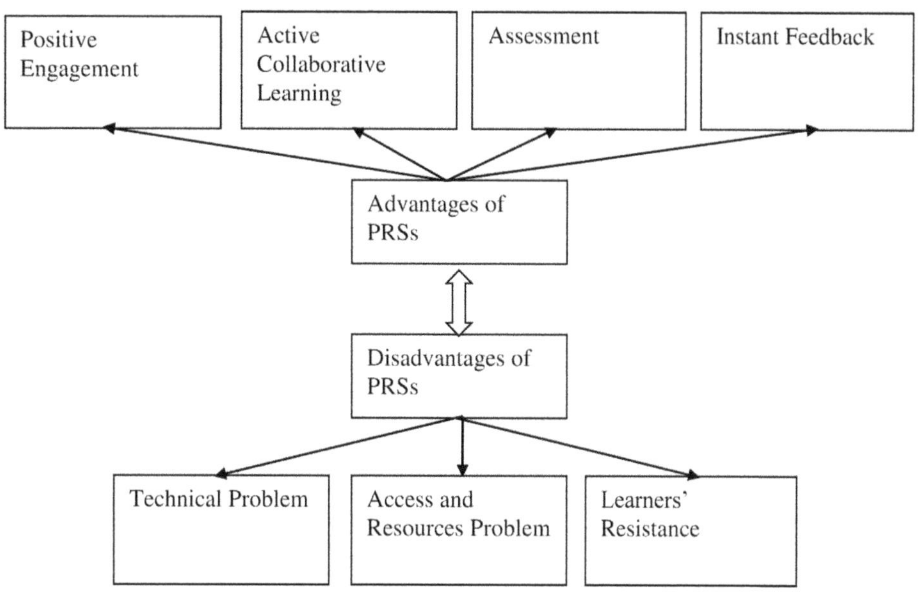

Positive Engagement

PRS functionality can promote positive student engagement during lessons (Barnett, 2006; Preszler, Dawe, Shuster, & Shuster, 2007; Simpson & Oliver, 2007). If instructors can pose well-chosen or well-crafted questions that appeal to the higher levels of Blooms Taxonomy to students during their lectures, students can better reflect on and absorb the course content. The integration of formal questions in class can facilitate students' engagement (Sun, 2014). In classroom settings, PRSs provide students with innovative ways to respond to review questions. Students can respond to questions asked during a lecture anonymously, which may increase their engagement (Voelkel & Bennett, 2014).

Blasco-Arcas, Buil, Hernandez-Ortega, and Sese (2013) found that the high level of student interactivity with instructors resulting from PRS usage is more likely to enhance students' engagement. Students' interactions with the environment influence their perceptions and engagement (Tloaele, Hofman, Naidoo, & Winnips, 2014). PRSs can enable instructors to check individual learners' attendance based on their responses. They can thus be used to ensure students' attendance because they have a motive to attend regularly. The use of PRSs would enhance students' motivation for learning which acts like a supplement to help teachers for constructing unstructured content in the classrooms.

Active Collaborative Learning

PRSs can enable students to engage in active collaborative learning because it is an individual response device that allows students to respond anonymously to questions presented during the lesson. Students can read, respond, and reflect on their learning with their instructors more actively. PRSs greatly increase students' ability to respond critically and encourage them to give and share their opinions (Kay & LeSage, 2009). The use of PRSs provide an opportunity for students to discuss, and teachers to follow-up on any interesting debates or topics.Students play active roles in creating new knowledge that contributes to the learning process.

The growth of PRSs use has created an opportunity to promote active collaborative learning in the classroom. The increase in the ability foster and facilitate discussion during lessons can lead to more interactive discussions between the instructors and students during the learning process. Although some argue that PRSs detract from real genuine engagement in class, this enhancement is supported by researchers (Barnett, 2006; Preszler, Dawe, Shuster, & Shuster, 2007; Simpson & Oliver, 2007; Siau, Sheng, & Nah, 2006) who have noted that creating a positive communication channel between instructors and individual learners which is a critical component of teaching and learning . PRSs can facilitate active collaborative learning, which allows students to appreciate their level of learning relative to their peers during class. PRSs are an effective tool for encouraging students' thinking and provide an active collaborative learning opportunity by engaging in the discussion of the results. PRSs facilitate the active learning mode and encourage students to participate in the creation of knowledge by responding the questions. The change of learning mode from face-to-face to the use of technology in the classroom would foster more constructive ideas and feedback for collaboration among the peers. Students are more collaborative with teachers, peers, and even creation of materials used for discussion so as to enhance their learning experience (Blasco-Arcas, Buil, Hernandez-Ortega, & Sese, 2013). For example, after receiving the responses in the classroom, students can involve in discussion for some surprising results.

Assessment

PRSs provide a means to assess students in the process of collecting responses, knowledge, and responses. Students generally appreciate anonymity and ease of use, and participate more when PRSs are used in class (Kay & LeSage, 2009). The assessments can be used to evaluate students' learning and determine whether they are performing well during lessons. PRSs can also be used to grade students' responses and tests at their learning levels during assessments.

PRSs can be assessed by having students give responses, which provides real-time information flow. Instructors can use these responses in class as assessments to make changes to the learning process. PRSs are important to maintain attention and identify the learning needs of students, and they allow students to focus longer on the content and enable instructors to determine whether they fully understand the important points during the lessons.

Instant Feedback

PRSs are effective at engaging students in class, providing a platform for interactions among students, providing immediate feedback on their understanding of the lessons, and facilitating their active participation in the learning process through the discussion of answers. This instant feedback is beneficial to teaching and learning (Lantz & Stawiski, 2014; Simpson & Oliver, 2007). For example, instructors immediately receive responses regarding the students' responses (Chen, Whittinghill, & Kadlowec, 2010), and students can check whether their responses were right or wrong. Chen, Whittinghill, and Kadlowec (2010) identified the benefits of timely and specific feedback on a task or concept by PRSs.

PRSs allow instructors to teach and discuss the course materials differently. Instructors can use the immediate feedback to dynamically alter the class based on the nature and distribution of students' answers class (Yourstone, Kraye, & Albaum, 2008). Instructors can use PRSs as a medium for obtaining overall feedback, as they can obtain a full picture of students who understand and follow the teaching materials.

Disadvantages

Some of the disadvantages of PRSs should also be discussed (Kay & LeSage, 2009). Although the problems of technology have decreased in recent years, other significant problems with their use remain: technical problems, access and resource problems, and learners' resistance.

Technical Problems

Although PRSs are easy to use in general, there are technical concerns related to Internet service and device unreliability. Response systems often have connection problems or require long loading times when students respond to polls (Stowell, 2015). PRSs must be connected to and recognized by the computer before the software is started (Caldwell, 2007). To support the use of PRSs, students are responsible for bringing the appropriate devices to the lessons.

PRSs can lead to unsatisfactory or poor situations due to technical problems, such as slow loading speeds and Wi-Fi connections. The software system and Internet support (i.e. Wi-Fi) must function properly. On some occasions, devices do not function properly or the signal is too weak for the instructor's computer to connect. Such problems can be avoided by ensuring good system maintenance, stable content delivery, and reliable loading of software.

Access and Resource Problems

Despite the benefits gained, instructors may be reluctant to use PRSs because of access and resource problems. Accessibility becomes an issue with information system solutions. Internet (i.e. Wi-Fi) support, Internet service, mobile or computer speed, and availability must be compatible with the PRS package. Another issue is the availability of adequate hardware, such as computers and mobile devices for students. Instructors who use PRSs must set up the response systems before their lessons. Students may encounter difficulties in accessing and connecting to the Internet using their mobile phones or computers to respond.

The resources required to set up PRSs and the pre-class preparation are important as it requires hardware, systems, software, design of the instructors, etc (Barnett, 2006). Most are password-protected and require students to log into the system. Instructors may feel that students need to master certain material before further explaining a particular topic. Class-wide discussion takes too much time; institutions must support instructors by providing relevant support.

Learners' Resistance

As the format of teaching and learning has changed with PRSs, students may resist adopting them. Such students may experience stress, frustration, and resistance to review questions in class. These negative reactions should be monitored to ensure individual learners' participation (Caldwell, 2007). Requiring students to learn to use PRSs creates challenges to and complaints about these new polling systems.

For example, in a large class setting, lectures tend to require the instructor to provide information and sharing experiences with the students in a one-way approach (Trees & Jackson, 2007). Instructors must be aware of individual learners who are resistant to adopting PRSs and find ways to ensure that they use them properly (Carnaghan, Edmonds, Lechner, & Olds, 2011).

THEORETICAL MODEL OF PERSONAL RESPONSE SYSTEMS

Most PRS research has concentrated on the features of the organization, with a lack of empirical research on other factors (e.g. Blasco-Arcas et al., 2013; Han, 2014). There is a need to identify the antecedents on students' learning attitudes in a virtual learning and academic environment. More studies have examined the impact of antecedents of PRSs on students' learning performance (Lantz & Stawiski, 2014; Rana, Dwivedi, & Al-Khowaiter, 2016; Rana & Dwivedi, 2015). Whereas empirical studies have identified the effects of the interactivity, collaboration, and engagement of the response system on student learning performance, it is also important to offer high-potential learning experiences (Choi, Kim, & Kim, 2007).

Three main characteristics (i.e., learner characteristics, instructor characteristics, and contextual factors) influence students' learning in the context of the electronic PRS assessment and learning system (Blasco-Arcas, Buil, Hernandez-Ortega, & Sese, 2013; Choi, Kim, & Kim, 2007). Choi, Kim & Kim (2007) examined the factors include learner interface, interaction, instructor attitude towards students, instructor technical competence and content on attitude towards e-learning. The results explained the significance of learner characteristics, instructor characteristics, and contextual factors in the system usage. The interactive abilities of the system allow for timely feedback for teaching and learning. These factors are the some of the key determinants of students' performance. Instructors play positive roles in learners' PRS experience (Kay & LeSage, 2009). Instructors' attitudes and technical competence affect learners' attitudes and the effectiveness of the learning process (Hunsu, Adesope, & Bayly, 2016; Simpson & Oliver, 2007). This chapter attempts to fill this gap by examining learners' preferences, instructors' interactions, and other contextual factors to explain learning outcomes. Individual learners and instructors can benefit from the use of PRSs (Simpson & Oliver, 2007).

A theoretical framework for PRSs is developed and depicted in Figure 2. The theoretical framework ties together six factors representing three main categories: learner characteristics (learner interface and interaction), instructor characteristics (instructor attitudes toward students and instructor technical competence), and contextual factors (content and types of questions).

Learner Characteristics

PRSs have increasingly important to offer high-potential characteristics to individual learners. The effects of interactivity, collaboration, and engagement of PRSs on learning performance have been examined (Blasco-Arcas et al., 2013). The learner interface and interaction are two main factors that influence students' learning performance (Choi, Kim, & Kim, 2007; Rana, Dwivedi, & Al-Khowaiter, 2016).

Learner Interface

Learner interface, an important component for promoting the use of PRSs, is defined as the extent to which learners can use them to connect. A learner's preference for PRSs can be encouraged by involve-

Figure 2. Theoretical framework

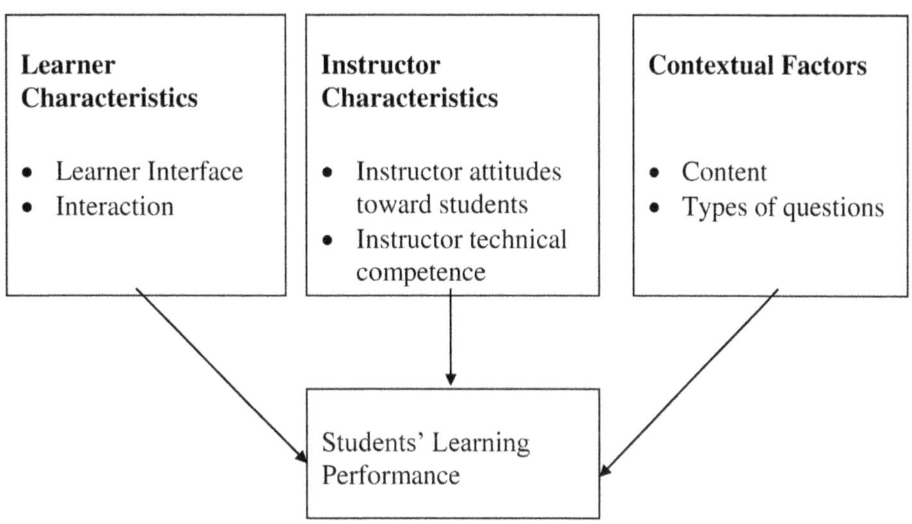

ment with the system (Lai, Wang, & Lei, 2012). Individual learners are expected to have enough PRS experience to overcome the interface (Simpson & Oliver, 2007). Easy-to-use learner interfaces can help to save time in learning. Such PRSs are thus user-friendly for individual learners, making them more comfortable with their use.

PRSs enable individual learners to share their experiences by demonstrating their knowledge in class activities. In the learning environment, PRSs facilitate as a way for communication between learners and instructors, and they can help spread new knowledge, keep users informed about the latest information, help users engage in new learning experiences, and facilitate monitoring of the learning experience.

Interaction

Interaction refers to the ability of the instructor and students to interact via PRSs. PRSs can engage students by providing them with rapid and compelling interactions and feedback. It is important to identify whether students can interact with instructors in class through the PRS. If so, they can easily contact the instructor and receive timely feedback (Latham & Hill, 2014).

PRSs provide opportunities for students to be actively involved and engaged in lectures. They can share their views and comment on review questions, topics of content delivery, and materials provided in class (Yourstone, Kraye, & Albaum, 2008). The results can help change students' misunderstandings and facilitate discussions with instructors. For example, in a class of business and management, the collaboration between instructors and individual learners can enable better communication in the discussion of current business environments.

Instructor Characteristics

Individual learners and instructors are actively interactive among themselves. Instructor characteristics play a significant role in the application of PRSs in the classroom (Choi, Kim, & Kim, 2007) because instructors can change their teaching and learning practices based on their students' attitudes and technical competence.

Instructor Attitude Toward Students

Instructor attitude toward students is defined as how instructors respond to the use of PRSs. The teaching and learning approach of an instructor can help students learn through the response systems. Instructors can increase the interest of students who seek advice on PRSs. With a positive instructor attitude, the classroom environment will be more open, and with more direct responses, students will be better able to recognize clear goals and receive instant feedback (Kay & LeSage, 2009).

In learning environments, instructors can use different teaching styles with the help of technology. In a more open environment, instructors can use more interactive teaching styles with the help of PRSs to engage individual learners (Choi, Kim, & Kim, 2007). With positive instructor attitudes, PRSs provide more opportunities for immediate feedback, different teaching methods, and interaction between instructors and learners. Instructors should continue to seek new technologies to meet students' learning needs.

Instructor Technical Competence

Instructor technical competence refers to instructors' knowledge in handling information technology or skill in using PRSs in classes. Many instructors who are interested in using PRSs may avoid them because they do not know how to implement them effectively (Choi, Kim, & Kim, 2007; Rana, Dwivedi, & Al-Khowaiter, 2016). Thus, the use of PRSs improves the learning experience of students in educational settings.

Competent instructors can provide learners with clear goals and the necessary technical competence in advanced technology and systems (Lantz & Stawiski, 2014; Volery & Lord, 2000), and they can effectively explain how to use PRSs. Instructors' technical competence in PRSs is likely to be related to students' learning performance. Instructors should thus have the ability to handle PRSs effectively and attentively and strategically develop relevant questions.

Contextual Factors

Contextual factors play crucial roles in learners' attitudes and learning performance (Choi, Kim, & Kim, 2007). Content and question type are two factors to be discussed.

Content

Content relevant for their use can motivate individuals to learn and enhance learning performance. The use of PRSs can transform the content and format of learning in large introductory lecture courses and in medium-sized and small group discussion courses. The content of the questions presents challenges to individual learners to respond.

For students to have a clear picture of what they are supposed to learn, content should be useful and fit individual needs. The content of the questions can increase the learning interest and performance of the students. For example, if an instructor wants to discuss thought-provoking questions in a large class, students should think of the content for their responses. This learning action can enhance their learning performance.

Question Type

In addition, types of questions in PRSs can affect students' learning (Gier and Kreiner, 2009). The most effective questions should be specific to avoid confusion, should give direction in a new context, and should provoke a wide range of responses. Questions of various kinds, from yes–no and true–false to multiple-choice and open-ended, are offered in PRSs.

Results indicate that students with experience using PRSs in the classroom obtain higher scores on examinations than those without such experience. Active learning by asking relevant questions was responsible for the increase in grades. PRSs are particularly efficient for deriving responses to questions at lower levels at which questions concerning facts, such as remembering and understanding, are involved.

STUDENTS' LEARNING PERFORMANCE

Research by Keough, 2012 as well as Scott & Walczak, 2009 has indicated that students' engagement with PRSs improves learning performance (Shapiro, Sims-Knight, O'Rielly, Capaldo, Pedlow, Gordon, & Monteiro, 2017). Hedgcock & Rouwenhorst (2014) examined how to use PRSs to provide feedback to enhance student learning performance. Student learning performance is a crucial part of the evaluation and impact of PRSs (Nelson & Hauck, 2008; Zhu, 2012).

The effective use of PRSs facilitates positive learning attitudes and engenders learner engagement and learning performance (Yourstone, Kraye, & Albaum, 2008). The use of PRSs integrated with peer instructions significantly influenced students' performance in class settings. They are also useful for instructors to understand what proportion of students are not participating in the discussion. Students who participate in classes with PRSs have better academic results. By using PRSs, classes have made students feel valuable, which may enhance their learning performance.

DISCUSSION

The growing popularity of PRSs has substantially improved the quality of teaching and learning performance among students (e.g. Blasco-Arcas et al., 2013; Moss & Crowley, 2011; Chien, Chang, & Chang, 2016). Through the literature review, this chapter explored the effectiveness of PRSs in classroom learning. A careful examination of the antecedent of PRSs using a theoretical model and empirically examined their effects on students' learning performance (Blasco-Arcas, Buil, Hernandez-Ortega, & Sese, 2013; Choi, Kim, & Kim, 2007). A model of the impact of PRSs and students' learning performance is discussed and developed a promising theoretical foundation of PRSs. PRSs can be useful for understanding the learning styles of individual learners. From the theoretical perspective, this chapter explained their relevant factors to enhance both learner characteristics and instructor characteristics in students' learning performance. PRSs can increase engagement in the learning experience, helping instructors provide adequate feedback and interface for learning.

Second, the effects of the contextual factors of PRSs on learning performance as an innovative means to deliver questions in education settings were examined. One of their main functions is to provide direct delivery of the content and the types of questions. It is important to identify which content is offered to learners. PRSs can help provide an innovative way for students to participate in lessons. The real-time feedback can benefit both instructors and students. Future research could seek guidance from experts at various levels of study to resolve challenges on when and how to set the content and types of questions as supplements in the learning process.

Third, PRSs can improve individual learners' attention in the classroom. Students' participation in the classroom is shown to enhance learning performance. Academics and students enjoy the innovative education experience with effective usage of PRSs. PRSs are also useful in curriculum design of to capture students' attention. This would increase learning opportunities for students with different learning styles. PRSs can help instructors design balanced teaching approaches that address students' needs and direct new teaching and learning visions by increasing the focus on students. Students are encouraged to develop problem-solving skills with flexible applications in classes. Students' responses can effectively serve as a roadmap for teaching and learning.

Finally, as implications for practitioners, this chapter promotes and encourages academics and practitioners to adopt PRSs to satisfy the knowledge gaps and learning needs of instructors and students. Their use can lead to greater success and understanding for students and can complement traditional teaching and learning tools as an innovative means of communication between instructors and students. This chapter provides implications for educators based on a theoretical model developed in the current competitive environment. Given the strong linkage in the field of teaching and learning, academic educators could seek to ascertain precisely how the benefits of PRSs can be achieved.

In addition, PRSs would help practitioners to enhance the learning condition in corporate training. The use of PRSs is good for practitioners to share their work experiences and to devote extra effort in the real life practices. The interactivity of PRSs allows employees to have better learning experience. It is particular important for corporate training to receive what participants' have learned and reinforce the learning expectations through feedback. PRSs provide thought-provoking learning experiences and motivate participants to learn instantly in real world scenarios.

FUTURE RESEARCH

This chapter explored and investigated the main categories of antecedents of PRSs on students' learning performance (i.e., learners' characteristics, instructors' characteristics, and other contextual factors). Their learner interface and instructor characteristics are two of the main factors that affect the students' attitudes and learning performance. Instructors should follow up from the point at which students use the technological devices to interact with them. Instructors can then create appropriate strategies to explain concepts clearly and then create student groups to debate the questions that arise. Other factors, such as disciplinary differences and course levels, may also affect students' learning performance (Han, 2014). While the references and examples cited in this chapter are at times dated, it must be noted there are numerous contexts in the USA and elsewhere where these PRS practices persists, ignoring mobile phone developments and mobile learning replacements in this regard. Future research could seek to identify the key success factors of PRSs. What are the other potential variables that influence students' learning performance? What are the conditions that make PRSs more effective? These questions are left to further study.

This chapter is significant for future research because it suggests a way to improve students' learning performance. The unique characteristics of the PRSs are more restrictive for delivery knowledge. Individual learners' characteristics can be more important in an interactive learning environment, and their experience can be improved by changing lecture content, adopting an inclusive method of teaching that involves everyone in the class, and enhancing class interactivity. Future research can examine the individual differences in the use of PRSs, such as gender (Kang, Lundeberg, Wolter, delMas, & Herreid, 2012), learning styles, and experience with PRSs (Lantz, 2010; Rana, & Dwivedi, 2015).) Further research is required to understand whether gender differences affect the overall learning experience of students.

Moreover, the characteristics of individual learners and instructors are important for enhancing students' learning performance in different education settings (Lai, Hill, & Ma, 2015). The findings may be generalizable to a range of levels of business students at other universities (Rana & Dwivedi, 2015). Rana and Dwivedi (2015) reviwered that the literature has compared these findings with reviews presented for other multidisciplinary research on PRSs (Farag, Park, & Kaupins, 2015). Future research could examine how instructors create review questions in PRSs, how they share information or results in class, and how PRSs lead to the creation of new learning practices.

Finally, the chapter offers important insights for the formulation of effective PRSs, which can be aligned with different learning philosophies to maintain competitive advantage. The use of PRSs is a new trend in higher education. Their characteristics, advantages, and disadvantages have been discussed. The effects of specific types of questions can create student-oriented learning. PRSs can thus deliver effective learning to students that use it. Future research should evaluate their effectiveness in the classroom (Hunsu, Adesope, & Bayly, 2016; Lantz, & Stawiski, 2014; Masikunis et al., 2008; Shaffer & Collura, 2009). Students can thus evaluate types of lectures as more interesting, entertaining, learning, and interactive (Masikunis, Panayiotidis, & Burke, 2009).

CONCLUSION

As mobile phones, educational technologies and information systems continue to change, the role of PRSs as an innovative means to implement interactive polling experiences between instructors and students in the classroom has grown. PRSs have received increasing acceptance among academics as a supplement that can support student learning. Academic institutions have benefited from their use for years because they offer great promise for engaging students' learning. PRSs promote learning platforms based on sound pedagogical practices. Students are expected to have high satisfaction with their use and to appreciate the instant feedback provided by the instructors.

This chapter fills the gap in the literature about the effects of PRSs in academic institutions. Practically, PRSs are effective tools for enhancing students' learning performance across disciplines and class sizes in various academic contexts.

REFERENCES

Addison, S., Wright, A., & Milner, R. (2009). Using clickers to improve student engagement and performance in an introductory biochemistry class. *Biochemistry and Molecular Biology Education*, *37*(2), 84–91. doi:10.1002/bmb.20264 PMID:21567711

Anthis, K. (2011). Is it the clicker, or is it the question? Untangling the effects of student response use. *Technology Teacher*, *38*(3), 189–193.

Barnett, J. (2006). Implementation of personal response units in very large lecture classes: Student perceptions. *Australasian Journal of Educational Technology*, *22*(4), 474–494. doi:10.14742/ajet.1281

Blasco-Arcas, L., Buil, I., Hernandez-Ortega, B., & Sese, F. J. (2013). Using clickers in class. The role of interactivity, active collaborative learning, and engagement in learning performance. *Computers & Education*, *32*, 102–110. doi:10.1016/j.compedu.2012.10.019

Blood, E., & Neel, R. (2008). Using student response systems in lecture-based instruction: Does it change student engagement and learning? *Journal of Technology and Teacher Education*, *16*(3), 375–383.

Bonaiuti, G., Calvani, A., & Piazza, D. (2015). Increasing classroom engagement and student comprehension through the use of clickers: An Italian secondary school experience. *Research on Education and Media*, *5*(1), 95–108.

Boyle, J. T., & Nicol, D. J. (2003). Using classroom communication systems to support interaction and discussion in large class settings. *Association of Learning Technology Journal*, *11*(3), 43–57. doi:10.3402/rlt.v11i3.11284

Buli, I., Catalan, S., & Martinez, E. (2016). Do clickers enhance learning? A control-value theory approach. *Computers & Education*, *103*, 170–182. doi:10.1016/j.compedu.2016.10.009

Caldwell, J. E. (2007). Clickers in the large classroom: Current research and best-practice tips. *Life Sciences Education*, *6*(1), 9–20. doi:10.1187/cbe.06-12-0205 PMID:17339389

Carnaghan, C., Edmonds, T. P., Lechner, T. A., & Olds, P. R. (2011). Using student response systems in the accounting classroom: Strengths, strategies, and limitations. *Journal of Accounting Education*, *29*(4), 265–283. doi:10.1016/j.jaccedu.2012.05.002

Chen, J. C., Whittinghill, D. C., & Kadlowec, J. A. (2010). Classes that click: Fast, rich feedback to enhance students' learning and satisfaction. *Journal of Engineering Education*, *99*(2), 158–169. doi:10.1002/j.2168-9830.2010.tb01052.x

Chien, Y. T., Chang, Y. H., & Chang, C. Y. (2016). Do we click in the right way? A meta-analytic review of clicker-integrated instruction. *Educational Research Review*, *17*, 1–18. doi:10.1016/j.edurev.2015.10.003

Choi, D. H., Kim, J., & Kim, S. H. (2007). ERP training with a web-based electronic learning system: The flow theory perspective. *International Journal of Human-Computer Studies*, *65*(3), 223–243. doi:10.1016/j.ijhcs.2006.10.002

Eastman, J. K., Iyer, R., & Eastman, K. L. (2011). Business students' perceptions, attitudes, and satisfaction with interactive technology: An exploratory study. *Journal of Education for Business*, *86*(1), 36–43. doi:10.1080/08832321003774756

Farag, D. M., Park, S., & Kaupins, G. (2015). Faculty perceptions of the adoption and use of clickers in the legal studies in business classroom. *Journal of Education for Business*, *90*(4), 208–216. doi:10.1080/08832323.2015.1014459

Gier, V. S., & Kreiner, D. S. (2009). Incorporating active learning with PowerPoint-based lectures using content-based questions. *Teaching of Psychology*, *36*(2), 134–139. doi:10.1080/00986280902739792

Han, J. H. (2014). Unpacking and repacking the factors affecting students' perceptions of the use of classroom communication systems (CCS) technology. *Computers & Education*, *79*, 159–176. doi:10.1016/j.compedu.2014.07.011

Hedgcock, W. H., & Rouwenhorst, R. M. (2014). Clicking their way to success: Using student response systems as a tool for feedback. *Journal for Advancement of Marketing Education*, *22*(2), 16–25.

Hunsu, N. J., Adesope, O., & Bayly, D. J. (2016). A meta-analysis of the effects of audience response systems (clicker-based technologies) on cognition and affect. *Computer Education*, *94*, 102–119. doi:10.1016/j.compedu.2015.11.013

Kang, H., Lundeberg, M., Wolter, B., delMas, R., & Herreid, C. F. (2012). Gender differences in student performance in large lecture classrooms using personal response systems ('clickers') with narrative case studies. *Learning, Media and Technology*, *37*(1), 53–76. doi:10.1080/17439884.2011.556123

Kay, B. H., & LeSage, A. (2009). Examining the benefits and challenges of using audience response systems: A review of the literature. *Computers & Education*, *53*(3), 819–827. doi:10.1016/j.compedu.2009.05.001

Keough, S. M. (2012). Clickers in the classroom: A review and a replication. *Journal of Management Education*, *36*(6), 822–847. doi:10.1177/1052562912454808

Lai, C., Wang, Q., & Lei, J. (2012). What factors predict undergraduate students' use of technology for learning? A case from Hong Kong. *Computers & Education*, *59*(2), 569–579. doi:10.1016/j.compedu.2012.03.006

Lai, G., Hill, V., & Ma, Y. (2015). Clickers in the classroom: A business professor's adoption of a classroom response system. *International Journal of Innovation and Learning*, *18*(4), 451–470. doi:10.1504/IJIL.2015.072458

Lantz, M. E. (2010). The use of 'clickers' in the classroom: Teaching innovation or merely an amusing novelty? *Computers in Human Behavior*, *26*(4), 556–561. doi:10.1016/j.chb.2010.02.014

Lantz, M. E., & Stawiski, A. (2014). Effectiveness of clickers: Effect of feedback and the timing of questions on learning. *Computers in Human Behavior*, *31*, 280–286. doi:10.1016/j.chb.2013.10.009

Latham, A., & Hill, N. S. (2014). Preference for anonymous classroom participation: Linking student characteristics and reactions to electronic response systems. *Journal of Management Education*, *38*(2), 192–215. doi:10.1177/1052562913488109

MacGeorge, E. K., Homan, S. R., Dunning, J. B., Elmore, D., Bodie, G. D., Evans, E., ... Lichti, S. M. (2008). The influence of learning characteristics on evaluation of audience response technology. *Journal of Computing in Higher Education*, *19*(2), 25–46. doi:10.1007/BF03033425

Masikunis, G., Panayiotidis, A., & Burke, L. (2009). Changing the nature of lectures using a personal response system. *Innovations in Education and Teaching International*, *46*(2), 199–212. doi:10.1080/14703290902843935

Mayer, R. E., Stull, A., DeLeeuw, K., Almeroth, K., Bimber, B., Chun, D., ... Zhang, H. (2009). Clickers in college classrooms: Fostering learning with questioning methods in large lecture classes. *Contemporary Educational Psychology*, *34*(1), 51–57. doi:10.1016/j.cedpsych.2008.04.002

Moss, J., & Crowley, M. (2011). Effective learning in science: The use of personal response systems with a wide range of audiences. *Computers & Education*, *56*(1), 36–43. doi:10.1016/j.compedu.2010.03.021

Nelson, M. K., & Hauck, R. V. (2008). Clicking to learn: A case study of embedding radio-frequency based clickers in an introductory management information systems course. *Journal of Information Systems Education*, *19*(1), 55–64.

Preszler, R. W., Dawe, A., Shuster, C. B., & Shuster, M. (2007). Assessment of the effects of student response systems on student learning and attitudes over a broad range of biology courses. *CBE Life Sciences Education*, *6*(1), 29–41. doi:10.1187/cbe.06-09-0190 PMID:17339392

Rana, N. P., & Dwivedi, Y. K. (2015). Using clickers in a large business class: Examining use behavior and satisfaction. *Journal of Marketing Education*, *38*(1), 47–64. doi:10.1177/0273475315590660

Rana, N. P., Dwivedi, Y. K., & Al-Khowaiter, W. A. A. (2016). A review of literature on the use of clickers in the business and management discipline. *International Journal of Management Education*, *14*(2), 74–91. doi:10.1016/j.ijme.2016.02.002

Roblyer, M. D., & Wiencke, W. R. (2003). Design and use of a rubric to access and encourage interactive qualities in distance courses. *American Journal of Distance Education*, *17*(2), 77–98. doi:10.1207/S15389286AJDE1702_2

Scott, J. E., & Walczak, S. (2009). Cognitive engagement with a multimedia ERP training tool: Assessing computer self-efficacy and technology acceptance. *Information & Management*, *46*(4), 221–232. doi:10.1016/j.im.2008.10.003

Shaffer, D. M., & Collura, M. J. (2009). Evaluating the effectiveness of a personal response system in the classroom. *Teaching of Psychology*, *36*(4), 273–277. doi:10.1080/00986280903175749

Shapiro, A. M., Sims-Knight, J., O'Rielly, G. V., Capaldo, P., Pedlow, T., Gordon, K., & Monteiro, K. (2017). Clickers can promote fact retention but impede conceptual understanding: The effect of the interaction between clicker use and pedagogy on learning. *Computers & Education*, *111*, 44–59. doi:10.1016/j.compedu.2017.03.017

Siau, J., Sheng, H., & Nah, F. F. H. (2006). Use of a classroom response system to enhance classroom interactivity. *IEEE Transactions on Education*, *49*(3), 398–403. doi:10.1109/TE.2006.879802

Simpson, V., & Oliver, M. (2007). Electronic voting systems for lectures then and now: A comparison of research and practice. *Australasian Journal of Educational Technology*, *23*(2), 187–208. doi:10.14742/ajet.1264

Sprague, E. W., & Dahl, D. W. (2010). Learning to click: An evaluation of the personal response system clicker technology in introductory marketing courses. *Journal of Marketing Education*, *32*(1), 93–103. doi:10.1177/0273475309344806

Stowell, J. R. (2015). Use of clickers vs mobile devices for classroom polling. *Computers & Education*, *82*, 329–334. doi:10.1016/j.compedu.2014.12.008

Stowell, J. R., & Nelson, J. M. (2007). Benefits of electronic audience response systems on student participation, learning, and emotion. *Teaching of Psychology*, *34*(4), 253–258. doi:10.1080/00986280701700391

Stuart, S., Brown, M., & Draper, S. (2004). Using an electronic voting system in logic lectures: One practitioner's application. *Journal of Computer Assisted Learning*, *20*(2), 95–102. doi:10.1111/j.1365-2729.2004.00075.x

Sun, J. C. Y. (2014). Influence of polling technologies on student engagement: An analysis of student motivation, academic performance, and brainwave data. *Computers & Education*, *72*, 80–89. doi:10.1016/j.compedu.2013.10.010

Tloaele, M., Hofman, A., Naidoo, A., & Winnips, K. (2014). Using clickers to facilitate interactive engagement activities in a lecture room for improved performance by students. *Innovations in Education and Teaching International*, *51*(5), 497–509. doi:10.1080/14703297.2013.796725

Trees, A. R., & Jackson, M. H. (2007). The learning environment in clicker classrooms: Student processes of learning and involvement in large university-level courses using student response systems. *Learning, Media and Technology*, *32*(1), 21–40. doi:10.1080/17439880601141179

Trew, J. L., & Nelsen, J. L. (2012). Getting the most out of audience response systems: Predicting stu dent reactions. *Learning, Media and Technology*, *37*(4), 379–394. doi:10.1080/17439884.2011.621957

Voelkel, S., & Bennett, D. (2014). New uses for a familiar technology: Introducing mobile phone polling in large classes. *Innovations in Education and Teaching International*, *51*(1), 46–58. doi:10.1080/14703297.2013.770267

Volery, T., & Lord, D. (2000). Critical success factors in online education. *International Journal of Educational Management*, *14*(5), 216–223. doi:10.1108/09513540010344731

Yourstone, S. A., Kraye, H. S., & Albaum, G. (2008). Classroom questioning with immediate electronic response: Do clickers improve learning? *Decision Sciences Journal of Innovative Education*, *6*(1), 75–88. doi:10.1111/j.1540-4609.2007.00166.x

Zhu, C. (2012). Student satisfaction, performance, and knowledge construction in online collaborative learning. *Journal of Educational Technology & Society*, *15*(1), 127–136.

Chapter 13
Conceptualizing Experiential Learning for Computer-Mediated Engagement

Himanshu Joshi
Teachers College, Columbia University, USA

Dominic Mentor
Teachers College, Columbia University, USA

ABSTRACT

This chapter aims to equip readers with a conceptual understanding to help them leverage experience-based learning in electronic (e) and mobile (m) learning environments. We are in times where learning goals needs to cater to increasingly complex scenarios that require non-didactic methods. Experiential learning emerges as a promising way to deliver such outcomes. David Kolb's experiential learning model emerges as a popular model to conceptualize such learning. E-learning and m-learning cater to the needs of an increasingly mobile learner who seeks situated and personalized learning. There is a need to incorporate experiential features in e/m-learning in a workplace informed manner. The authors weave the learnings from a pilot research and from real world examples to conceptualize a model of experiential learning for e/m-learning environments. The purpose of the models is to help learners critically evaluate other learning applications in digital environments or even design their own.

INTRODUCTION

Computers and mobile technologies have revolutionized multiple aspects of life in the 21st century. Hardly any aspect of life is left untouched by technology, and more so now with the ubiquity of mobile phones. From virtually hailing a cab, ordering groceries, interpersonal communication, entertainment, to managing our exercise and health. No wonder that mobile and computer-based learning is finding increased application in education (Naismith, 2004) and continues to offer revolutionary learning opportunities in multiple environments (Mentor, 2018). Using technology, universities and workplaces across the world are incorporating technology based learning to enhance their reach to learners, irrespective of boundaries

DOI: 10.4018/978-1-5225-4111-0.ch013

of time and geography. For profit companies like Coursera and Udacity offer Massively Open Online Courses (MOOCs), at times in partnerships with universities, and are looking at ways to disrupt the global education industry (Verger, Lubienski & Steiner-Khamsi, 2016). EdX founded by Massachusetts Institute of Technology (MIT) and Harvard in partnership with 90 universities and schools is also engaged in a similar pursuit through their offerings with free preview courses (Lazaroiu, Popescu, & Nica, 2016). Udemy, another player, claims more than 50,000 online courses on its platform where professors and instructors can design and offer courses at a rate of their own choosing (Epelboin, 2017; Gaebel, 2014; Täuscher & Kietzmann, 2017). Some claim that workers are increasingly adopting MOOCs as a route to further their professional development (Yuan, Powell & Cetis, 2013; Savino, 2014). Learning and development departments of workplaces across the globe are increasingly incorporating digital learning in order to provide on-demand, ongoing, self-directed learning aligned to the needs of rapidly changing workforce needs, often at lesser costs compared to face-to-face delivery models (KPMG: Wildi-Yune,J., Cordero, C. (2015), Deloitte: Pelster, Haims, Stempel & Vyver, 2016).

This chapter also acknowledges that many adult educators in schools and workplaces try to comprehend e/m- learning through models based in traditional education models. We believe that conceptualizing computer-mediated learning rooted in experiential learning, will be of great support to educators who are looking to embrace computer based learning from a lens they philosophically align with.

This chapter shows how computer-based learning is evolving beyond the conventional presentation of a static, mono-directional content approach. It also presents theories, research, as well as practical pedagogical and andragogical examples. The chapter will highlight that despite the evolution of mobile and eLearning, computer-based learning still needs to become more versatile in a multi-screen world.

Digital Learning a Norm Requiring Quality Informed by Educational Objectives

Online learning can take various forms. In its earliest forms, it started as converted correspondence courses (Ko, & Rossen, 2017) and others such as those delivered over the internet by for profit universities. Today, online learning is popular in the form of Massive Online Open Courses (MOOCs) and Small Private Online Courses (SPOCs) (Kaplan & Michael, 2016). According to data collected by Class-Central, a leading MOOC aggregator, over 700 universities across the world offered 6850 MOOCs in 2016 with 48.3 million users across the top five providers- Coursera, edX, XuetangX, FutureLearn and Udacity (Shah, 2016). This is over and above Small Private Online courses (SPOCs) delivered by universities and organizations over learning management systems (for example: Canvas, Moodle and Schoology to name but a few) for closed groups of learners in the form of stand-alone, blended and/or flipped learning programs (Bonk & Graham, 2006; Friesen, Norm, 2012) where learners can interact with course material before a course or class commences. Apart from computers, handheld computers, MP3 players, notebooks, mobile phones and tablets are also part of this learning revolution and are usually classified under m-learning (Trentin & Repetto, 2013). The terms online learning, digital learning and e-learning are used interchangeably in this chapter as the differences aren't significant to the research in this chapter. M-learning is considered unique given the element interactivity based on form factors (Chen, Woolcott, & Sweller, 2017; Mentor, 2011; 2016). Thus, while m-learning falls under the same umbrella of online learning, but with full appreciation for its uniqueness.

However, there are also educators who are skeptical about technology aided learning because they believe it de-humanizes education, leads to excess stimulation, borders on entertainment, and comes at the cost of *authentic* and *natural* ways of learning (Buckingham, 2013; Okan, 2003; Selwyn, 2016).

The educational technology skeptics seem to suggest criticisms of education technology and can be described as solutions in search of a problems (Selwyn, 2016). Based on the above, we surmise that as digital learning starts to become a norm, these criticisms call for deploying approaches that addresses concerns pertaining to quality of learning on digital platforms.

Impactful Digital Solutions Informed by Principles of Experiential Learning

Experiential learning can simply be defined by the two words it consists of, hence as a process of learning through experience, but as Patrick (2011) more accurately unpacks it, experiential learning requires reflection as part of that 'doing'. Experiential learning has been considered a superior learning method that fosters better understanding since Dewey (1938) extolled its virtues. Reflective experiential also addresses learning preferences of diverse learners as compared to a one size fits all-didactic way of teaching. This form of learning has been adopted in various settings such as K-12, higher education, workplace learning as well as self-directed learning (Andersen, Boud, & Cohen, 2000). Sometimes experiential learning and e/m-learning are considered opposites, especially when they are associated with hands-on learning and virtual learning respectively. However, there are several examples that enable us to see that these can be the same or complementary. For example, medical practitioners often learn complex surgical procedures that are both technology enabled and experiential. McKinsey Capability Centers use such experiential simulations to build new skills and attitudes as they prepare their clients for transformations (Benkert, Dam, 2015). According to a Deloitte study (Pelster et al., 2016), learning and development departments of tomorrow will provide learning that "is experiential, relies on case studies and flipped classrooms…from today's expert delivered lecture-based models" (p.60). With proven success of experiential models in classrooms and inevitability of digitization of learning, there is an opportunity to see how e/m- learning can be informed by principles of experiential learning or rather how technology solutions can help enhance the impact of experiential learning.

This chapter aims to equip educators, curriculum designers and instructional technologists with a perspective that they can use to foster experiential learning in a computer aided learning environment. We start by offering definitions of learning from various lenses notably experiential learning and e/m learning. We subsequently present examples of computer-based learning that integrate elements of experiential learning. We have chosen examples where we have been personally involved as designers, researchers and users, even though we realize that there could be better examples of computer based experiential learning. We tie our learnings from these to present a model derived from David Kolb's experiential learning model (Kolb, 1984, Kolb & Kolb, 2012). We hope that by the end of the chapter, learners, teachers, corporate trainers or professional developers will be able to start critically evaluating e/m-learning designs and be better equipped to develop experiential e/m- based learning programs.

There are several important and related questions, notably regarding the suitability of certain contexts for technology in learning and conceptualization of experiential learning itself. For reasons of scope and focus we have chosen to stay away from these based on our assumptions that e/m- learning is here to stay and experiential learning method, of which Kolb's conceptualization is a good representation, works in the classroom. And of course, with mobile learning, it works outside of the classroom and breaks down the literal and artificial divide between learning institutions and the outside world. We hope the reader will be able to gain perspectives on the unanswered questions elsewhere, inside or outside of this book, and finds the theoretical framework useful for their own computer mediated learning contexts.

CONCEPTUAL THEORETICAL FRAMEWORK

In this section we offer concepts and theories to help the frame the understanding of experiential learning in computer-based learning systems. Goals and objectives sometime falls by the wayside when integrating computer mediated learning Goals and methods of learning have evolve over time to cater to changing needs of the society, economy and polity. One way to look at this evolution is through the theoretical schools of learning: Behaviorist, Humanist, Progressive and Radical (Elias & Merriam,1995). Behaviorists such as Guthrie, Hull, Pavlov, Skinner, Thorndike, Tolman and Watson looked at deploying the process of stimuli and response to encourage desired behaviors (Merriam, Caffarella, & Baumgartner, 2012). Taylor's model of learning (Ralph, 1949) based on behaviorism found much use in workplace learning when organizations were responding to the needs of an industrializing society that needed to create a large workforce performing standard tasks and behaviors required by the employer. As society and industry evolved, need was felt for a citizenry and workforce that could think for itself, be innovative and go beyond a regimen of drill and practice. Consequently, organizations moved beyond principles of scientific management, also called Taylorism, and started shifting to knowledge and team-based work (Kasworm, Rose, & Ross-Gordon, 2010). Considering this shift, human resource development started to incorporate ideas from other schools, notably the humanist and progressive schools (Watkins & Marsick, 2014).

Emancipatory Educational Approaches to Promote Agency

As adults develop and live in a more complex world, they need learning that goes beyond the mere acquisition of knowledge, towards critical processing or transferability of that knowledge to multiple domains and the ability to critically reflect on self in such systems (Kegan, 2009). Hence, as the world continues to change at a faster pace catalyzed by the advent and rapid development of technology, the nature of learning goals will continue to evolve. In some ways, this is already the need of the hour as the challenges we face in the twenty-first century are getting increasingly complex and adaptive (Heifetz,1994; Heifetz & Linsky, 2014). In ambiguity filled scenarios it is often difficult to define learning goals with the specificity when educators using models such as Blooms Taxonomy (Bloom, Krathwohl & Masia, 1984). The Glossary of Education Reform defined Bloom's taxonomy as "a classification system used to define and distinguish different levels of human cognition—i.e., thinking, learning, and understanding" (Great School Partnership, 2014). Given the huge differences in contexts and challenges faced across a diversity of communities in an interconnected world, it is also reasonable to argue that we cannot have one school of thought that can guide us to define learning goals and processes. While we will continue to have leadership-education based on humanist and progressive tradition (Watkins & Marsick, 2014) that prepares leaders for workplaces, we will also see emancipatory education (Freire, 2000) based on deconstructive education models (Custer, Deutscher, & Haddad, 2016; Dahlgren, 2017; Peters, 2016) to empower impoverished communities and intellectually impoverished individuals. This means that an educator in today's world will need to not only identify the learning goals, but also consider the culture while choosing methods, tools and techniques to deliver learning. Likewise, educators, corporate trainers, and researchers should be mindful of the variety of contexts within which their customers operate while proposing learning models and defining the scope and boundaries of their model. With this backdrop, we humbly submit that our chapter has been written in the current digital milieu of higher education settings and workplace learning. We would be delighted if educators outside these domains find our ideas meaningful. Therefore, this chapter describes a conceptual framework that should help develop a better latitude on understanding learning and in particular, computer mediated learning for workforce development.

There are several definitions of Learning. Jack Mezirow (2000) classified learning into three types: Instrumental or technical, communicative and transformative. Technical learning focuses on gaining knowledge, information and skills. For example, learning a language, plumbing, carpentry and so forth. Communicative learning focuses on building understanding of self and others, and transformative learning focuses on becoming aware of one's own limiting assumptions and developing more permeable meaning making schemes. Kegan (2009) uses the words technical and transformational learning to differentiate learning into two types. The latter pertains to helping people identify their assumptions and be better prepared to deal with complexity, an increased feature of our times. According to Kegan (2009), Technical learning is about gathering more information, while Transformational learning is about how we see the information that we have. In the practical world learning situations are a blend of technical and transformational goals. Watkins, Marsick and Faller (2012) make a case for transformative learning approaches in workplaces to better address today's challenges that are rooted in greater ambiguity and complexity. They suggest mechanisms such as action research, action science, action learning, and cooperative inquiry, which have a common dynamic of "collective identification, sharing, and examination of different points of view" (Taylor & Cranton, 2012, p.380). It's not difficult to argue that learning goals—which fall into the realm of dealing with complexity and ambiguity—require methods that go beyond facilitating a one-way broadcast of information. Even if the learning goals are largely technical, the learning can be strengthened if the process provides for the learner to engage in activities, and exchange ideas with other learners rather than simply accepting information from an expert. In this way the learner will be better placed to understand the knowhow and create his/her own solutions. From the perspective of cognitive capacities, learning gets enhanced if multiple modes are used as the cognitive load gets distributed across various modes so that learners get to engage in their learning in their preferred way (Plass, Moreno & Brünken, 2010; Barbe, Milone, & Swassing, 1979). We believe that these ideas of using experiential learning and multiple modes should hold or pursue to hold to be relevant whether we are using a face-to-face methods or e/m- learning environments.

In the western world, John Dewey is considered one of the earliest and foremost proponents of experiential learning. He posited that, "…all genuine education comes about through experience" (Dewey, 1938, p. 25). Various models of learning from experience, have subsequently evolved that capture a host of issues involved in learning. Andersen, Boud & Cohen (2000), consider the Montessori method of teaching as essentially an experiential learning method that "supports a more participative and learner-centered approach. (p.225)". They highlight the importance of reflection and attending to feelings during and after the experience (Boud, Keogh & Walker, 1985, 2013; Kolb,1984, 2014). In light of these more dynamic, participative and learner centered approaches, experiential learning has been extensively deployed to achieve goals that seem technical, communicative and possibly even transformative.

Traxler (2007), states that e-learning and m-learning are different. He refers to the traditional virtual learning solutions that use online computer-based platforms as e-learning. Such learning is seen as *structured*, *media-rich*, *broadband*, *interactive*, *intelligent*, and *usable*. On the other hand, he sees m- learning as that which is comparatively more "personal, contextual, and situated" (p. 1). He further mentions that advancement of technology will blur these differences, which will unbundle e-learning and bring it closer to descriptions associated with m-learning.

In that vein, Sharples, Taylor and Vavoula (2007) propose a wider conception of mobile learning as that which is not only supported by mobile devices, but also has mobility as a characteristic of learning as the user today is mobile. They talk about the emergence of *new learning* and *new technology* while describing mobile learning. They also suggest that such learning embeds ideas of personalization of learning, anytime-anywhere learning as well as collaboration and connection opportunities between learners. They define mobile learning as "the processes of coming to know through conversations across multiple contexts amongst people and personal interactive technologies" (p.224).

The Case and Concerns About E/M-Learning Rooted in Experience

Carrington (2013) re-conceptualized the pedagogical wheel and laid out ways in which computer-based learning has benefits beyond enhancing content presentation or automating processes of learning delivery. He emphasizes the fact that mobile devices offer opportunities to redefine learning beyond substitution, modification, and augmentation. For example, personalization of learning could be one such opportunity. Researchers, and educators are seeing increased integration and convergence across platforms, often enabled by cloud-based platforms. Cloud computing platforms provide cost efficient access to the critical infrastructure needed for these types of solutions. With both computer and mobile device technological advancements, the lines of separation are growingly overlapping and learns switch back and forth between computer and mobile devices. And with wireless advancements, there are overlying spaces where conversations between educators and learners are facilitated through the use of technology in a manner where learning is more personalized, situated, and connected to the context, experience and learner. In this conception, both historically and futuristically, technology facilitates access, conversations, interactivity and enhances experiences (Sharples, Taylor & Vavoula, 2007).

With this futuristic view of blurring boundaries between e-learning and m-learning as well as the emergence of integrated platforms, we wish to define e- learning and m- learning under the same umbrella of e/m- learning for the purposes of this chapter. Historically, computer-based learning started with applications to achieve technical learning goals. Today a large part of technical learning in organizations is being offered via e-learning modules which could facilitate more than transferring information, but also involves teaching skills, and testing knowledge. However, there is still an increased need to research and figure out how computer-based technologies can lend themselves to purposes beyond transferring or testing the absorption of content (Richards, 2012) and increase the transferability of knowledge to other contexts (Mentor, 2016). As an example, workplaces are using e/m-based learning for teaching leadership; a process that moves into communicative and transformative realms. Several collaborative tools (e.g. Slack, Zoom, Skype, Webex) are also helping practice effective team-ship and leadership. However, we believe that there is still a huge potential of e/m- learning to address communicative and transformative goals lying untapped. We believe incorporation of experiential elements will be a key to realize this potential.

Mobile phone devices today are loaded with a host of functionalities. With today's computers and phone screens, you can use touch, sight, and sound at the same time, run multiple applications, navigate in non-linear fashion, and connect to a community of learners and educators. At the same time the multifunctional ties of these devices are being leveraged to enhance learning experiences. As mentioned earlier in the chapter, there are also concerns about the potential of these devices to distract learners. The possibility of distraction and over stimulation at one end and the opportunity to offer an experience that may not be possible in traditionally designed teaching, training and learning settings, ups the ante for UI/UX professionals, instructional designers, and educators using these platforms. We illustrate a few examples where e/m- learning, in our view, has been meaningfully used to create experiences in the service of learning that is personalized, situated, contextual and blurs the line between formal and informal learning. We hope these will seed some ideas for our readers before we present our own conceptualization of e/m- based experiential learning cycle.

It is encouraging that educators, UX/UI designers, technologists and instructional designers are incorporating tools and techniques to make learning more hands on and experiential. Models such as Rogers' diffusion of innovation (Rogers, 2010), CCNMTL's (2013) learning design model informed by Bereiter (2002), Edelson, (2002) and, others (for example: Bannan, Cook, & Pachler, 2016; Obrenović, 2011; Van der Stappen & Zitter, 2016) theorize putting into practice, how an educator or instructional designer could conceptualize and design such learning. At the same time, many adult educators in schools and workplaces try to comprehend e/m- learning through models based in traditional education models. We believe that conceptualizing computer-based learning rooted in experiential learning will be of great support to educators who are looking to embrace computer-based learning from a lens they philosophically align with.

As previously indicated, this chapter provides a vignette of some examples that build a case for experiential learning model for computer-based learning. The next portion of the chapter elaborates how computer-based learning is evolving beyond the conventional presentation of a static, mono-directional content approach. We also present research to show that despite the long-term evolution of e-Learning, computer-based learning still needs to become more versatile in a multi-screen world. This is of consequence as contemporary learners are prone to even more hypermedia distraction, and at the same time prefer to use micro-moments of learning. We believe these ideas can be explored using Kolb's learning cycle as an anchor.

Creating Transformative Experiences Through Knowledge Generation and Transfer

David Kolb's cyclical learning model is one of the most widely used concepts of experiential learning. As per Kolb (1984), "Learning is the process whereby knowledge is created through the transformation of experience (p.38)." David Kolb describes experiential learning cycle that comprises of the steps of *concrete experience(CE), reflective observation (RO), active reflection (AE), abstract conceptualization (AC),* and *active experimentation (AE).* In other words, a cycle of learning requires that a learner reflects on an experience, which can lead to meaning making and conceptualization of that meaning. This is followed by active experimentation where one might explore possible applications of an idea which informs further experiences and meaning making.

Figure 1 depicts the four steps in Kolb's Experiential Learning Cycle (Kolb, 1984). Kolb also mentions that learners have their own preferred learning styles formed because of an interplay of preferences along the CE-AC and AC-RO axes. Hence, when they learn, learners go through the entire cycle, but each has her preferred step or location in the cycle. Kolb names four unique learning styles and hence four types of learners: Divergers, Assimilators, Convergers and Accommodators. We believe that Kolb's learning cycle and the idea of diverse learners can be a useful idea to guide the design of computer mediated learning for workforce development. If we are talking about a conversational approach facilitated in mobile learning, then Kolb's idea of meaningful experiences reflected upon in conversation with others seems a powerful idea to leverage in e/m-learning design.

Figure 1. A representation of Kolb's (1984) Experiential Learning Cycle

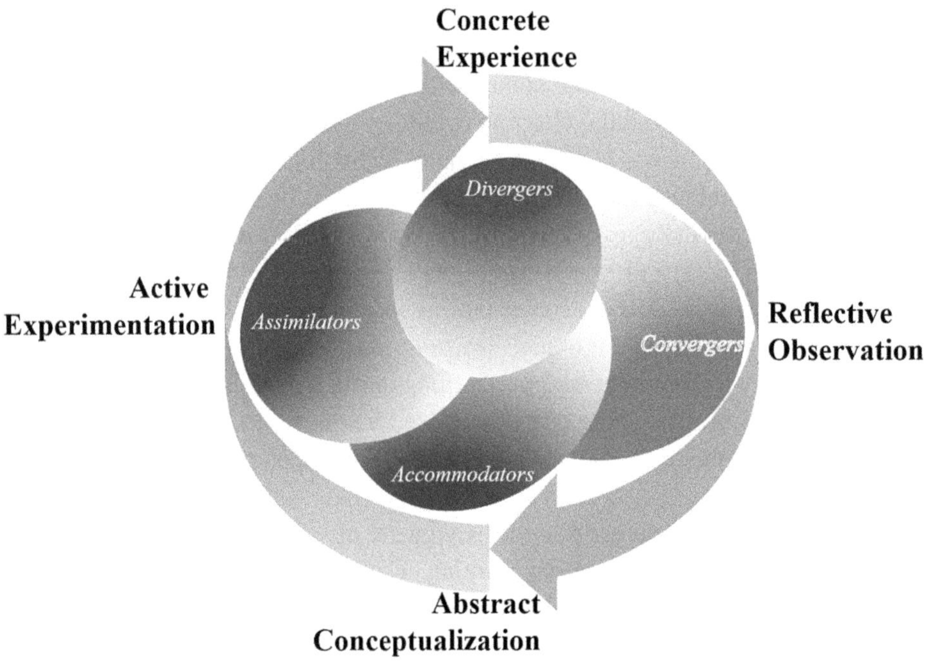

EXAMPLES OF COMPUTER BASED EXPERIENTIAL LEARNING

We now present three examples of computer mediated learning that we believe are exemplars of experiential learning. The first example pertains to an online leadership development course offered at a leading business school, the second explains a few mindfulness teaching apps developed for mobile devices, while the last example focusses on employability skills development as part of a workforce training organization. The first two examples provide a micro picture of experiential learning design while the last example provides a macro picture of a program that used experiential elements.

Online Executive Education for Leadership Development

A leading business school offers online leadership courses targeted at working professionals. The courses already have face-to-face instruction-based versions that continue to be popular with MBA students and corporate executives. The online courses were developed to expand reach of these courses to audiences who couldn't attend these courses for reasons of time, distance and money. Later on, it emerged that online modules could also enhance impact of the face-to-face versions and today, many face-to-face programs have an embedded online component as well. The courses are exemplars of how the positive experience of an in-person program can be retained when adapted to a digital format.

A typical online course gives the learner access to an e/m platform. Thus, the learner can access the content on both a smart phone or a computer. There are various aspects of the course that ensure that the instruction is not reliant on one-way communication. The professionally filmed lecture videos are the centerpiece of experience. These videos have the professor share interesting stories, cases, thought provoking information, situations, and questions. The questions posed at the end of videos become reflective stems. Participants post their reflections on a shared forum, thus getting an opportunity to learn from each other's perspectives. Key concepts are shared through videos or Portable Document Format (PDF) files that participants can download for future reference. Post learning the key concepts, the professor poses application-based questions and/or assignments that help participants apply learning in real life and facilitators assess impact of their teaching. One remarkable example of such application is role-play assignments. Participants watch video role play situations, built around their work situations, and are asked to respond as one of the actors. They video record their responses and share them with computer assigned peers for assessment and feedback. The peers provide quantitative feedback on a ten-point scale on pre-defined dimensions such as clarity of communication, strength of arguments and so forth. They can add Qualitative comments that comprise open ended feedback. The application pushes notifications to let participants know when they have received feedback. Notifications are also used to remind participants of upcoming tasks, new posts on the forum that might interest them and the release of new modules. You can probably recollect that typical leadership programs conducted in a face-to-face format has presentations by facilitators, exercises, reflective pauses, assignments, and role plays. These online courses mirror such elements using a judicious blend of experiential elements in an e/m learning environment. In addition, the format offers the benefit of anytime-anywhere access, opportunities for learners to study at their own pace and review content multiple times, something which becomes challenging in a face to face program. Despite having a pre-designed structure and flow, the online course offers greater flexibility and choice to the learner compared to a face-to-face program. Another aspect of the application is that the entire learning interaction is recorded on the shared platform for post program reference and reflection. This is a rich source that can also be used to inform follow-up assessments and exercises in three or six-month time to take steps to ensure that the program meets its goals.

Meditation and Mindfulness Training App

Mindfulness training has become popular in the west as a means of beating stress and training attention control. While earlier influences and applications emerged from yogic and meditation practices of the east, it gradually entered the world of scholarship in mental health notably through the work of Kabat-Zinn (2003). A meta-analysis by Khoury, Sharma, Rush, and Fournier (2015) examined 29 studies and concluded that mindfulness-based stress reduction is moderately effective in "reducing stress, depression, anxiety and distress and in ameliorating the quality of life of healthy individuals" (p.1). Today, meditation and mindfulness applications have applications beyond mental health to address needs of healthy individuals. While workshops and in person classes have been the popular mechanisms to deliver mindfulness learning and practices, mobile mindfulness applications (apps) are increasingly becoming popular. In a preliminary review of a few applications, the authors found that applications to teach mindfulness practices rely heavily on guided instructions, or videos making it a very passive and didactic process. This experience is validated in a review of various mindfulness apps by Plaza, Demarzo, Herrera-Mercadal, and García-Campayo, (2013) where they mention that these applications are mostly instructional and non-interactive and that there were opportunities to leverage social learning and to track data regarding practice and health benefits.

However, a leading mindfulness app has several features that address issues raised above. While it continues to have recorded animated videos that teach basic concepts, in contrast to several other applications that offer a one-size fits all model of meditation, this app offers an array of options based on the meditator's needs. It asks the learner a series of questions about her mood and feelings and based on responses, suggests a few meditation options to pick from. This supports personalization of the mindfulness practice. The guided meditation provides the learner detailed instructions and options to vary the length of time devoted to the practice. For advanced learners, it offers options to self-time and meditate without guidance. The application helps the learner to keep a record of their meditation sessions and pushes daily reminders to promote a regularity in one's practice. It also integrates with a tracking application in the phone to record health data by capturing active use of the app and subsequent activity related to the app. In another application with a similar approach, the learner has the option to regulate length of one's breathing cycles of inhalation and exhalation. This is aided by the timed dimming and brightening of the screen, which is based on learner inputs. Use of this feature avoids a disturbance one might experience in applications based on vocal guiding instructions. Such an application helps the one learn meditation in ways that mimic face-to-face methods or even surpass them in certain outcomes. Such applications, while not replacement for a personal trainer or a group class, have several features that help the learner experience meditation, track their progress, and seek expert inputs in their own contexts outside the classroom. While there is still a long way to improve functionalities, like communication with other learners to seek support, these apps have surpassed earlier versions of computer mediated meditation training, which were simply recorded videos of an expert with scrolling text—and/or accompanying booklets—summarizing key steps. They are also more visually appealing with digital enhancements such as high pixel graphics, clear sound and seamless video.

Employability Skills for the 21st Century Economy

An innovative social enterprise Empowered Professionals for Internship and College Credits (EPICC) secures internship opportunities with corporate partners for young adults from predominantly disadvantaged backgrounds. Its educators have believed in the power of experiential learning to foster meaningful skill building to empower these young adults. While undergoing the training, the interns get trained in professionalism and career development, and are held to an employability skills development contract. Earlier, the training program ran for six months conducted through in-person training and followed by six months of internship. The learning and development (L&D) courses were redesigned to solidify EPICC's high expectations and high support model and to go beyond traditional higher education objectives to include actual experiences that trainees would encounter during the internship. The practical exercises of the courses developed an experiential knowledge that would help convey possible career trajectories with opportunities to hone their employability. Digitization was an important aspect of the redesign process and technology was leveraged to support the trainees, interns and alumni. Social connectedness formed a part of the experiential learning and theoretical framework to cultivate communities of practice (Mentor, 2017). Prior to digitization, experiential elements of internship training were offered in a cyclical fashion of modules and programmatic elements. In that model, students were highly reliant on physical textbooks and were time bound at the physical location to print and submit their work.

After digitization of the training program, the trainees could also earn college credit recommendations for the training and the internship experience. For example, as with all the courses, a professional skills course was rebuilt for college credit recommendations and designed for e/m- blended learning delivery and engagement. Workplace scenarios were embedded and integrated throughout the training and students were trained. For example, on how to overcome challenges of punctuality, meeting work assignment deadlines, committing to workplace level professionalism, delivering work at appropriate quality levels, and develop confidence during their L&D for the internship phase of the program and beyond. Online access, interaction, and virtual learning communities of support offered real time support and data. The latter strengthened the visibility and transparency of the trainers' and trainees' work. Leveraging mobile technology in hand, both instructors and trainees were also provided a network of supportive relationships. The digitization of the learning community further fostered a powerful sense of agency for the students—encouraging them to be active agents in their workforce learning and development, internship and preparing them for a career path. The virtual communities of practice were formed through career track sorting after a foundational phase of training as well as through in-person interactive exercise. The sense of belonging and community affiliation was further promoted through an 'academic facebook' style learning management system. Trainees could message one another and their instructors to bond, query a concept, or just ask a question in general.

Since the adoption of the academic vision and strategy, with e/m-learning as an integral tool in support of those efforts, the academic program could now offer valuable data learning analytics faster than it could before, and be replicated and scaled faster as a model to other city's training centers. This tracking also helped the educators assess where learners' levels were in knowledge, skills building and transference of knowledge and skills, and offer them appropriate support to further their learning. The addition of the digital engagement enhanced the organization's traditional model and helped better prepare the trainees by offering multimodal stimulation which could be revisited, shared and offer real time and asynchronous collaborative learning opportunities.

The L&D organization won an Optimas Gold Award from Workforce Magazine for enacting this e/m-learning vision in 2014, and an award for best blended learning from the International E-Learning Association for a number of its own built curricula and courses.

Across the three examples above, the underlying learning elements closely connect with experiential learning. Table 1 lists certain connections we saw between elements in the above three examples and Kolb's (1984) learning cycle organized by underlying steps.

If we look across the above three examples, we can conclude that the learning goals and context of the learner contributed to the choice of the learning elements, nature of conversations and process. Each of the examples demonstrated how learning design moved beyond knowledge transfer to one in which there was an opportunity for the participant to engage with the subject, experience personalized learning, and engage outside classrooms in situated contexts. These could be construed as cases of Communicative Learning if not Transformative (Mezirow, 2000). While these apps encouraged learners to construct their own meaning, there remained certain aspects of learning that stayed constant and at an informational level. These include the specific techniques in meditation app, the knowledge for the ideas about professionalism such as punctuality in the EPICC examples, and key principles and tools on which the leadership

Table 1. Experiential elements in examples of e/m learning

	Concrete Experience	Reflective Observation	Abstract Conceptualization	Active Experimentation
Online Executive Education Program	Video cases and stories, Feedback	Discussion boards	Video lectures, Tool, Tool kit	Quizzes, Role plays
Mindfulness Applications	Guided meditation	Mood prompts	Guided meditations, Video lectures and animations	Self-timed meditation
Employability Skills	Workplace scenarios	Virtual Learning communities, Feedback	Online content, Messaging feature	Practical exercises

development program was based. Across examples, assessment and tracking was a key feature. Learners were helped to become aware of their progress and use that to guide further learning and improvement. Additionally, even though the applications used e/m-technologies, they strongly incorporated *doing* or *experimentation* connected to real life situations as a learning element with examples ranging from customer service trouble shooting, offering mock interviews with people from industry, and unpacking real world business examples. The meditation practice, role plays, and assignments are examples of learning by doing. The applications also differed from each other in many ways. In the meditation app and leadership training app, the entire course was offered over the e/m- platform whereas in the L&D example, the e/m- features were blended with face-to-face elements of the internship. The differences between the applications can be attributed to difference in technical capability available to the designers, their own philosophy of teaching, and the context for which they were developing the applications.

Hopefully, the above examples help to see emergent themes that will inform our conceptual model for learning programs using e/m- methods that incorporate experiential elements. We now move on to describe our pilot research before showcasing our conceptual model mentioned above.

MOBILE ENHANCED APPLICATION PILOT TO TEACH MINDFUL BREATHING

We tested the concept of our research using a pilot model of a mobile app that teaches the regulation of breathing because breathing is one of the core processes across various mindfulness techniques. Breathing has been identified over millennia by ancient cultures ranging from Africa's indigenous knowledge, to India and Chines ancient cultural practices as is evident in such practices as chanting, yoga, tai-chi, body-oriented psychotherapy, body awareness therapy, mindfulness-based therapies, meditation, Alexander technique, and breathing therapy (Wahbeh, Elsas, & Oken, 2008; Ospina et al., 2008). According to Bennett (2011), "although diverse in their specific orientation and application, these fields all share the common belief that the body, mind, and spirit are inherently interconnected and utilize the body as a resource for learning, spirituality, wellness, personal growth and development, and diverse methods of healing" (p.128). Learning to regulate attention focused on breath as part of mindfulness techniques can help us achieve psychological well-being (Wielgosz, Schuyler, Lutz, & Davidson, 2016; Fernros, Furhoff, & Wändell, 2008; Krygier et al., 2013).

Since our primary goal was to test experiential learning in an e/m-learning environment, our learning design incorporated all aspects of Kolb (1984)'s experiential learning cycle-experience, reflection, analysis and experimentation. We also incorporated elements of assessment that could serve the purpose of tracking and motivation.

The goal translated to the following objectives for the process:

1. Leverage experiences to teach aspects of breathing.
2. Facilitate reflections through technology enabled conversations (conversation theory; Pask, 1976).
3. Limit instructions to the conceptualization process. (this was both an objective and a parameter of the exercise)
4. Enhance learning by assessment:
    a. Making learning goals explicit to learners (Richards, 2012)
    b. Appraise one's breathing patterns to inform corrective action

The design was tested with graduate school students and their responses analyzed through qualitative thematic analysis (Braun & Clarke, 2006). The research questions concerned their views about efficacy of the process, ways in which the process was superior or inferior to conventional face-to-face or linear instructor lead designs, and design elements they saw as most useful in furthering their learning. Overall, the learner-users preferred the proposed design as compared to conventional and linear approaches. No single feature or element emerged as a winner. Based on the following observations, we interpret that learner-users latched on to different elements (videos, reflections, or practice) based on their individual learning preferences.

They mentioned, as other key benefits of the approach:

- Flexibility of navigation an exploration
- The choice of skipping uninteresting elements.
- Learning though different elements appealing
- NO need or coercion to follow all the elements in the sequence presented in the application.
- They could spend greater time in elements that looked meaningful.

For example, some learner-users did not see meaning in trying to understand how others experienced the meditation process. Instead, they were keen to learn the tips and tricks shared by the expert and practice. Others wanted to practice and reflect on what difference it made and how they could enhance their practice. These learner-users mentioned shortage of time as one of the reasons why they would not go through all elements, but rather pick the one they thought were most useful. This phenomenon is a small example of how an experiential learning cycle creates the space for learners (who are also users) to step into the learning using methods they are more comfortable using, prior to stretching themselves in another part of the cycle.

ELUCIDATING THE JOSHI-MENTOR MODEL

This next section will tie the concepts together and distill this discussion down to a few key ideas. Our separate research of implementing experiential learning cycles, and collaborative reflection, led to the creation of the Joshi-Mentor model, which is depicted in Figure 2, and was developed based on concepts used in the chapter. The model has several features of Kolb (1984)'s experiential learning model. It has tasks that facilitate active experimentation and/or concrete experience. The learner's experience often captured through assessments or reflection stems, facilitates RO. The reflection component could take the form of individual or community-based reflection in online or face-to-face contexts. The conceptualization phase could happen through content transfer using videos from experts, and subsequent discourse amongst peers. In addition, the computer-based tasks often incorporate assessment, feedback and/or tracking. We also introduce the added dimension of time as a global factor surrounding all elements in this model as there are many underlying time dependent variables: Access to the device, availability of time and synchronicity, or a-synchronicity of learning design that guide the app.

E/m environments also facilitate hypermedia technology, i.e. flexibility of not following a rigid process but navigating based on interests, which becomes difficult in classroom experiential workshops. E/m learning makes it feasible to allow the learner to enter anywhere in the learning cycle rather than necessarily start with an experience. The authors of the chapter believe that the freedom of choice depending on preferences, temporality, a/synchronous choices of engagement, or other considerations and constraints, is a definite advantage of experiential learning in computer-based environment based on, but not limited to the options listed in this sentence. Further, the cost of providing such learning is less than providing face-to-face learning especially if one needs to transcend significant boundaries of time and space. Hence this becomes appropriate for deployment where high quality has to be provided on scale.

A brief description of the individual components follows:

- **Learning Goals and Context:** Experiential learning can rely on past experiences of the learner, use structured experiences designed to achieve specific goals, or a combination. With the current stage of technological development, most e/m- learning is likely to focus on specific knowledge, skills and attitudinal goals. These goals will influence the nature of the components placed in the periphery. Similarly, the context of the learner will influence all the peripheral elements.
- **Time:** Learning or the facilitation of learning could be synchronous or asynchronous. Learning design needs to be informed by how the tech-educator envisions or anticipates the learner will use the application. Would the learner be able to undertake bite-sized learning in small chunks or would the learner need to carve out extensive durations of time? These aspects too will influence all the components placed in the periphery. It is also possible to imagine that time is a vertical axis placed perpendicular to the figure. This makes it possible to conceive that the learning goals and hence the nature of E/m-aided tasks will change with time as the learner progresses to advanced levels of understanding.
- **E/m-Aided Tasks:** These tasks facilitate engagement of the learner with the subject matter. They could be considered akin to experiences that are designed to facilitate achievement of learning goals. Depending on the purpose of e/m- learning, these could take various forms such as guided

Figure 2. E/m based Experiential Learning Model developed based on mindful breathing application pilot study and Kolb's experiential cycle

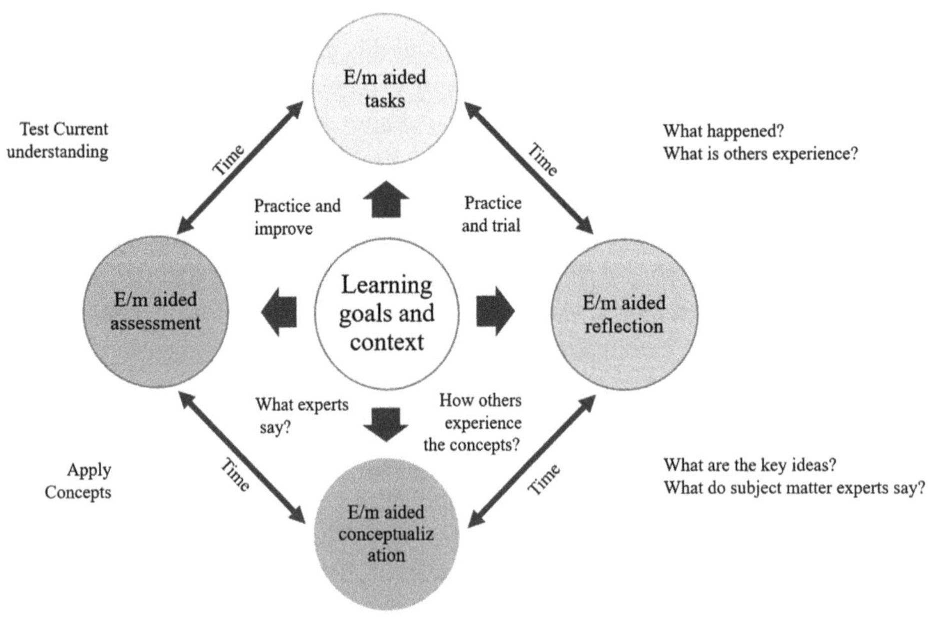

meditation, puzzles, games, simulations, audio-video cases or challenges. The tasks are closely connected to other elements in the periphery and often the integration may be seamless. For example, a game (the task) and its scoring (assessment) may be experienced by the learner as one and the same. However, it is important for the tech-educator and the designer to be informed of the sub-component purposes (assessment, experience, reflection, conceptual input) underlying an integrated task.

- **E/m-Aided Reflection:** At this stage, a learner is trying to make meaning of the task and/or the conceptual input. Social connectivity through computers and mobile devices offer the fascinating possibility of this meaning making being informed by experiences of fellow learners in the community. As an example, this could take forms such as a discussion thread to share how a suggested guided meditation played out for learners, what learners think of a communication technique offered by a professor in an instructional video, or learners responding to video recordings of each other posted on the e/m- learning platform. While the reflection can be individual or group, in case of communicative and transformative learning where questioning and challenging of one's perspective in dialogue with others is involved, social reflection becomes a necessity.
- **E/m-Aided Conceptual Inputs:** It emerged in our pilot study and examples that given the situated nature of such learning, learners expect expert inputs and concise presentation of key concepts and ideas that they can immediately apply. This often takes forms of instructional videos, audio-visual presentations, or text and visual models summarizing key ideas, rules and concepts. These conceptual inputs can also become the triggers for subsequent reflection or assessment.
- **E/m-Aided Assessment:** Learners wish to be informed about their progress. Health management applications are great examples wherein they track and present key statistics which act as motivators and metrics. E/m- learning applications not only capture the digital footprint of the learner within the app (time taken to accomplish tasks, accuracy etc.) but also the context in which learning took place (what was the learner doing while learning, which places was the learner in) using the telemetry-based affordances of mobile devices such as Global Positioning Systems (GPS) and other location features like Near Field Communication (NFC) for situated cognition). These possibilities can help draw fascinating insights on aspects such as learner preferences and styles that can help individual learners and tech-educators. Assessment also serves to motivate the learner. Educators can incorporate reward and recognition elements to enhance motivation.
- **Connections and Flow:** Our pilot revealed to us that e/m- learning offer the possibility to not be stuck with any sequence of the various components. Such flexibility can be a distinct advantage of e/m- based learning as compared to non-digital formats of experiential learning. A learner should be provided options to enter the learning cycle at places according to her preference, spend greater time or even repeat components that support better learning. Each of these components can create bi-directional triggers towards other components as depicted in Figure 2.

CONCLUSION

Educators adapted experiential learning approaches to earlier models of learning to enhance the teaching-learning process and reach more learners. The authors of this chapter believe that computer-mediated learning also needs to become more experiential in a judicious and educationally informed manner. Flexibility, lower cost of experimentation (failure) and responsiveness to learners' learning preferences seem to be clear advantages over experiential learning in a classroom environment. Such an approach is being used for technical as well as communicative learning goals. We would be keen to see how this can extend to transformative learning goals. For the latter goals, it is imperative that technology help promote inter-connection between perspectives of different learners as that is at the heart of facilitating communicative and transformative learning (Mezirow, 2000) and help build the capabilities of hosting multiple perspectives beyond binary views of the world.

We explored examples of e/m learning from a micro and macro perspective. The workforce development, leadership teaching application and mindfulness application focused on experiential learning embedded in one application and interfacing with the learner while the employability skills example focused on a macro perspective beyond a single application in a blended manner with virtual distance learning components interacting with other processes in the organization. The examples mentioned in this chapter will need to be revisited as technological innovations will continuously impact these target audiences and influence the contexts. From these examples we concluded that models rooted in basics such as Kolb's model can be effective frameworks to inform design approaches and shared our conceptualization of one such model as a possible takeaway for practical application. However, with rapid pace of change, these models will need to continuously evolve, and will require further research.

FUTURE RESEARCH DIRECTIONS

Technological advances continue to facilitate greater diffusion between the virtual-real boundary. We are going to see greater integration of mobile devises in all aspects of our life. Augmented and Virtual Reality are already seeing increased application in education settings, even as their efficacy is being debated as one would expect to happen whenever changes occur (Merchant, Goetz, Cifuentes, Keeney-Kennicutt, & Davis, 2014; Bower, Howe, McCredie, Robinson, & Grover, 2014). The learning machines will become more intelligent and this will lead to enhancement of personalization features. Greater convergence between machines and aspects of life, more availability of learner data too, will inform mobile learning. It might be feasible to identify unique aspects of a learning cycle that suit a learner, thereby letting one completely minimize or eliminate the other aspects. However, deep personalization also comes at a possible cost of learners getting used to learn in their preferred way but not developing the abilities to learn in ways outside their preference. This could lead to efficient learning in the short run but narrow learning approaches in the longer run that could filter out new learning possibilities or become limiting in certain situations. How flexible should such learning be and how can designers and educators balance the twin aims of greater personalization in learning methods yet retaining the stretch element is an issue we foresee requiring research. On almost the counter side of deep personalization of learning, social connectedness through computers and mobile devices require further research into the building of abilities to entertain multiple truths, discern tropes from truths, and further meaning making being informed by experiences of fellow learners in the digital community As mentioned earlier in the chapter, our conceptualization is based on our experiences limited to the contexts of our higher education, research and workplaces. Further research could validate the model and verify its applicability in other learning contexts. Finally, from our research and pilot we see that such models can facilitate technical and communicative learning. It still needs to be researched as to what extend and in what ways can e/m learning support transformative learning.

REFERENCES

Andersen, L., Boud, D., & Cohen, R. (2000). Experience-based learning. In Understanding adult education and training (2nd ed.; pp. 225-239). Sydney: Allen & Unwin.

Bannan, B., Cook, J., & Pachler, N. (2016). Reconceptualizing design research in the age of mobile learning. *Interactive Learning Environments*, *24*(5), 938–953. doi:10.1080/10494820.2015.1018911

Barbe, W. B., Milone, M. N., & Swassing, R. H. (1979). *Teaching Through Modality Strenghts: Concepts and Practices*. Zaner-Bloser.

Benkert, C., & Dam, N. (2015). *Experiential learning: What's missing in most change programs*. Retrieved from http://www.mckinsey.com/business-functions/operations/our-insights/experiential-learning-whats-missing-in-most-change-programs

Bennett, C. C. (2012). *The role of the body in leading and learning: A case study of a somatic leadership development program* (Doctoral dissertation). Teachers College, Columbia University.

Bereiter, C. (2002). Design research for sustained innovation. *Cognitive Studies*, *9*(3), 321–327.

Bloom, B. S., Krathwohl, D. R., & Masia, B. B. (1984). *Bloom taxonomy of educational objectives*. Boston, MA: Allyn and Bacon.

Boud, D., Keogh, R., & Walker, D. (1985). Promoting reflection in learning: A model. *Reflection: Turning experience into learning*, 18-40.

Boud, D., Keogh, R., & Walker, D. (Eds.). (2013). *Reflection: Turning experience into learning*. Routledge.

Bower, M., Howe, C., McCredie, N., Robinson, A., & Grover, D. (2014). Augmented Reality in education–cases, places and potentials. *Educational Media International*, *51*(1), 1–15. doi:10.1080/09523987.2014.889400

Braun, V., & Clarke, V. (2006). Using thematic analysis in psychology. *Qualitative Research in Psychology*, *3*(2), 77–101. doi:10.1191/1478088706qp063oa

Buckingham, D. (2013). *Beyond technology: Children's learning in the age of digital culture*. John Wiley & Sons.

CCNMTL. (2013, June 25). *Design Research at CCNMTL | About Design Research at CCNMTL*. Retrieved August 05, 2017, from http://ccnmtl.columbia.edu/dr/about/

Chen, O., Woolcott, G., & Sweller, J. (2017). Using cognitive load theory to structure computer-based learning including MOOCs. *Journal of Computer Assisted Learning*, *33*(4), 293–305. doi:10.1111/jcal.12188

Custer, O., Deutscher, P., & Haddad, S. (Eds.). (2016). *Foucault/Derrida Fifty Years Later: The Futures of Genealogy, Deconstruction, and Politics*. Columbia University Press. doi:10.7312/cust17194

Dahlgren, R. L. (2017). Education and Popular Culture Narratives. In *From Martyrs to Murderers* (pp. 13–30). SensePublishers.

Dewey, J. (1938). *Experience and Education*. New York: Collier Books.

Edelson, D. C. (2002). Design research: What we learn when we engage in design. *Journal of the Learning Sciences, 11*(1), 105–121. doi:10.1207/S15327809JLS1101_4

Elias, J. L., & Merriam, S. B. (1995). *Philosophical foundations of adult education*. Melbourne, FL: Krieger Publishing.

Epelboin, Y. (2017). MOOCs: A Viable Business Model? In *Open Education: from OERs to MOOCs* (pp. 241–259). Springer Berlin Heidelberg. doi:10.1007/978-3-662-52925-6_13

Felicia, P. (Ed.). (2011). *Handbook of Research on Improving Learning and Motivation through Educational Games: Multidisciplinary Approaches: Multidisciplinary Approaches*. IGI Global. doi:10.4018/978-1-60960-495-0

Fernros, L., Furhoff, A. K., & Wändell, P. E. (2008). Improving quality of life using compound mind-body therapies: Evaluation of a course intervention with body movement and breath therapy, guided imagery, chakra experiencing and mindfulness meditation. *Quality of Life Research: An International Journal of Quality of Life Aspects of Treatment, Care and Rehabilitation, 17*(3), 367–376. doi:10.100711136-008-9321-x PMID:18324479

Friesen, N. (2012). *Report: Defining blended learning*. Retrieved from http://learningspaces. org/papers/ Defining_Blended_Learnin g_NF. pdf

Gaebel, M. (2014). *MOOCs: Massive open online courses*. EUA.

Great School Partnership. (2014, March 05). *Bloom's Taxonomy Definition by The Glossary of Education Reform*. Retrieved March 06, 2018, from https://www.edglossary.org/blooms-taxonomy/

Heifetz, R. A. (1994). *Leadership without easy answers* (Vol. 465). Harvard University Press.

Heifetz, R. A., & Linsky, M. (2014). *Adaptive Leadership: The Heifetz Collection (3 Items)*. Harvard Business Review Press.

Kabat-Zinn, J. (2003). Mindfulness-based interventions in context: Past, present, and future. *Clinical Psychology: Science and Practice, 10*(2), 144–156. doi:10.1093/clipsy.bpg016

Kaplan Andreas, M., & Michael, H. (2016). Higher education and the digital revolution: About MOOCs. *SPOCs, social media, and the Cookie Monster, Business Horizon, 59*.

Kasworm, C. E., Rose, A. D., & Ross-Gordon, J. M. (Eds.). (2010). *Handbook of adult and continuing education*. Sage.

Kegan, R., & Lahey, L. L. (2009). *Immunity to change: How to overcome it and unlock potential in yourself and your organization*. Harvard Business Press.

Khoury, B., Sharma, M., Rush, S. E., & Fournier, C. (2015). Mindfulness-based stress reduction for healthy individuals: A meta-analysis. *Journal of Psychosomatic Research, 78*(6), 519–528. doi:10.1016/j.jpsychores.2015.03.009 PMID:25818837

Ko, S., & Rossen, S. (2017). *Teaching online: A practical guide*. Taylor & Francis.

Kolb, A. Y., & Kolb, D. A. (2012). Experiential learning theory. In Encyclopedia of the Sciences of Learning (pp. 1215-1219). Springer US.

Kolb, D. A. (1984). *Experience as the source of learning and development. Prentice-Hall.*

Kolb, D. A. (2014). *Experiential learning: Experience as the source of learning and development.* FT Press.

Krygier, J. R., Heathers, J. A., Shahrestani, S., Abbott, M., Gross, J. J., & Kemp, A. H. (2013). Mindfulness meditation, well-being, and heart rate variability: A preliminary investigation into the impact of intensive Vipassana meditation. *International Journal of Psychophysiology*, *89*(3), 305–313. doi:10.1016/j.ijpsycho.2013.06.017 PMID:23797150

Lazaroiu, G., Popescu, G. H., & Nica, E. (2016, July). Democratizing education: the potential of EdX in revolutionizing learning. In *The International Scientific Conference eLearning and Software for Education* (*Vol. 3*, p. 34). "Carol I" National Defence University.

Mentor, D. (2016). EMxC3= e&mLearning Cultivating Connected Communities: Sustainable Workforce Talent Development. Handbook of Research on Mobile Learning in Contemporary Classrooms, 240-259.

Mentor, D. (2017, April 25). Cultivating Digital TLC – Teaching and Learning Communities. *ModSim World 2017*. Retrieved July 23, 2017, from http://www.modsimworld.org/2017/documents/Working-Agenda-17-April.pdf

Mentor, D. (2018). Micro to Macro Social Connectedness Through Mobile Phone Engagement. In *Encyclopedia of Information Science and Technology* (4th ed.; pp. 6184–6194). IGI Global. doi:10.4018/978-1-5225-2255-3.ch537

Mentor, D. J. (2011). *Exploring social connectedness via mobile phone texting* (Ph.D. thesis). Teachers College, Columbia University. Retrieved December 25, 2017 from https://www.learntechlib.org/p/117314/

Merchant, Z., Goetz, E. T., Cifuentes, L., Keeney-Kennicutt, W., & Davis, T. J. (2014). Effectiveness of virtual reality-based instruction on students' learning outcomes in K-12 and higher education: A meta-analysis. *Computers & Education*, *70*, 29–40. doi:10.1016/j.compedu.2013.07.033

Merriam, S. B., Caffarella, R. S., & Baumgartner, L. M. (2012). *Learning in adulthood: A comprehensive guide*. John Wiley & Sons.

Mezirow, J. (2000). Learning as Transformation: Critical Perspectives on a Theory in Progress. The Jossey-Bass Higher and Adult Education Series. Jossey-Bass Publishers.

Naismith, L., Sharples, M., Vavoula, G., & Lonsdale, P. (2004). *Literature review in mobile technologies and learning*. Academic Press.

Obrenović, Ž. (2011). Design-based research: what we learn when we engage in design of interactive systems. *Interactions, 18*(5), 56-59.

Okan, Z. (2003). Edutainment: Is learning at risk? *British Journal of Educational Technology*, *34*(3), 255–264. doi:10.1111/1467-8535.00325

Ospina, M. B., Bond, K., Karkhaneh, M., Buscemi, N., Dryden, D. M., Barnes, V., & Shannahoff Khalsa, D. (2008). Clinical trials of meditation practices in health care: Characteristics and quality. *Journal of Alternative and Complementary Medicine (New York, N.Y.)*, *14*(10), 1199–1213. PubMed

Pask, G. (1976). *Conversation theory: Applications in education and epistemology*. Elsevier Publishing Company.

Pelster, B., Haims, J., Stempel, J., & Vyver, B. (2016). Learning: Employees take charge. Global Human Capital Trends 2016. *The new organization: Different by design*. Retrieved from https://www2.deloitte.com/us/en/pages/human-capital/articles/introduction-human-capital-trends-2016.html

Peters, M. A. (2016). The postcolonial university. *Linguistic and Philosophical Investigations*, *15*, 77.

Plass, J. L., Moreno, R., & Brünken, R. (2010). *Cognitive load theory*. Cambridge University Press. doi:10.1017/CBO9780511844744

Plaza, I., Demarzo, M. M. P., Herrera-Mercadal, P., & García-Campayo, J. (2013). Mindfulness based mobile applications: Literature review and analysis of current features. *JMIR mHealth and uHealth*, *1*(2), e24. doi:10.2196/mhealth.2733 PMID:25099314

Ralph, T. W. (1949). Basic principles of curriculum and instruction. Syllabus for Education.

Richards, R. (2012). Exploring Formative Assessment with a Project Using Mobile Phones. *Ubiquitous Learning. International Journal (Toronto, Ont.)*, *4*(2).

Rogers, E. M. (2010). *Diffusion of innovations*. Simon and Schuster.

Savino, D. M. (2014). The impact of MOOCs on human resource training and development. *Journal of Higher Education Theory and Practice*, *14*(3), 59.

Selwyn, N. (2016). *Education and technology: Key issues and debates*. Bloomsbury Publishing.

Shah, D. (2016, December 25). *By the numbers: MOOCS in 2016*. Retrieved from https://www.class-central.com/report/mooc-stats-2016/

Sharples, M., Taylor, J., & Vavoula, G. (2007). A Theory of Learning for the Mobile Age. In R. Andrews & C. Haythornthwaite (Eds.), *The Sage Handbook of Elearning Research* (pp. 221–247). London: Sage. doi:10.4135/9781848607859.n10

Täuscher, K., & Kietzmann, J. (2017). Learning from Failures in the Sharing Economy. *MIS Quarterly Executive*, *16*(4).

Taylor, E. W., & Cranton, P. (2012). *The handbook of transformative learning: Theory, research, and practice*. John Wiley & Sons.

Traxler, J. (2007). Defining, Discussing and Evaluating Mobile Learning: The moving finger writes and having writ..... *The International Review of Research in Open and Distributed Learning*, *8*(2). doi:10.19173/irrodl.v8i2.346

Van der Stappen, E., & Zitter, I. (2016). Exploring Design Principles for Technology-Enhanced Workplace Learning. In Bled eConference (p. 14). Academic Press.

Verger, A., Lubienski, C., & Steiner-Khamsi, G. (2016). The emergence and structuring of the global education industry: Towards an analytical framework. In A. Verger, C. Lubienski, & G. Steiner-Khamsi (Eds.), *World Yearbook of Education 2016: The Global Education Industry*. New York: Routledge.

Wahbeh, H., Elsas, S.-M., & Oken, B. S. (2008). Mind–body interventions: Applications in neurology. *Neurology*, *70*(24), 2321–2328. doi:10.1212/01.wnl.0000314667.16386.5e PMID:18541886

Watkins, K. E., & Marsick, V. J. (2014). Adult education & human resource development: Overlapping and disparate fields. *New Horizons in Adult Education and Human Resource Development*, *26*(1), 42–54. doi:10.1002/nha3.20052

Watkins, K. E., Marsick, V. J., & Faller, P. G. (2012). Transformative learning in the workplace: Leading learning for self and organizational change. The handbook of transformative learning, 373-387.

Wielgosz, J., Schuyler, B. S., Lutz, A., & Davidson, R. J. (2016). Long-term mindfulness training is associated with reliable differences in resting respiration rate. *Scientific Reports*, 6. PMID:27272738

Wildi-Yune, J., & Cordero, C. (2015). *Corporate Digital Learning: How to Get It "Right"*. Retrieved from https://www.imd.org/globalassets/publications/working-papers/docs/working-paper_corporate-digital-learning-final--05-05-15.pdf

Yuan, L., Powell, S., & Cetis, J. (2013). *MOOCs and open education: Implications for higher education*. Academic Press.

KEY TERMS AND DEFINITIONS

E-Learning: Conventionally refers to learning through online platforms and computers. Could be a tethered computer to a desk or un-tethered in the form of a laptop.

Experiential Learning Cycle: Developed by David Kolb, an American educational theorist credited with a popular model of learning through experience and learning styles.

Instrumental Learning: Also called technical learning. Refers to learning leading to getting information that helps carry out tasks in a pre-defined manner using a set of instructions and skills.

Learning Style: Refers to the idea that learners differ from each other in the way they prefer to learn or learn better. There are quite a few conceptualizations of learning styles and preferences including Kolb's learning styles.

Mindfulness: The phenomenon of being present to ones here and now thoughts, feelings, sensations, and actions.

M-Learning: Learning offered via mobile devices and/or learning happening for a mobile learner.

Transformative Learning: Learning that requires going beyond information and knowledge and challenging the assumptions based on which knowledge is based leading to new perspectives and approaches.

Compilation of References

Abas, Z. W. (2015). 21st century education: strategies to ignite and engage students. Special Edition on Teaching and Learning in South East Asia, 13(2), 5-16.

Abeles, P., Verduyn, C., Robinson, A., Smith, P., Yule, W., & Proudfoot, J. (2009). Computerized CBT for adolescent depression ("Stressbusters") and its initial evaluation through an extended case series. *Behavioural and Cognitive Psychotherapy*, *37*(02), 151–165. doi:10.1017/S1352465808005067 PMID:19364416

Abouzeid, R. (2011, January 21). Bouazizi: The man who set himself and Tunisia on fire. *Time*. Retrieved from http://content.time.com/time/magazine/article/0,9171,2044723,00.html

Abulrub, A., Attridge, A., & Williams, M. (2011). Virtual Reality in Engineering Education: The Future of Creative Learning. *International Journal Of Emerging Technologies In Learning*, *6*(4), 4–11. doi:10.3991/ijet.v6i4.1766

Accenture. (2016). *Contributing solutions to youth unemployment: Perspectives from a skills to succeed insight to action virtual event*. New York, NY: Author.

Adbusters. (2011). #OccupyWallStreet. *Adbusters blog*. Retrieved from https://www.adbusters.org/occupywallstreet/

Addison, S., Wright, A., & Milner, R. (2009). Using clickers to improve student engagement and performance in an introductory biochemistry class. *Biochemistry and Molecular Biology Education*, *37*(2), 84–91. doi:10.1002/bmb.20264 PMID:21567711

Akanegbu, A. (2013, April 30). Does the Khan Academy pass the MOOC "Duck Test"? *Ed Tech*. Retrieved from https://edtechmagazine.com/higher/article/2013/04/does-khan-academy-pass-mooc-duck-test

Albrecht, U. V., & von Jan, U. (2017). Safe, sound and desirable: Development of mHealth apps under the stress of rapid life cycles. *mHealth*, 3. PMID:28828374

Aldrich, C. (2005). *Learning by doing*. San Francisco, CA: Pfeiffer.

Alessi, M., Giangreco, E., Mainetti, L., Mighali, V., Patrono, L., Pinnella, M., ... Storelli, D. (2017). Abstracting IoT Complexity Through an Innovative Virtual Environment. *Recent Advances in Communications and Networking Technology*, *6*(1), 2–14. doi:10.2174/2215081106666170601123935

Alexander, B. (2004). Going nomadic: Mobile learning in higher education. *EDUCAUSE Review*, *39*(5), 29–35.

Allen, E.; Seaman, J. (2008). *Staying the Course: Online Education in the United States, Survey report.* Sloan Consortium (Sloan-C).

Allen, E., & Seaman, J. (2014). *Grade change: Tracking online education in the United States*. The Sloan Consortium.

Ally, M. (2008). Foundations of Educational Theory for Online Learning. In T. Anderson & F. Elloumni (Eds.), Theory and Practice of Online Learning (pp. 15-44). Athabasca, Canada: Athabasca University Press.

Andersen, L., Boud, D., & Cohen, R. (2000). Experience-based learning. In Understanding adult education and training (2nd ed.; pp. 225-239). Sydney: Allen & Unwin.

Anderson, J. R. (2015). *Cognitive psychology and its implications*. New York, NY: Worth Publishers.

Anderson, T. (2008). Towards a theory of online learning. In T. Anderson & F. Elloumni (Eds.), *Theory and practice of online learning* (pp. 45–74). Athabasca, Canada: Athabasca University Press.

Andriessen, J., Baker, M., & Suthers, D. (Eds.). (2013). *Arguing to learn: Confronting cognitions in computer-supported collaborative learning environments* (Vol. 1). Springer Science & Business Media.

Anonymous. (2011). #Anonymous #OccupyWallStreet call to action. *Anonymous blog*. Retrieved from http://anonops.blogspot.com/2011/09/anonymous-occupywallstreet-call-to.htm

Anthis, K. (2011). Is it the clicker, or is it the question? Untangling the effects of student response use. *Technology Teacher*, *38*(3), 189–193.

An, Y., & Reigeluth, C. (2012). Creating technology-enhanced, learner-centered classrooms: K-12 teachers' beliefs, perceptions, barriers, and support needs. *Journal of Digital Learning in Teacher Education*, *28*(2), 54–62. doi:10.1080/21532974.2011.10784681

Archbold Hufty Alegría, D., Boscardin, C., Poncelet, A., Mayfield, C., & Wamsley, M. (2014). Using tablets to support self-regulated learning in a longitudinal integrated clerkship. *Medical Education Online*, *19*(1), 23638. doi:10.3402/meo.v19.23638 PMID:28440172

Arnott, J. A., & Planey, S. L. (2017). Flipped classroom approaches lead to no improvement in learning outcomes or student perceptions. *The FASEB Journal*, *31*(1Supplement), 751–14. PMID:27871064

Ashman, A. F., & Conway, R. N. (2017). Using cognitive methods in the classroom. New York, NY: Routledge.

Askew, S. (Ed.). (2000). *Feedback for learning*. London: RoutledgeFalmer.

Association of American Medical Colleges. (2012). *Faculty Roster, Table 5*. Retrieved from https://www.aamc.org/data/facultyroster/reports/

Association of American Medical Colleges. (2017). *MD-PhD Summer Undergraduate Research Programs*. Retrieved from https://students-residents.aamc.org/applying-medical-school/article/md-phd-summer-undergrad-research-programs/

Astatke, Y., Ladeji-Osias, J. O., James, P., Moazzami, F., Scott, C., Connor, K., & Saka, A. (2016). Improving and Expanding Engineering Education in the Middle East and Africa Using Mobile Learning Technology and Innovative Pedagogy. In *Advances in Engineering Education in the Middle East and North Africa* (pp. 235–260). Springer International Publishing. doi:10.1007/978-3-319-15323-0_10

AT&T and Udacity to Offer Scholarships to Underserved Students for New Online Tech Training Program. (n.d.). *AT&T People | Planet | Possibilities*. Retrieved December 10, 2017, from http://about.att.com/content/csr/home/blog/2014/09/at_t_and_udacityto.html#sthash.U0r03eEW.dpuf

AT&T Aspire. (n.d.). *AT&T People | Planet | Possibilities*. Retrieved December 10, 2017, from http://about.att.com/content/csr/home/people/at-t-aspire.html

AT&T. (2014, September 14). *Att_and_MOOC service provider_launch_online_training_program_nanodegree*. Retrieved from http://about.att.com/story/att_and_MOOC service provider _launch_online_training_program_nanodegree.html

Atkinson, D., Neville, H., & Casas, A. (1991). The mentorship of ethnic minorities in professional psychology. *Professional Psychology, Research and Practice*, *22*(4), 336–338. doi:10.1037/0735-7028.22.4.336

Aydin, H., Ozfidan, B., & Carothers, D. (2017). Meeting the Challenges of Curriculum and Instruction in School Settings in the United States. *Journal of Social Studies Education Research*, *8*(3), 76–92.

Azuma, R. T. (1997). A survey of augmented reality. *Presence (Cambridge, Mass.)*, *6*(4), 355–385. doi:10.1162/pres.1997.6.4.355

Baber, C. (2003). *Cognition and tool use*. New York, NY: Taylor & Francis. doi:10.1201/9781420024203

Bacharach, S., & Bacharach, Y. (2012). The Coaching Mindset for Engaging and Developing Others. *eCornell.*

Bakhtin, M. (1981). *The dialogic imagination: Four essays* (C. Emerson & M. Holquist, Trans.). Austin, TX: University of Texas Press.

Bakker, A. B., & Demerouti, E. (2008). Towards a model of work engagement. *Career Development International*, *13*(3), 209–223. doi:10.1108/13620430810870476

Balfanz, R., Bridgeland, J. M., Hornig Fox, J., DePaoli, J. L., Ingram, E. S., & Maushard, M. (2014). *Building a grad nation: Progress and challenge in ending the high school dropout epidemic*. Baltimore, MD: Civic Enterprises, Everyone Graduates Center at the School of Education at Johns Hopkins University, America's Promise Alliance, Alliance for Excellent Education, AT&T, Target.

Bannan, B., Cook, J., & Pachler, N. (2016). Reconceptualizing design research in the age of mobile learning. *Interactive Learning Environments*, *24*(5), 938–953. doi:10.1080/10494820.2015.1018911

Barbe, W. B., Milone, M. N., & Swassing, R. H. (1979). *Teaching Through Modality Strenghts: Concepts and Practices*. Zaner-Bloser.

Barley, S. R. (1997). Competence without credentials: The promise and potential problems of computer-based distance education. Paper presented at the workshop on Competence Without Credential, Washington, DC.

Barnett, J. (2006). Implementation of personal response units in very large lecture classes: Student perceptions. *Australasian Journal of Educational Technology*, *22*(4), 474–494. doi:10.14742/ajet.1281

Barondess, J. (1994). A brief history of mentoring. *Transactions of the American Clinical and Climatological Association*, *106*, 1–24. PMID:7483164

Barton, A. H. (1968). The Columbia crisis: Campus, Vietnam, and the ghetto. *Public Opinion Quarterly*, *32*(3), 333–351. doi:10.1086/267619

Bastos, M. T., Mercea, D., & Charpentier, A. (2015). Tents, tweets, and events: The interplay between ongoing protests and social media. *Journal of Communication*, *65*(2), 320–350. doi:10.1111/jcom.12145

Bates, M. S., Phalen, L., & Moran, C. (2016). Online professional development: A primer. *Phi Delta Kappan*, *97*(5), 70–73. doi:10.1177/0031721716629662

Battou, A. (2017). Designing an Adaptive Learning System Based on a Balanced Combination of Agile Learner Design and Learner Centered Approach. *American Scientific Research Journal for Engineering, Technology, and Sciences*, *37*(1), 178–186.

Beaudouin-Lafon, M., & Mackay, M. (2000). Research directions in situated computing. In CHI'00 Extended Abstracts on Human Factors in Computing Systems (pp. 369-369). ACM. doi:10.1145/633292.633516

Beaumont, R., Rotolone, C., & Sofronoff, K. (2015). The Secret Agent Society social skills program for children with high-functioning autism spectrum disorders: A comparison of two brief versions for schools. *Psychology in the Schools*, *52*(4), 290–402. doi:10.1002/pits.21831

Beetham, H., & Sharpe, R. (2013). *Rethinking pedagogy for a digital age: Designing for 21st century learning*. Routledge.

Bell, B., & Federman, J. (2013). E-learning in postsecondary education. *The Future of Children*, *23*(1), 165–185. doi:10.1353/foc.2013.0007 PMID:25522650

Benkert, C., & Dam, N. (2015). *Experiential learning: What's missing in most change programs*. Retrieved from http://www.mckinsey.com/business-functions/operations/our-insights/experiential-learning-whats-missing-in-most-change-programs

Bennett, C. C. (2012). *The role of the body in leading and learning: A case study of a somatic leadership development program* (Doctoral dissertation). Teachers College, Columbia University.

Bereiter, C. (2002). Design research for sustained innovation. *Cognitive Studies*, *9*(3), 321–327.

Bersin, J., Agarwal, D., Pelster, B., & Schwartz, J. (n.d.). *Global Human Capital Trends 2015*. Retrieved from https://www2.deloitte.com/insights/us/en/focus/human-capital-trends/2015.html

Besser, H. (2012). Archiving media from the "Occupy" movement: Methods for archives trying to manage large amounts of user-generated audiovisual media. *Imatge I Recerca, 12es Jornades Antoni Varés 2012, Ponencies, Experiencies I Communicacions: Proceedings of conference in Girona*, pp. 106–110.

Bicer, A., & Carpraro, R. (2017). Logitudinal effects of technology integration and teacher professional development on students' mathematics achievement. *EURASIA Journal of Mathematics Science and Technology Education*, *13*(3), 815–833. doi:10.12973/eurasia.2017.00645a

Bierema, L. L., & Hill, J. R. (2005). Virtual Mentoring and HRD. *Advances in Developing Human Resources*, *7*(4), 556–568. doi:10.1177/1523422305279688

Bierema, L., & Merriam, S. (2002). E-mentoring: Using computer mediated communciation to enhance the mentoring process. *Innovative Higher Education*, *26*(3), 211–227. doi:10.1023/A:1017921023103

Bilgiç, H. G., Duman, D., & Seferoğlu, S. S. (2011). Dijital yerlilerin özellikleri ve çevrim içi ortamların tasarlanmasındaki etkileri. *Akademik Bilişim*, 2-4.

Black, D. W. (2017). *Preferencing Educational Choice: The Constitutional Limits*. Academic Press.

Blakely, E. J., Hadi, S., & Johnson, P. (1995). *Information city and the ghetto: The LA experience*. Los Angeles, CA: Lusk Center Research Institute.

Blank, G. (2013). Who creates content? stratification and content creation on the internet. *Information Communication and Society*, *16*(4), 590–612. doi:10.1080/1369118X.2013.777758

Blasco-Arcas, L., Buil, I., Hernandez-Ortega, B., & Sese, F. J. (2013). Using clickers in class. The role of interactivity, active collaborative learning, and engagement in learning performance. *Computers & Education*, *32*, 102–110. doi:10.1016/j.compedu.2012.10.019

Blood, E., & Neel, R. (2008). Using student response systems in lecture-based instruction: Does it change student engagement and learning? *Journal of Technology and Teacher Education*, *16*(3), 375–383.

Bloom, B. S., Krathwohl, D. R., & Masia, B. B. (1984). *Bloom taxonomy of educational objectives*. Boston, MA: Allyn and Bacon.

Bonaiuti, G., Calvani, A., & Piazza, D. (2015). Increasing classroom engagement and student comprehension through the use of clickers: An Italian secondary school experience. *Research on Education and Media*, *5*(1), 95–108.

Bonilla, Y., & Rosa, J. (2015). #Ferguson: Digital protest, hashtag ethnography, and the racial politics of social media in the United States. *American Ethnologist*, *42*(1), 4–16. doi:10.1111/amet.12112

Bonsón, E., Royo, S., & Ratkai, M. (2017). Facebook practices in Western European municipalities: An empirical analysis of activity and citizens' engagement. *Administration & Society*, *49*(3), 320–347. doi:10.1177/0095399714544945

Boud, D., Keogh, R., & Walker, D. (1985). Promoting reflection in learning: A model. *Reflection: Turning experience into learning*, 18-40.

Boud, D., Keogh, R., & Walker, D. (Eds.). (2013). *Reflection: Turning experience into learning*. Routledge.

Boulos, M. N. K., Lu, Z., Guerrero, P., Jennett, C., & Steed, A. (2017). From urban planning and emergency training to Pokémon Go: Applications of virtual reality GIS (VRGIS) and augmented reality GIS (ARGIS) in personal, public and environmental health. *International Journal of Health Geographics*, *16*(1), 7. doi:10.118612942-017-0081-0 PMID:28219378

Bower, M., Howe, C., McCredie, N., Robinson, A., & Grover, D. (2014). Augmented Reality in education–cases, places and potentials. *Educational Media International*, *51*(1), 1–15. doi:10.1080/09523987.2014.889400

Boykin, W. A., & Noguera, P. (2011). *Creating the opportunity to learn moving from research to practice to close the achievement gap*. Alexandria, VA: ASCD.

Boyle, J. T., & Nicol, D. J. (2003). Using classroom communication systems to support interaction and discussion in large class settings. *Association of Learning Technology Journal*, *11*(3), 43–57. doi:10.3402/rlt.v11i3.11284

Bradley, S. (2003). "Gym Crow Must Go!" Black student activism at Columbia University, 1967–1968. *Journal of African American History*, *88*(2), 163–181. doi:10.2307/3559064

Brandzel, B. (2010, November 15). What Malcolm Gladwell missed about online organizing and creating big change. *The Nation*. Retrieved from https://www.thenation.com/article/what-malcolm-gladwell-missed-about-online-organizing-and-creating-big-change/

Bransford, J. D., & Schwartz, D. L. (1999). Rethinking transfer: A simple proposal with multiple implications. *Review of Research in Education*, 61–100. doi:10.3102/0091732X024001061

Brass,, L.F., Akabas,, M.H., Burnley,, L.D., Engman,, D.M., & Wilsey,, C.A., & Os, A. (2010). Are MD-PhD programs meeting their goals? An analysis of career choices made by graduates of 24 MD-PhD programs. *Academic Medicine*, *85*, 692–701. PMID:20186033

Braun, V., & Clarke, V. (2006). Using thematic analysis in psychology. *Qualitative Research in Psychology*, *3*(2), 77–101. doi:10.1191/1478088706qp063oa

Breslin, M., & Buchanan, R. (2008). On the case study method of research and teaching in design. *Design Issues*, *24*(1), 36–40. doi:10.1162/desi.2008.24.1.36

Breslow, L., Pritchard, D. E., DeBoer, J., Stump, G. S., Ho, A. D., & Seaton, D. T. (2013). Studying learning in the worldwide classroom: Research into edX's first MOOC. *Research & Practice in Assessment, 8.*

Briz-Ponce, L., Juanes-Mendez, J. A., & García-Peñalvo, F. J. (Eds.). (2016). *Handbook of Research on Mobile Devices and Applications in Higher Education Settings*. IGI Global. doi:10.4018/978-1-5225-0256-2

Brookfield, S. (1993). Self-directed learning, political clarity, and the critical practice of adult education. *Adult Education Quarterly*, *43*(4), 227–242. doi:10.1177/0741713693043004002

Brown, S., & Adler, R. P. (2008). Minds on fire. Open education, the long trail, and learning 2.0. *Educause*, *43*, 17–32.

Bruns, A. (2009). *Blogs, Wikipedia, second life and beyond: From production to produsage*. New York, NY: Peter Lang.

Buckingham, D., (2003) Media Education and the End of the Critical Consumer. *Harvard Educational Review, 73*(3).

Buckingham, D. (2013). *Beyond technology: Children's learning in the age of digital culture*. John Wiley & Sons.

Buczynski, S., & Mathews, K. (2016). An urban school district's 21st century teaching vision: Integration and readiness to incorporate technology. *Annual International Conference On Education & E-Learning*, 34. 10.5176/2251-1814_EeL16.12

Buddeberg-Fischer, B., & Herta, K. (2006). Formal mentoring programmes for medical students and doctors-a review of the Medline literature. *Medical Teacher*, *28*(3), 248–257. doi:10.1080/01421590500313043 PMID:16753724

Buli, I., Catalan, S., & Martinez, E. (2016). Do clickers enhance learning? A control-value theory approach. *Computers & Education*, *103*, 170–182. doi:10.1016/j.compedu.2016.10.009

Buraphadeja, V., & Dawson, K. (2008, July). Content Analysis in Computer-Mediated Communication: Analyzing Models for Assessing Critical Thinking Through the Lens of Social Constructivism. *American Journal of Distance Education*, *22*(3), 130–145. doi:10.1080/08923640802224568

Burbach, M., Matkin, G., & Fritz, S. (2004). Teaching critical thinking in an introductory leadership course utilizing active learning strategies: A confirmatory study. *College Student Journal*, *38*(3), 482–493.

Burke, J., & Tumbleson, E. (2016). Choosing Content-Creation Products. *Library Technology Reports*, *52*(2), 23-27.

Burke, W., & Litwin, G. H. (1992). A causal model of organizational performance and change. *Journal of Management*, *18*(3), 532–545. doi:10.1177/014920639201800306

Byrne, J. (2012, April 30). 9 moments in the Occupy movement history. *The Huffington Post*. Retrieved from http://www.huffingtonpost.com/janet-byrne/occupy-movement_b_1464717.html?slideshow=true#gallery/223338/0

Caldwell, J. E. (2007). Clickers in the large classroom: Current research and best-practice tips. *Life Sciences Education*, *6*(1), 9–20. doi:10.1187/cbe.06-12-0205 PMID:17339389

Calhoun, C. (2013). Occupy Wall Street in perspective. *The British Journal of Sociology*, *64*(1), 26–38. doi:10.1111/1468-4446.12002 PMID:23488697

Candy, P. (1991). Four Dimensions of Self-Direction. In Self-direction for lifelong learning (pp. 97-126). Jossey-Bass.

Candy, P. C. (2004). *Linking thinking: Self-directed learning in the digital age*. Canberra, Australia: Department of Education, Science and Training.

Carliner, S. (2017). Informal learning. In R. A. Reiser & J. V. Dempsey (Eds.), *Trends and Issues in Instructional Design and Technology* (pp. 142–151). Boston, MA: Pearson.

Carlson, D. (2017). Teachers and crisis: Urban school reform and teachers' work culture (Vol. 1). New York, NY: Routledge.

Carnaghan, C., Edmonds, T. P., Lechner, T. A., & Olds, P. R. (2011). Using student response systems in the accounting classroom: Strengths, strategies, and limitations. *Journal of Accounting Education*, *29*(4), 265–283. doi:10.1016/j.jaccedu.2012.05.002

Caro-López, H. (2010). Ecuadorians in New York City 1990—2008. *Center for Latin American, Caribbean & Latino Studies*. Retrieved from http://clacls.gc.cuny.edu/files/2013/10/Ecuadorians-in-New-York-1990-2008.pdf

Carter, T. J., & Adkins, B. (2017). Situated learning, communities of practice, and the social construction of knowledge. *Theory and Practice of Adult and Higher Education*, 113.

CCNMTL. (2013, June 25). *Design Research at CCNMTL | About Design Research at CCNMTL*. Retrieved August 05, 2017, from http://ccnmtl.columbia.edu/dr/about/

CEO Forum on Education and Technology. (2001). *Key building blocks for student achievement in the 21" century: Assessment, alignment, accountability, access and analysis. Year 4 report.* Retrieved May 24, 2017 from http://www.ceoforum.org

Chamerland, K., Sanchez, M., Panahi, S., Provencher, V., Gagnon, J., & Drapeau, V. (2017). The impact of an innovative web-based school nutrition intervention to increase fruits and vegetables and milk and alternatives in adolescents: A clustered randomized trial. *The International Journal of Behavioral Nutrition and Physical Activity*, *14*(1), 140. doi:10.118612966-017-0595-7 PMID:29037203

Champion, R. (2017). *Digitally Sound? Teachers' Use of Digital Literacies in Predominantly African American Classrooms in a Low SES Urban School Setting*. Academic Press.

Chang, C.-C. (2007). A study of the quantitative analysis of the development and implementation of Electronic Performance Support Systems. *International Journal of Instructional Media*, *34*(3), 275–284.

Chapman, B. (2008). Learning Technology Primer. In ASTD Handbook for Workplace Learning Professionals. Baltimore, MD: United Book Press.

Cheema, J. R., & Hang, B. (2013). Quantity and quality of computer use and academic achievement: Evidence from a large-scale international test program. *International Journal of Education and Development Using Information and Communication Technology*, *9*(2), 95–106.

Cheng, Y., & Wang, S. H. (2011). Applying a 3D virtual learning environment to facilitate student's application ability–The case of marketing. *Computers in Human Behavior*, *27*(1), 576–584. doi:10.1016/j.chb.2010.10.008

Chen, H.-T., & Rossi, P. H. (1980). The multi-goal, theory-driven approach to evaluation: A model linking basic and applied social science. *Social Forces*, *59*(1), 106–122. doi:10.1093f/59.1.106

Chen, J. C., Whittinghill, D. C., & Kadlowec, J. A. (2010). Classes that click: Fast, rich feedback to enhance students' learning and satisfaction. *Journal of Engineering Education*, *99*(2), 158–169. doi:10.1002/j.2168-9830.2010.tb01052.x

Chen, O., Woolcott, G., & Sweller, J. (2017). Using cognitive load theory to structure computer-based learning including MOOCs. *Journal of Computer Assisted Learning*, *33*(4), 293–305. doi:10.1111/jcal.12188

Chien, Y. T., Chang, Y. H., & Chang, C. Y. (2016). Do we click in the right way? A meta-analytic review of clicker-integrated instruction. *Educational Research Review*, *17*, 1–18. doi:10.1016/j.edurev.2015.10.003

Choi, D. H., Kim, J., & Kim, S. H. (2007). ERP training with a web-based electronic learning system: The flow theory perspective. *International Journal of Human-Computer Studies*, *65*(3), 223–243. doi:10.1016/j.ijhcs.2006.10.002

Choy, S. C., & Cheah, P. K. (2009). Teacher Perceptions of Critical Thinking Among Students and its Influence on Higher Education. *International Journal on Teaching and Learning in Higher Education*, *20*(2), 198–206.

Chronicle Research Services (2009). *The College Of 2020: Students Executive Summary*. Author.

Chu, S. K. W., Reynolds, R. B., Tavares, N. J., Notari, M., & Lee, C. W. Y. (2017). *Twenty-First Century Skills Education in Switzerland: An Example of Project-Based Learning Using Wiki in Science Education. In 21st Century Skills Development Through Inquiry-Based Learning*. Springer.

Ciampa, K. (2017). Building bridges between technology and content literacy in special education: Lessons learned from special educators' use of integrated technology and perceived benefits for students. *Literacy Research and Instruction*, *56*(2), 85–113. doi:10.1080/19388071.2017.1280863

Clancey, W. J. (1995). A tutorial on situated learning. In J. Self (Ed.), *Proceedings of the International Conference on Computers and Education* (pp. 49-70). Charlottesville, VA: AACE.

Clark, A. (2003). *Natural-born cyborgs: why minds and technologies are made to merge*. Oxford University Press.

Clark, K. (2005). Serving underserved communities with instructional technologies: Giving them what they need, not what you want. *Urban Education*, *40*(4), 430–445. doi:10.1177/0042085905276388

Cobb, J. (2016, March 14). The matter of black lives. *The New Yorker*. Retrieved from http://www.newyorker.com/magazine/2016/03/14/where-is-black-lives-matter-headed

Coleman, L., Cotten, S., Gibson, P., Stringer, K., & Howell-Moroney, M. (2016). Integrating computing across the curriculum. *Journal of Educational Computing Research*, *54*(2), 275–294. doi:10.1177/0735633115616645

Collins, L. J., & Liang, X. (2015). Examining high quality online teacher professional development: Teachers' voices. *International Journal of Teacher Leadership*, *6*(1), 18–34.

Comi, S. L., Argentin, G., Gui, M., Origo, F., & Pagani, L. (2017). Is it the way they use it? Teachers, ICT, and student achievement. *Economics of Education Review*, *56*, 24–39. doi:10.1016/j.econedurev.2016.11.007

CommLabIndia. (2016, February 24). *Top Mobile Learning Trends*. Retrieved December 02, 2017, from http://blog.tristit.com/top-mobile-learning-trends/271

Common Sense Media. (2011). *Zero to eight: Children's media use in America*. Retrieved September 2017 from http://www.commonsensemedia.org/sites/default/files/research/zerotoeightfinal2011.pdf

Conchas, G. Q., Gottfried, M., Hinga, B. M., & Oseguera, L. (Eds.). (2017). *Educational Policy Goes to School: Case Studies on the Limitations and Possibilities of Educational Innovation*. Routledge.

Connors-Kellgren, A., Parker, C. E., Blustein, D. L., & Barnett, M. (2016). Innovations and challenges in project-based STEM education: Lessons from ITEST. *Journal of Science Education and Technology*, *25*(6), 825–832. doi:10.100710956-016-9658-9

Conover, M. D., Ferrara, E., Menczer, F., & Flammini, A. (2013). The digital evolution of Occupy Wall Street. *PLoS One*, *8*(5), e64679. doi:10.1371/journal.pone.0064679 PMID:23734215

Corcino, C., & Mentor, D. (2018). *Computer programming for under resourced communities - from training, to internship, to employment*. Presented at the UNESCO Mobile Learning Week's Skills for a connected world 2018, UNESCO Headquarters, Paris, France.

Costa, C., Murphy, M., Pereira, A. L., & Taylor, Y. (2018). Higher education students' experiences of digital learning and (dis) empowerment. *Australasian Journal of Educational Technology*, *34*(3), 140–152.

Costley, K.C., Bell, D., & Legett, T. (2014). *More feet hitting the road: Ten ways to get impoverished childrens' test scores up*. Academic Press.

Cotten, S. R., Hale, T. M., Moroney, M. H., O'Neal, L., & Borch, C. (2011). Using affordable technology to decrease digital inequality: Results from birmingham's one laptop per child XO laptop project. *Information Communication and Society*, *14*(4), 424–444. doi:10.1080/1369118X.2011.559266

Cottle, S. (2011). Media and the Arab uprisings of 2011: Research notes. *Journalism*, *12*(5), 647–659. doi:10.1177/1464884911410017

Cox, J. (2015, December). *Benefits of technology in the classroom*. Retrieved from http://www.teachhub.com/benefits-technology-classroom

Cranton, P. (2006). *Understanding and Promoting Transformative Learning: A Guide for Educators of Adults* (2nd ed.). San Francisco, CA: Jossey-Bass.

Cross, J. (2005). *Informal learning: Rediscovering the natural pathways that inspire innovation and performance*. San Francisco, CA: Pfeiffer.

Cruz-Neira, C., Sandin, D. J., DeFanti, T. A., Kenyon, R. V., & Hart, J. C. (1992). The CAVE: Audio visual experience automatic virtual environment. *Communications of the ACM*, *35*(6), 64–73. doi:10.1145/129888.129892

Cuban, L. (2001). *Oversold and underused: Computers in the classroom*. Cambridge, MA: Harvard University Press.

Cubbage, A. (2017, January 20). Northwestern university forms new partnership with lake view high school. *Northwestern Now*. Retrieved from https://news.northwestern.edu/stories/ 2017/01/northwestern-university-new-partnership-lake-view-high-school-cps/

Curinga, M. X., & Saravanos, A. (2016). Mobile First E-Learning. In D. Mentor (Ed.), *Handbook of Research on Mobile Learning in Contemporary Classrooms* (pp. 23–36). Hershey, PA: IGI Global. doi:10.4018/978-1-5225-0251-7.ch002

Custer, O., Deutscher, P., & Haddad, S. (Eds.). (2016). *Foucault/Derrida Fifty Years Later: The Futures of Genealogy, Deconstruction, and Politics*. Columbia University Press. doi:10.7312/cust17194

Dahlgren, R. L. (2017). Education and Popular Culture Narratives. In *From Martyrs to Murderers* (pp. 13–30). SensePublishers.

Daloz, L. A. (1999). Mentor (2nd ed.). San Francisco: Jossey-Bass.

Daloz, L. A. (1986). *Effective Teaching and Mentoring* (A. B. Knox, Ed.). San Francisco, CA: Jossey-Bass.

Darling-Hammond, L. (1995). Inequality and access to knowledge. In J. A. Banks & C. A. Banks (Eds.), *The handbook of multicultural education* (pp. 465–483). New York: MacMillan.

Darling-Hammond, L. (2015). *The flat world and education: How America's commitment to equity will determine our future*. Teachers College Press.

Daskal, L. (2016, January 28). 6 social media skills every leader must learn. *Inc*. Retrieved from https://www.inc.com/lolly-daskal/6-social-media-skills-you-need-to-lead.html

Davidson, S. P., Cain, S. M., McGinnis, R. S., Vitali, R. R., Perkins, N. C., & McLean, S. G. (2016). Quantifying warfighter performance in a target acquisition and aiming task using wireless inertial sensors. *Applied Ergonomics*, *56*, 27–33. doi:10.1016/j.apergo.2016.03.001 PMID:27184308

Davis, N., Hansmeyer, E., Minic, B., Prakash, S., & Rangan, S. (2013). TEN youth: Unlocking enterprise growth by focusing on the fortune at the bottom of the talent pyramid. *Innovations*, *8*(1/2), 141–157.

Davis, A., Little, P., & Stewart, B. (2008). Developing an Infrastructure for Online Learning. In T. Anderson & F. Elloumni (Eds.), *Theory and Practice of Online Learning*. Athabasca, Canada: Athabasca University Press.

Davis, D., O'Brien, M. A. T., Freemantle, N., Wolf, F. M., Mazmanian, P., & Taylor-Vaisey, A. (1999). Impact of formal continuing medical education: Do conferences, workshops, rounds, and other traditional continuing education activities change physician behavior or health care outcomes? *Journal of the American Medical Association*, *282*(9), 867–874. doi:10.1001/jama.282.9.867 PMID:10478694

Davis, F. D. (1993). User acceptance of information technology: System characteristics, user perceptions and behavioral impacts. *International Journal of Man-Machine Studies*, *38*(3), 475–487. doi:10.1006/imms.1993.1022

Davis, K. K., Ambrose, A., & Orand, M. (2017). Identity and agency in school and afterschool settings: Investigating digital media's supporting role. *Digital Culture & Education*, *9*, 31–47.

Davis, M. C., Can, D. D., Pindrik, J., Rocque, B. G., & Johnston, J. M. (2016). Virtual interactive presence in global surgical education: International collaboration through augmented reality. *World Neurosurgery*, *86*, 103–111. doi:10.1016/j.wneu.2015.08.053 PMID:26342783

Davison, E., & Cotten, S. R. (2010). Connection disparities: The importance of broadband connections in understanding today's digital divide. In E. Ferro (Ed.), *Handbook of research on overcoming the digital divides: constructing an equitable and competitive information society* (Vol. 1). IGI Global. doi:10.4018/978-1-60566-699-0.ch019

Dawson, M. R. W. (2013). *Mind, body, world: Foundations of cognitive science*. Edmonton, Canada: AU Press.

Day, E. (2015, July 19). #BlackLivesMatter: The birth of a new civil rights movement. *The Guardian*. Retrieved from https://www.theguardian.com/world/2015/jul/19/blacklivesmatter-birth-civil-rights-movement

Day, V. (2017). *Understanding the Relationship Between K-12 Teachers' Perceptions of Their Levels of Teaching Innovation and Their Experiences with Technology-Driven Professional Development*. Drexel University.

de Aquino, C. T. E., Robertson, R. W., Allen, P., & Withey, P. (2017). A global learning-centered approach to higher education: Workplace development in the 21st century. *Revista Tecnología Ciência & Educação (Bauru)*, 6.

de Block, L., Buckingham, D., & Banaji, S. (2005). *Final project report: Children in communication about migration*. London: Institute of Education, University of London.

Deci, E. L., Koestner, R., & Ryan, R. M. (2001). Extrinsic rewards and intrinsic motivation in education: Reconsidered once again. *Review of Educational Research*, *71*(1), 1–27. doi:10.3102/00346543071001001

Degennaro, D., & Brown, T. L. (2009). Youth voices: Connections between history, enacted culture and identity in a digital divide initiative. *Cultural Studies of Science Education*, *4*(1), 13–19. doi:10.100711422-008-9108-y

Deiser, R., & Newton, S. (2013, February). Six social media skills every leader needs. *McKinsey Quarterly*. Retrieved from http://www.mckinsey.com/industries/high-tech/our-insights/six-social-media-skills-every-leader-needs

Delgado, R. (1989). Storytelling for oppositionists and others: A plea for narrative. *Michigan Law Review*, *87*(8), 2411–2441. doi:10.2307/1289308

DeLozier, S. J., & Rhodes, M. G. (2017). Flipped classrooms: A review of key ideas and recommendations for practice. *Educational Psychology Review*, *29*(1), 141–151. doi:10.100710648-015-9356-9

Demby, G. (2014, December 31). The birth of a new civil rights movement. *Politico*. Retrieved from http://www.politico.com/magazine/story/2014/12/ferguson-new-civil-rights-movement-113906

DeMers, J. (2015, June 8). 5 skills your social media manager must have. *Forbes*. Retrieved from https://www.forbes.com/sites/jaysondemers/2015/06/08/5-skills-your-social-media-manager-must-have/#3577b703ac56

Demir, K., & Akpınar, E. (2016). Mobil öğrenmeye yönelik tutum ölçeği geliştirme çalışması. *Eğitim Teknolojisi Kuram ve Uygulama, 6*(1).

Desimone, L. M. (2009). Improving impact studies of teachers' professional development: Toward better conceptualizations and measures. *Educational Researcher, 38*(3), 181–199. doi:10.3102/0013189X08331140

DeVega, C. (2017). From "All Lives Matter" to the terror in Charlottesville: How the media's phony fairness got us here. *Salon*. Retrieved from http://www.salon.com/2017/08/14/from-all-lives-matter-to-the-terror-in-charlottesville-how-the-medias-phony-fairness-got-us-here/

Dewey, J. (1913). *Interest and effort in education*. Houghton Mifflin. doi:10.1037/14633-000

Dewey, J. (1916). *Democracy and education: An introduction to the philosophy of education*. New York: Macmillan.

Dewey, J. (1938). *Experience and education*. New York, NY: Macmillan/Collier.

Dewey, J. (1938). *Experience and Education*. New York: Collier Books.

Dexter, S., Doering, A. H., & Riedel, E. S. (2006). Content area specific technology integration: A model for educating teachers. *Journal of Technology and Teacher Education*, (2): 325.

Dillahunt, T. R., Wang, B. Z., & Teasley, S. (2014). Democratizing higher education: Exploring MOOC use among those who cannot afford a formal education. *The International Review of Research in Open and Distributed Learning, 15*(5). doi:10.19173/irrodl.v15i5.1841

Domine, V. (2010). Chapter ten: How important is technology in urban education? *Counterpoints, 215*, 145–158.

Driscoll, M. P. (2017). Psychological foundations of instructional design. In R. A. Reiser & J. V. Dempsey (Eds.), *Trends and Issues in Instructional Design and Technology* (pp. 52–60). Boston, MA: Pearson.

Drossel, K., & Eickelmann, B. (2017). Teachers' participation in professional development concerning the implementation of new technologies in class: A latent class analysis of teachers and the relationship with the use of computers, ICT self-efficacy and emphasis on teaching ICT skills. *Large-Scale Assessments in Education, 5*(1), 1–13. doi:10.118640536-017-0053-7

Du, J., Sansing, W., & Yu, C. (2004). The impact of technology use on low-income and minority students' academic achievements: Educational longitudinal study of 2002. *Association For Educational Communications and Technology*, 274-283.

Dumford, A. D., & Miller, A. L. (2017). Assessing alumni success: Income is NOT the only outcome! *Assessment & Evaluation in Higher Education, 42*(2), 195–207.

Durak, H. İ. (2016). Eğitimde Temel Kavramlar ve Kuramlar. In Tıp Eğiticinin El Kitabı (pp. 15-22). Ankara: Güneş Tıp Kitabevleri.

Earle, E. (2012). *A brief history of Occupy Wall Street*. New York, NY: Rosa Luxemburg Stiftung.

Eastman, J. K., Iyer, R., & Eastman, K. L. (2011). Business students' perceptions, attitudes, and satisfaction with interactive technology: An exploratory study. *Journal of Education for Business, 86*(1), 36–43. doi:10.1080/08832321003774756

Eby, L. T., Allen, T. D., Evan, S. C., Ng, T., & DuBois, D. (2008). Does Mentoring Matter? A Multidiciplinary Meta-Analysis Comparing Mentored and Non-Mentored Individuals. *Journal of Vocational Behavior*, *72*(2), 254–267. doi:10.1016/j.jvb.2007.04.005 PMID:19343074

Economou, D., Mitchell, W. L., & Boyle, T. (2000). Requirements elicitation for virtual actors in collaborative learning environments. *Computers & Education*, *34*(3), 225–239. doi:10.1016/S0360-1315(99)00047-0

Edelson, D. C. (2002). Design research: What we learn when we engage in design. *Journal of the Learning Sciences*, *11*(1), 105–121. doi:10.1207/S15327809JLS1101_4

Edinger, M. J. (2017). Online teacher professional development for gifted education: Examining the impact of a new pedagogical model. *Gifted Child Quarterly*, *61*(4), 300–312. doi:10.1177/0016986217722616

Education Week. (2007, March 30). *Technology counts 2007: A digital decade*. Retrieved May 20, 2017, from http://www.edweek.org/ew/toc/2007/03/29/

Edyburn, D., Higgins, K., & Boone, R. (2005). *Handbook of special education technology research and practice*. Whitefish Bay, WI: Knowledge by Design, Inc.

Egloffstein, M., & Ifenthaler, D. (2017). Employee perspectives on MOOCs for workplace learning. *TechTrends*, *61*(1), 65–70.

Eickelmann, B., Gerick, J., & Koop, C. (2016). ICT use in mathematics lessons and the mathematics achievement of secondary school students by international comparison: Which role do school level factors play? *Education and Information Technologies*, 1–25.

Einfeld, S., Beaumont, R., Clark, T., Clarke, K., Costley, D., Gray, K. M., ... Howlin, P. (in press). School-based social skills training for young people with Autism Spectrum Disorders. *The Journal of Special Education*.

Elam, C. L., Stratton, T. D., Hafferty, F. W., & Haidet, P. (2009). Identity, Social Networks, and Relationships: Theoretical Underpinnings of Critical Mass and Diversity. *Academic Medicine*, *84*(10Supplement), S135–S140. doi:10.1097/ACM.0b013e3181b370ad PMID:19907377

El-Hussein, M. O. M., & Cronje, J. C. (2010). Defining mobile learning in the higher education landscape. *Journal of Educational Technology & Society*, *13*(3), 12–21.

Elias, J. L., & Merriam, S. B. (1995). *Philosophical foundations of adult education*. Melbourne, FL: Krieger Publishing.

Eligon, J. (2015, November 18). One slogan, many methods: Black Lives Matter enters politics. *The New York Times*. Retrieved from https://www.nytimes.com/2015/11/19/us/one-slogan-many-methods-black-lives-matter-enters-politics.html?_r=0

Elvezio, C., Sukan, M., Oda, O., Feiner, S., & Tversky, B. (2017, July). Remote collaboration in AR and VR using virtual replicas. In ACM SIGGRAPH 2017 VR Village (p. 13). ACM. doi:10.1145/3089269.3089281

Epelboin, Y. (2017). MOOCs: A Viable Business Model? In *Open Education: from OERs to MOOCs* (pp. 241–259). Springer Berlin Heidelberg. doi:10.1007/978-3-662-52925-6_13

Epstein, I. (2015). No turning back, neoliberalism exposed: Youth protest in Chile, Spain and the U.S. In I. Epstein (Ed.), *The whole world is texting: Youth protest in the information age* (pp. 81–114). Rotterdam: Sense Publishers. doi:10.1007/978-94-6300-055-0_5

Eraut, M. (2004). Informal learning in the workplace. *Studies in Continuing Education*, *26*(2), 247–273. doi:10.1080/158037042000225245

Ericsson, K. A., Krampe, R. T., & Tesch-Romer, C. (1993). The role of deliberate practice in the acquisition of expert performance. *Psychological Review, 100*(3), 363–406. doi:10.1037/0033-295X.100.3.363

Ertmer, P. A., & Newby, T. J. (1993). Behaviorism, cognitivism, constructivism: Comparing critical features from an instructional design perspective. *Performance Improvement Quarterly, 6*(4), 50–72. doi:10.1111/j.1937-8327.1993.tb00605.x

Ertmer, P. A., & Ottenbreit-Leftwich, A. (2013). Removing obstacles to the pedagogical changes required by Jonassen's vision of authentic technology-enabled learning. *Computers & Education, 64*, 175–182. doi:10.1016/j.compedu.2012.10.008

Evans, B. J., & Baker, R. B. (2016). MOOCs and persistence: Definitions and predictors. *New Directions for Institutional Research, 2015*(167), 69–85. doi:10.1002/ir.20155

Evans, M. S., & Johri, A. (2008). Facilitating guided participation through mobile technologies: Designing creative learning environments for self and others. *Journal of Computing in Higher Education, 20*(2), 92–105. doi:10.100712528-008-9004-1

Facebook. (2016). *FbStart Reveals Apps of the Year Winners and Introduces New Program Updates*. Retrieved May 25, 2017, from https://developers.facebook.com/blog/post/2016/05/31/fbstart-apps-of-the-year-program-updates/

Fairlie, R. (2007). Explaining differences in access to home computers and the Internet: A comparison of Latino groups to other ethnic and racial groups. *Electronic Commerce Research, 7*(3/4), 265–291. doi:10.100710660-007-9006-5

Fairlie, R. W. (2012). Academic achievement, technology and race: Experimental evidence. *Economics of Education Review, 31*(5), 663–679. doi:10.1016/j.econedurev.2012.04.003

Farag, D. M., Park, S., & Kaupins, G. (2015). Faculty perceptions of the adoption and use of clickers in the legal studies in business classroom. *Journal of Education for Business, 90*(4), 208–216. doi:10.1080/08832323.2015.1014459

Felicia, P. (Ed.). (2011). *Handbook of Research on Improving Learning and Motivation through Educational Games: Multidisciplinary Approaches: Multidisciplinary Approaches*. IGI Global. doi:10.4018/978-1-60960-495-0

Fenn, J., & Linden, A. (2005). *Gartner's Hype Cycle Special Report for 2005*. Academic Press.

Fernros, L., Furhoff, A. K., & Wändell, P. E. (2008). Improving quality of life using compound mind-body therapies: Evaluation of a course intervention with body movement and breath therapy, guided imagery, chakra experiencing and mindfulness meditation. *Quality of Life Research: An International Journal of Quality of Life Aspects of Treatment, Care and Rehabilitation, 17*(3), 367–376. doi:10.100711136-008-9321-x PMID:18324479

Fine, M. (1991). *Framing dropouts: notes on the politics of an urban public high school*. New York: State University of New York Press.

Fischer, F., Kollar, I., Stegmann, K., & Wecker, C. (2013). Toward a script theory of guidance in computer-supported collaborative learning. *Educational Psychologist, 48*(1), 56–66. doi:10.1080/00461520.2012.748005 PMID:23378679

Fisher, J. B., Schumaker, J. B., Culbertson, J., & Deshler, D. D. (2010). Effects of a computerized professional development program on teacher and student outcomes. *Journal of Teacher Education, 61*(4), 302–312. doi:10.1177/0022487110369556

Fishman, B. J., Konstantopoulos, S., Kubitskey, B. W., Vath, R., Park, G., Johnson, H., & Edelson, D. C. (2013). Comparing the impact of online and face-to-face professional development in the context of curriculum implementation. *Journal of Teacher Education, 64*(5), 426–438. doi:10.1177/0022487113494413

Frank, T. J., & Castek, J. (2017). From digital literacies to digital problem solving: Expanding technology-rich learning opportunities for adults. *Journal of Research and Practice for Adult Literacy, Secondary, and Basic Education, 6*(2), 66–70.

Frederick, R., & Shockley, K. (2008). Culturally responsive uses of computer technology: A portrait of three teachers working in urban schools. *Electronic Journal for Instructional Technology*, *7*(3), 3–21.

Friedman, T. L. (2014). It takes a mentor. *New York Times*, p. A12.

Friesen, N. (2012). *Report: Defining blended learning*. Retrieved from http://learningspaces. org/papers/Defining_ Blended_Learnin g_NF. pdf

Fullam, J. (2017). Becoming a youth activist in the internet age: A case study on social media activism and identity development. *International Journal of Qualitative Studies in Education: QSE*, *30*(4), 406–422. doi:10.1080/09518398 .2016.1250176

Gaebel, M. (2014). *MOOCs: Massive open online courses*. EUA.

Galkandage, C., Calic, J., Dogan, S., & Guillemaut, J. Y. (2017). Stereoscopic video quality assessment using binocular energy. *IEEE Journal of Selected Topics in Signal Processing*, *11*(1), 102–112. doi:10.1109/JSTSP.2016.2632045

Gao, Y., Liu, T. C., & Paas, F. (2016). Effects of mode of target task selection on learning about plants in a mobile learning environment: Effortful manual selection versus effortless QR-code selection. *Journal of Educational Psychology*, *108*(5), 694–704. doi:10.1037/edu0000080

Garrison, D. R., & Kanuka, H. (2004). Blended learning: Uncovering its transformative potential in higher education. *The Internet and Higher Education*, *7*(2), 95–105. doi:10.1016/j.iheduc.2004.02.001

Garrison, H. H. B., & Prudence, W. (1985). *Minority Access to Research Careers: An Evaluation of the Honors Undergraduate Research Training Program*. National Academies Press.

Garza, A. (2014, October 7). A herstory of the #BlackLivesMatter movement. *The Feminist Wire*. Retrieved from http://www.thefeministwire.com/2014/10/blacklivesmatter-2/

Gephart, M. A., & Marsick, V. J. (2016). *Strategy organizational learning: Using system dynamics for innovation and sustained performance*. Springer-Verlag. doi:10.1007/978-3-662-48642-9

Gershman, A. V., McCarthy, J. F., & Frano, A. E. (1999). Situated computing: Bridging the gap between intention and action. In *Wearable Computers, 1999. Digest of Papers. The Third International Symposium on Wearable Computers* (pp. 3-9). IEEE.

Gery, G. J. (1991). *Electronic performance support systems*. Gery Associates.

Ge, X., Planas, L. G., & Er, N. (2010). A cognitive support system to scaffold students' problem-based learning in a web-based learning environment. *The Interdisciplinary Journal of Problem-Based Learning*, *4*(1), 30–56. doi:10.7771/1541-5015.1093

Gier, V. S., & Kreiner, D. S. (2009). Incorporating active learning with PowerPoint-based lectures using content-based questions. *Teaching of Psychology*, *36*(2), 134–139. doi:10.1080/00986280902739792

Gladwell, M. (2010). Small change: Why the revolution will not be tweeted. *The New Yorker*. Retrieved from http://www.newyorker.com/reporting/2010/10/04/101004fa_fact_gladwell?currentPage=all

Gladwell, M. (2000). *The tipping point: How little things can make a big difference*. Boston, MA: Little, Brown and Company.

Global Communities. (2016). *Partners for good, adapting and sustaining demand-driven training programs: A case study from the Middle East*. Retrieved from https://www.globalcommunities.org/publications/2016-MENA-YES.pdf

Global Human Capital Trends. (2016). *The New Organization: Different by Design*. Westlake, TX: Deloitte University Press.

Golkar, S. (2015). Student activism, social media, and authoritarian rule in Iran. In I. Epstein (Ed.), *The whole world is texting: Youth protest in the information age* (pp. 61–79). Rotterdam: Sense Publishers. doi:10.1007/978-94-6300-055-0_4

Goodman, S. (2010). Educating for democracy: An uncommon standard. *The New York Times,* pp. 1-6.

Goodman, S. (2003). *Teaching youth media*. New York, NY: Teachers College Columbia University.

Good, T. L., & Lavigne, A. L. (2017). *Looking in classrooms*. Routledge.

Goodyear, P., Jones, C., & Thompson, K. (2014). Computer-supported collaborative learning: Instructional approaches, group processes and educational designs. In *Handbook of research on educational communications and technology* (pp. 439–451). Springer New York. doi:10.1007/978-1-4614-3185-5_35

Goslee, S. (1998). *Losing ground bit by bit: low-income communities in the information age*. Washington, DC: Benton Foundation.

Gotian, R. (2017). *Optimizing Success of Physician-Scientists* (EdD). Teachers College Columbia University. ProQuest database. (10255977)

Gotian, R. (2016). Mentoring the Mentors: Just Because You Have the Title Doesn't Mean You Know What You Are Doing. *College Student Journal*, *50*(1), 1–4.

Gotian, R., Raymore, J., Rhooms, S.-K., Liberman, L., & Andersen, O. S. (2017). Gateways to the Laboratory: How an MD-PhD Program Increased the Number of Minority Physician-Scientists. *Academic Medicine*, *92*(5), 628–634. doi:10.1097/ACM.0000000000001478 PMID:28441673

Granovetter, M. S. (1973). The strength of weak ties. *American Journal of Sociology*, *78*(6), 1360–1380. doi:10.1086/225469

Grant, L. (2006). Using wikis in schools: A case study. *FutureLab*. Retrieved from https://www.nfer.ac.uk/publications/FUTL98

Gray, L., Thomas, N., & Lewis, L. (2010). *Teachers' use of educational technology in U.S. public schools: 2009 (NCES, 2010)*. Washington, DC: National Center for Education Statistics, Institute of Education Sciences, U.S. Department of Education.

Great School Partnership. (2014, March 05). *Bloom's Taxonomy Definition by The Glossary of Education Reform*. Retrieved March 06, 2018, from https://www.edglossary.org/blooms-taxonomy/

Greenwald, S., Kulik, A., Kunert, A., Beck, S., Frohlich, B., Cobb, S., ... Snyder, A. (2017). *Technology and applications for collaborative learning in virtual reality*. Academic Press.

Gruber, H., & Voneche, J. (1977). *The essential Piaget: An interpretive reference and guide*. New York, NY: Basic Books.

Gunkel, D. J. (2003). Second thoughts: Toward a critique of the digital divide. *New Media & Society*, *5*(4), 499–522. doi:10.1177/146144480354003

Guynn, J. (2015, March 4). Meet the woman who coined #BlackLivesMatter. *USA Today*. Retrieved from https://www.usatoday.com/story/tech/2015/03/04/alicia-garza-black-lives-matter/24341593

Guzdial, M. (2014). Limitations of MOOCs for Computing Education-Addressing our needs: MOOCs and technology to advance learning and learning research (Ubiquity symposium). *Ubiquity*. doi:10.1145/2591683

Han, J. H. (2014). Unpacking and repacking the factors affecting students' perceptions of the use of classroom communication systems (CCS) technology. *Computers & Education*, *79*, 159–176. doi:10.1016/j.compedu.2014.07.011

Hannaford, B., Barral, J. K., Rephaeli, E., Ching, C. D., & Bajaj, V. S. (2017). *U.S. Patent No. 9,645,785*. Washington, DC: U.S. Patent and Trademark Office.

Hanson, K., & Shelton, B. E. (2008). Design and Development of Virtual Reality: Analysis of Challenges Faced by Educators. *Journal of Educational Technology & Society*, *11*(1), 118–131.

Hargittai, E. (2002). Second-level digital divide: Differences in people's online skills. *First Monday*, *7*(4), 1–14. doi:10.5210/fm.v7i4.942

Hargittai, E. (2010). Digital na(t)ives? variation in internet skills and uses among members of the "net generation". *Sociological Inquiry*, *80*(1), 92–113. doi:10.1111/j.1475-682X.2009.00317.x

Hariman, R., & Lucaites, J. L. (2001). Dissent and emotional management in a liberal-democratic society: The Kent State iconic photograph. *Rhetoric Society Quarterly*, *31*(3), 5–31. doi:10.1080/02773940109391204

Harris, F. C. (2015). *The challenges of My Brother's Keeper*. New York, NY: The Brookings Institute.

Harris, M. (2017). *Kids These Days: Human Capital and the Making of Millennials*. Little, Brown.

Hathaway, R., Nagda, B., & Gregerman, S. (2002). The Relationship of Undergraduate Research Participation to Graduate and Professional Education Pursuit: An Empirical Study. *Journal of College Student Development*, *43*(5), 614–630.

Hawkins, J. B. (2011). *Bridging the knowledge gap: The effectiveness of compulsory computer-based training in federal employees' professional education* (Doctoral dissertation). Retrieved from https://search-proquest-com.ezproxy.cul.columbia.edu/docview/916240004?accountid=10226 (Order No. 3489818)

Hedgcock, W. H., & Rouwenhorst, R. M. (2014). Clicking their way to success: Using student response systems as a tool for feedback. *Journal for Advancement of Marketing Education*, *22*(2), 16–25.

Heifetz, R. A. (1994). *Leadership without easy answers* (Vol. 465). Harvard University Press.

Heifetz, R. A., & Linsky, M. (2014). *Adaptive Leadership: The Heifetz Collection (3 Items)*. Harvard Business Review Press.

Henderson, S., & Feiner, S. (n.d.). *Augmented Reality for Maintenance and Repair (ARMAR)*. Columbia University Computer Graphics & User Interfaces Lab. Retrieved from http://monet.cs.columbia.edu/projects/armar/

Hennessy, S., Haßler, B., & Hofmann, R. (2015). Challenges and opportunities for teacher professional development in interactive use of technology in African schools. *Technology, Pedagogy and Education*, *24*(5), 1–28. doi:10.1080/1475939X.2015.1092466

Henriksen, D., Mishra, P., & Fisser, P. (2016). Infusing Creativity and Technology in 21st Century Education: A Systemic View for Change. *Journal of Educational Technology & Society*, *19*(3), 27–37.

Heron, J. (1999). *The Complete Facilitator's Handbook*. Sterling, VA: Kogan Page.

Hewlett, S. A. (2013). *Forget a Mentor: Find a Sponsor (Thew New Way to Fast Track Your Career)*. Boston, MA: Harvard Business Review Press.

Holton, E. III, & Baldwin, T. T. (Eds.). (2003). *Improving learning transfer in organizations*. San Francisco: Jossey-Bass.

Hricko, M. (2017). Personal Learning Environments. In *Handbook of Research on Instructional Systems and Educational Technology* (pp. 236–248). IGI Global. doi:10.4018/978-1-5225-2399-4.ch021

Hsu, Y. C., Baldwin, S., & Ching, Y. H. (2017). Learning through making and maker education. *TechTrends*, 1–6.

Hughs, J. E., & Ooms, A. (2004). Content-focused technology inquiry groups: Preparing urban teachers to integrate technology to transform student learning. *Journal of Research on Technology in Education, 36*(4), 397–411. doi:10.1080/15391523.2004.10782422

Hull, R., Neaves, P., & Bedford-Roberts, J. (1997). Towards situated computing. In *Wearable Computers, 1997. Digest of Papers. The First International Symposium on Wearable Computers* (pp. 146-153). IEEE. 10.1109/ISWC.1997.629931

Hunsu, N. J., Adesope, O., & Bayly, D. J. (2016). A meta-analysis of the effects of audience response systems (clicker-based technologies) on cognition and affect. *Computer Education, 94*, 102–119. doi:10.1016/j.compedu.2015.11.013

Hwang, G. J., Tsai, C. C., & Yang, S. J. H. (2008). Criteria, strategies and research issues of context-aware ubiquitous learning. *Journal of Educational Technology & Society, 11*(2), 81–91.

Ifenthaler, D. (2012). Learning Management System. In Encyclopedia of the Sciences of Learning. Springer.

Ingram, M. (2011, February 3). Gladwell still missing the point about social media and activism. *GigaOm*. Retrieved from https://gigaom.com/2011/02/03/gladwell-still-missing-the-point-about-social-media-and-activism/

International Labor Organization. (2015). *Global employment trends for youth 2015: Scaling Up investments in decent jobs for youth*. Geneva: Author.

International Labor Organization. (2016). *World employment social outlook: Trends for youth 2016*. Geneva: Author.

International Labour Organization. (2013). *Global employment trends for youth 2013: A generation at risk*. Geneva: Author. Retrieved from http://www.ilo.org/wcmsp5/groups/public/---dgreports/--dcomm/documents/publication/wcms_212423.pdf,1.10.14

Irby, D. R. (2017). *Middle School Student and Teacher Perceptions About the Effectiveness of the Technology Integration in the Classroom*. Academic Press.

Istepanian, R., Laxminarayan, S., & Pattichis, C. S. (2014). *M-health*. John Wiley & Sons.

Jack, A. A. (2014). Culture Shock Revisited: The Social and Cultural Contingencies to Class Marginality. *Sociological Forum, 29*(2), 453–475. doi:10.1111ocf.12092

Jamu, J. T., Lowi-Jones, H., & Mitchell, C. (2016). Just in time? Using QR codes for multi-professional learning in clinical practice. *Nurse Education in Practice, 19*, 107–112. doi:10.1016/j.nepr.2016.03.007 PMID:27428702

Jenkins, H. (2007, December 7). *Reconsidering Digital Immigrants...* Retrieved December 09, 2017, from http://henryjenkins.org/blog/2007/12/reconsidering_digital_immigran.html

Johnson, L., Adams Becker, S., Estrada, V., & Freeman, A. (2014). *The NMC Horizon Report: 2014 Higher Education Edition*. Retrieved from http://cdn.nmc.org/media/2014-nmc-horizon-report-he-EN-SC.pdf

Johnson, L., Adams Becker, S., Cummins, M., Estrada, V., & Freeman, A. (2014). *NMC Horizon report: 2014 K- 12*. Austin, TX: The New Media Consortium.

Jones, V., & Jo, J. H. (2004). Ubiqutous learning environment: An adaptive teaching system using ubiquitous technology. In R. Atkinson, C. McBeath, D. Jonas-Dwyer, & R. Phillips (Eds.), *Beyond the comfort zone: Proceedings of the 21st ASCILITE Conference* (pp. 468-474). Perth, Australia: ASCILITE.

Kabat-Zinn, J. (2003). Mindfulness-based interventions in context: Past, present, and future. *Clinical Psychology: Science and Practice, 10*(2), 144–156. doi:10.1093/clipsy.bpg016

Kang, H., Lundeberg, M., Wolter, B., delMas, R., & Herreid, C. F. (2012). Gender differences in student performance in large lecture classrooms using personal response systems ('clickers') with narrative case studies. *Learning, Media and Technology, 37*(1), 53–76. doi:10.1080/17439884.2011.556123

Kanuka, H. (2008). Understanding e-Learning Technologies-in-Practice through Philosophies-in-Practice. In T. Anderson & F. Elloumni (Eds.), *Theory and Practice of Online Learning*. Athabasca, Canada: Athabasca University Press.

Kaplan Andreas, M., & Michael, H. (2016). Higher education and the digital revolution: About MOOCs. *SPOCs, social media, and the Cookie Monster, Business Horizon, 59*.

Kardash, C. M. (2000). Evaluation of an Undergraduate Research Experience: Perceptions of Undergraduate Interns and Their Faculty Mentors. *Journal of Educational Psychology, 92*(1), 191–201. doi:10.1037/0022-0663.92.1.191

Kartsonaki, E., Lawrie, G., Sharpe, P., & Bailey, C. (2017, August). Online, virtual, and adaptive learning environments: improving the journey through large first year chemistry courses. In *Proceedings of The Australian Conference on Science and Mathematics Education (formerly UniServe Science Conference)* (p. 74). Academic Press.

Kasworm, C. E., Rose, A. D., & Ross-Gordon, J. M. (Eds.). (2010). *Handbook of adult and continuing education*. Sage.

Kasworm, C., Rose, A. D., & Ross-Gordon, J. M. (2010). *Handbook of Adult and Continuing Education* (2010 ed.). SAGE Publications, Inc.

Kayapa, N., & Tong, T. (2011). Sanal gerçeklik ortamında algı. *Mühendislik ve Fen Bilimleri Dergisi, 3*, 348–354.

Kay, B. H., & LeSage, A. (2009). Examining the benefits and challenges of using audience response systems: A review of the literature. *Computers & Education, 53*(3), 819–827. doi:10.1016/j.compedu.2009.05.001

Kegan, R., & Lahey, L. L. (2009). *Immunity to change: How to overcome it and unlock potential in yourself and your organization*. Harvard Business Press.

Keough, S. M. (2012). Clickers in the classroom: A review and a replication. *Journal of Management Education, 36*(6), 822–847. doi:10.1177/1052562912454808

Kereluik, K., Mishra, P., Fahnoe, C., & Terry, L. (2013). What knowledge is of most worth: Teacher knowledge for 21st century learning. *Journal Of Digital Learning In Teacher Education, 29*(4), 127–140. doi:10.1080/21532974.2013.10784716

Kerr, S. (2011). High school online: Pedagogy, preferences, and practices of three online teachers. *Journal of Educational Technology Systems, 39*(3), 221–244. doi:10.2190/ET.39.3.b

Khaddage, F., Müller, W., & Flintoff, K. (2016). Advancing mobile learning in formal and informal settings via mobile app technology: Where to from here, and how? *Journal of Educational Technology & Society, 19*(3), 16.

Khanna, M. S., & Kendall, P. C. (2008). Computer-assisted CBT for child anxiety: The Coping Cat CD-ROM. *Cognitive and Behavioral Practice, 15*(2), 159–165. doi:10.1016/j.cbpra.2008.02.002

Khazraee, E., & Losey, J. (2016). Evolving repertoires: Digital media use in contentious politics. *Communication and the Public, 1*(1), 39–55. doi:10.1177/2057047315625076

Khoury, B., Sharma, M., Rush, S. E., & Fournier, C. (2015). Mindfulness-based stress reduction for healthy individuals: A meta-analysis. *Journal of Psychosomatic Research, 78*(6), 519–528. doi:10.1016/j.jpsychores.2015.03.009 PMID:25818837

Khurgin, A. (2017). Microlearning for Transformation, Not Information Transfer (Webinar video). *Training Magazine Network*. Retrieved from https://youtu.be/kGff91akTJ0

King, A., Crawford, P., San Miguel, C., Barrie, M., Spencer, S., Tyransky, A., & Kman, N. (2017). Novel Emergency Medicine Curriculum Utilizing Self-Directed Learning and the Flipped Classroom Method: Head, Eyes, Ears, Nose and Throat Emergencies Small Group Module. *Journal of Education and Teaching in Emergency Medicine*, *2*(4).

Kleiman, G. M. (2004). Myths and realities about technology in K-12 schools: Five years later. *Contemporary Issues in Technology & Teacher Education*, *4*(2), 248–253.

Knowles, M. S. (1970). *The modern practice of adult education* (Vol. 41). New York, New York: Association Press.

Knowles, M. S. (1984). *The adult learner: A neglected species* (3rd ed.). Houston, TX: Gulf.

Knox, J. (2017). How goes the revolution? Three themes in the shifting MOOC landscape. *Educação e Tecnologia: Abordagens Criticas*, *1*, 386–408.

Kohut, A. (2011, October 27). 'Haves' and 'have-nots.' *The New York Times*. Retrieved from http://www.nytimes.com/roomfordebate/2011/10/18/the-psychology-of-/occupy-wallstreet/occupy-wall-streets-taps-into-longstanding-concerns?scp=1&sq=kohut&st=cse

Kolb, A. Y., & Kolb, D. A. (2012). Experiential learning theory. In Encyclopedia of the Sciences of Learning (pp. 1215-1219). Springer US.

Kolb, D. (1984). Structural Foundations of the Learning Process. In Experiential learning: Experience as the source of learning and development. Englewood Cliffs, NJ: Prentice Hall.

Kolb, D. (1984). The Process of Experiential Learning. In Experiential learning: Experience as the source of learning and development (pp. 20-38). Englewood Cliffs, NJ: Prentice Hall.

Kolb, D. A. (2014). *Experiential learning: Experience as the source of learning and development*. FT Press.

Kolb, D. (1983). *Experiential Learning: Experience as the Source of Learning and Development*. Upper Saddle River, NJ: Prentice Hall.

Kolb, D. A. (1984). *Experience as the source of learning and development. Prentice-Hall.*

Kolesinski, M., Nelson-Weaver, E., & Diamond, D. (2013). Digital solidarity in education: promoting equity, diversity, and academic excellence through innovative instructional programs. In T. Bastiaens & G. Marks (Eds.), *proceedings of e-learn: world conference on e-learning in corporate, government, healthcare, and higher education 2013* (pp. 2228–2233). Association for the Advancement of Computing in Education.

Ko, S., & Rossen, S. (2017). *Teaching online: A practical guide*. Taylor & Francis.

Kounaves, S., Archer, L., King, H., & Pegram, E. (2016). Science Learning and Engagement in the Digital Age: Understanding the Effect of Mobile Technology on Adult Engagement Experiences at a Natural History Museum. *Mobile Learning Futures–Sustaining Quality Research and Practice in Mobile Learning*, 361.

Kozol, J. (1991). *Savage inequalities: children in America's schools*. New York: Harper-Collins.

KPMG. (2015). *Corporate digital learning*. Retrieved from https://assets.kpmg.com/content/dam/kpmg/pdf/2015/09/corporate-digital-learning-2015-KPMG.pdf

Kress, T. (2011). Going high tech under high surveillance: Technology integration, zero tolerance, and implications for access and equity. *Radical Teacher*, *90*(1), 15–24. doi:10.5406/radicalteacher.90.0015

Krygier, J. R., Heathers, J. A., Shahrestani, S., Abbott, M., Gross, J. J., & Kemp, A. H. (2013). Mindfulness meditation, well-being, and heart rate variability: A preliminary investigation into the impact of intensive Vipassana meditation. *International Journal of Psychophysiology*, *89*(3), 305–313. doi:10.1016/j.ijpsycho.2013.06.017 PMID:23797150

Kumar, S., Nilsen, W. J., Abernethy, A., Atienza, A., Patrick, K., Pavel, M., ... Hedeker, D. (2013). Mobile health technology evaluation: The mHealth evidence workshop. *American Journal of Preventive Medicine*, *45*(2), 228–236. doi:10.1016/j.amepre.2013.03.017 PMID:23867031

Kuzu Demir, E. B., & Demir, K. (2016). Enhancing Learning with Wearable Technologies In and Out of Educational Settings. *Digital Tools for Seamless Learning, 119.*

Kuzu, E. B. (2014). *Hizmet öncesi BT öğretmenleri arasında çevrimiçi sosyal ağların eğitim amacıyla kullanılması. Yayımlanmamış doktora tezi.* Eğitim Bilimleri Enstitüsü, Anadolu Üniversitesi, Türkiye.

Kvasny, L. (2006). Cultural (re) production of digital inequality in a US community technology initiative. *Information Communication and Society*, *9*(2), 160–181. doi:10.1080/13691180600630740

Kvasny, L., & Mark, K. (2006). The challenges of redressing the digital divide: A tale of two US cities. *Information Systems Journal*, *16*(1), 23–53. doi:10.1111/j.1365-2575.2006.00207.x

Ladson-Billings, G. (1995). Toward a theory of culturally relevant pedagogy. *Educational Research*, *32*(3), 465–491.

Lai, C., Wang, Q., & Lei, J. (2012). What factors predict undergraduate students' use of technology for learning? A case from Hong Kong. *Computers & Education*, *59*(2), 569–579. doi:10.1016/j.compedu.2012.03.006

Lai, G., Hill, V., & Ma, Y. (2015). Clickers in the classroom: A business professor's adoption of a classroom response system. *International Journal of Innovation and Learning*, *18*(4), 451–470. doi:10.1504/IJIL.2015.072458

Landsman, J., & Lewis, C. W. (2011). *White teachers, diverse classrooms: Creating inclusive schools, building on students' diversity, and providing true educational equity*. Sterling, VA: Stylus Pub.

Lantigua-Williams, J. (2016, July 13). How much can better training do to improve policing? *The Atlantic*. Retrieved from https://www.theatlantic.com/politics/archive/2016/07/police-training/490556/

Lantz, M. E. (2010). The use of 'clickers' in the classroom: Teaching innovation or merely an amusing novelty? *Computers in Human Behavior*, *26*(4), 556–561. doi:10.1016/j.chb.2010.02.014

Lantz, M. E., & Stawiski, A. (2014). Effectiveness of clickers: Effect of feedback and the timing of questions on learning. *Computers in Human Behavior*, *31*, 280–286. doi:10.1016/j.chb.2013.10.009

Latham, A., & Hill, N. S. (2014). Preference for anonymous classroom participation: Linking student characteristics and reactions to electronic response systems. *Journal of Management Education*, *38*(2), 192–215. doi:10.1177/1052562913488109

Lave, J., & Wenger, E. (1998). *Communities of practice.* Academic Press.

Lawless, K., & Pellegrino, J. W. (2007). Technology into teaching and learning: Knowns, unknowns, and ways to pursue better questions and answers. *Review of Educational Research*, *77*(4), 575–614. doi:10.3102/0034654307309921

Lazar, S. (2011, October 10). Occupy Wall Street: Interview with Micah White from Adbusters. *The Huffington Post.* Retrieved from http://www.huffingtonpost.com/shira-lazar/micah-white-adbusters-_b_996931.html

Lazaroiu, G., Popescu, G. H., & Nica, E. (2016, July). Democratizing education: the potential of EdX in revolutionizing learning. In *The International Scientific Conference eLearning and Software for Education* (*Vol. 3*, p. 34). "Carol I" National Defence University.

Lazaroiu, G., Popescu, G. H., & Nica, E. (2016, July). The sustainability of udacity's business model of producing first-rate online content and incorporating interactive learning aspects into an online course. In *The International Scientific Conference eLearning and Software for Education* (*Vol. 3*, p. 40). "Carol I" National Defence University.

Leffler, M. E. (2015). Digitally divided in Jackson: Are students getting the digital literacy skills they need to succeed? (Doctoral dissertation). Jackson State University.

LeHong, H., Fenn, J., & Leeb-du Toit, R. (2013). *Hype cycle for emerging technologies*. Stamford, CT: Gartner Inc.

Lemos, M. H., & Charles, G. U. E. (2017). *Patriotic Philanthropy?* Financing the State with Gifts to Government.

Lenhart, A., Purcell, K., Smith, A., & Zickuhn, K. (2010). Social media & mobile internet use among teens and young adults. *Pew Internet & American Life Project*. Pew Research Center. Retrieved 5/20/17 from http://www.pewinternet.org/w/media//Files/Reports/ 2010/PIP_Social_Media_and_Young_Adults_Report_Final_with_toplines.pdf

Lenhart, A., Madden, M., Smith, A., & Macgill, A. (2007). *Teens and social media. Pew Internet & American Life Project*. Pew Research Center.

Leue, M. C., Jung, T., & Dieck, D. (2015). Google glass augmented reality: Generic learning outcomes for art galleries. In *Information and Communication Technologies in Tourism 2015* (pp. 463–476). Cham: Springer. doi:10.1007/978-3-319-14343-9_34

Lewis, J. M., & Hensley, T. R. (1998). The May 4 shootings at Kent State University: The search for historical accuracy. *The Ohio Council for the Social Studies Review, 34*(1), 9–21. Retrieved from http://www.stetson.edu/law/conferences/highered/archive/media/higher-ed-archives-2009/ii-lewis-may-4-shooting-at-kent-state-pdf.pdf

Lewis, G. A. (2017). Cloud computing. *Computer*, *50*(5), 8–9. doi:10.1109/MC.2017.141

Lewis, O. (1959). *Five families: Mexican case studies in the culture of poverty*. New York: Basic Books.

Liebowitz, J., & Frank, M. (Eds.). (2016). *Knowledge management and e-learning*. CRC press.

Li, J., Snow, C., & White, C. (2015). Teen culture, technology and literacy instruction: Urban adolescent students' perspectives. *Canadian Journal of Learning and Technology*, *41*(3), 1–36. doi:10.21432/T2004H

Limniou, M., Roberts, D., & Papadopoulos, N. (2008). Full immersive virtual environment CAVE TM in chemistry education. *Computers & Education*, *51*(2), 584–593. doi:10.1016/j.compedu.2007.06.014

Lin, H. F. (2013). The effect of absorptive capacity perceptions on the context-aware ubiquitous learning acceptance. *Campus-Wide Information Systems*, *30*(4), 249–265. doi:10.1108/CWIS-09-2012-0031

Li, Q. (2017). Characteristics and social impact of the use of social media by Chinese Dama. *Telematics and Informatics*, *34*(3), 797–810. doi:10.1016/j.tele.2016.05.020

Littlejohn, A., Hood, N., Milligan, C., & Mustain, P. (2016). Learning in MOOCs: Motivations and self-regulated learning in MOOCs. *The Internet and Higher Education*, *29*, 40–48. doi:10.1016/j.iheduc.2015.12.003

Liu, D., Dede, C., Huang, R., & Richards, J. (2017). *Virtual*. Augmented, and Mixed Realities in Education.

Livet, M., Yannayon, M., Sheppard, K., Kocher, K., Upright, J., & McMillen, J. (2017). Exploring provider use of a digital implementation support system for school mental health: A pilot study. *Administration and Policy in Mental Health and Mental Health Services,* 1-19. DOI:10.100710488-017-0829-7

Loksa, D., Ko, A. J., Jernigan, W., Oleson, A., Mendez, C. J., & Burnett, M. M. (2016, May). Programming, Problem Solving, and Self-Awareness: Effects of Explicit Guidance. In *Proceedings of the 2016 CHI Conference on Human Factors in Computing Systems* (pp. 1449-1461). ACM. 10.1145/2858036.2858252

Lombard, J. (2016, August 10). Social media skills you need to hone now. *Media Bistro*. Retrieved from https://www.mediabistro.com/climb-the-ladder/social-media-skills/

Lombardo, M. M., & Eichinger, R. W. (1996). *The career architect development planner*. Minneapolis, MN: Lominger Limited.

Looi, C.-K., Seow, P., Zhang, B., So, H.-J., Chen, W., & Wong, L.-H. (2010). Leveraging mobile technology for sustainable seamless learning: A research agenda. *British Journal of Educational Technology*, *41*(2), 154–169. doi:10.1111/j.1467-8535.2008.00912.x

Lopatto, D. (2004). Survey of Undergraduate Research Experiences (SURE): First Findings. *Cell Biology Education*, *3*(4), 270–277. doi:10.1187/cbe.04-07-0045 PMID:15592600

Lotan, G., Graeff, E., Ananny, M., Gaffney, D., & Pearce, I., & boyd, d. (2011). The Arab Spring: The role of ICTs. The revolutions were tweeted: Information flows during the 2011 Tunisian and Egyptian revolutions. *International Journal of Communication*, *5*(2001), 1375–1405. Retrieved from http://ijoc.org/index.php/ijoc/article/view/1246/643

Loup-Escande, E., Frenoy, R., Poplimont, G., Thouvenin, I., Gapenne, O., & Megalakaki, O. (2017). Contributions of mixed reality in a calligraphy learning task: Effects of supplementary visual feedback and expertise on cognitive load, user experience, and gestural performance. *Computers in Human Behavior*, *75*, 42–49. doi:10.1016/j.chb.2017.05.006

Lua, A. (2017, February 23). 10 important skills your social media manager needs in 2017. *Entrepreneur*. Retrieved from https://www.entrepreneur.com/article/289363

Lubin, I. A. (2016). *Intentional ICT: Curriculum, education and development* (IBE Working Papers on Curriculum Issues No. 17). Geneva, Switzerland: IBE-UNESCO.

Luibrand, S. (2015, August 7). Black Lives Matter: How the events in Ferguson sparked a movement in America. *CBS News*. Retrieved from http://www.cbsnews.com/news/how-the-black-lives-matter-movement-changed-america-one-year-later/

Lu, J., Pan, Z., Lin, H., Zhang, M., & Shi, J. (2005). Virtual learning environment for medical education based on VRML and VTK. *Computers & Graphics*, *29*(2), 283–288. doi:10.1016/j.cag.2004.12.015

Lynch, B. (2008, May 1). Startup Poll Everywhere invented the technology to conduct real time polls through cell phones. *Boston Business Journal*.

Lynch, M. (2011, January 6). Obama's 'Arab Spring'? *Foreign Policy*. Retrieved from http://foreignpolicy.com/2011/01/06/obamas-arab-spring/

Maag, M. (2006). *iPod, uPod? An emerging mobile learning tool in nursing education and students' satisfaction*. Australasian Society for Computers in Learning in Tertiary Education (ASCILITE).

MacGeorge, E. K., Homan, S. R., Dunning, J. B., Elmore, D., Bodie, G. D., Evans, E., ... Lichti, S. M. (2008). The influence of learning characteristics on evaluation of audience response technology. *Journal of Computing in Higher Education*, *19*(2), 25–46. doi:10.1007/BF03033425

Magid, L. (2013, June 4). *A Third of Recently Married Couples Met Online and They're More Satisfied and Less Likely To Split-Up*. Retrieved December 09, 2017, from https://www.forbes.com/sites/larrymagid/2013/06/03/a-third-of-recently-married-couples-met-online-and-theyre-more-satisfied-and-less-likely-to-split-up/#2c62c61f595e

Majchrzak, A., Rice, R. E., King, N., Malhotra, A., & Ba, S. (2014). *Computer-mediated inter-organizational knowledge-sharing: Insights from a virtual team innovating using a collaborative tool.* Academic Press.

Malanga, D. F. (2017). Implementation of Mobile Health Initiatives in Malawi: Current Status, Issues, and Challenges. In Health Information Systems and the Advancement of Medical Practice in Developing Countries (pp. 115-128). IGI Global.

Manning, M. L., Baruth, L. G., & Lee, G. L. (2017). *Multicultural education of children and adolescents*. Taylor & Francis.

Marin, V. I., Jaakela, P., Hakkinen, P., & Juntunen, M. (2016). Seamless Learning Environments in Higher Education with Mobile Devices and Examples. *International Journal of Mobile and Blended Learning*, *8*(1), 51–68. doi:10.4018/IJMBL.2016010104

Marist College Institute for Public Opinion. (2015, September). Summary of National Survey findings. *PBS NewsHour/Marist Poll*. Retrieved from https://www.dropbox.com/s/j64c02ub48aipja/PBS-NEWSHOUR-MARIST-POLL-SEP2015.pdf

Marks, D. (2009). Literacy, instruction, and technology: Meeting Millennials on their own turf. *AACE Journal*, *17*(4), 363–377.

Marsick, V. J., & Neaman, A. (2017). Adult informal learning. In Informelles lernen. Eine Standortbestimmung. Wiesbaden: Springer VS.

Marsick, V. J., Watkins, K. E., Scully-Russ, E., & Nicolaides, A. (2017). Rethinking informal and incidental learning in terms of complexity and the social context. *Journal of Adult Learning, Knowledge and Innovation,* (0), 1-8.

Marsick, V. J. (1998). Transformative learning from experience in the knowledge era. *Daedalus*, *127*(4), 119–136.

Marsick, V. J., & Volpe, M. (1999). The nature and need for informal learning. *Advances in Developing Human Resources*, *1*(3), 1–9. doi:10.1177/152342239900100302

Marsick, V. J., Volpe, M., & Watkins, K. E. (1999). Theory and practice of informal learning in the knowledge era. *Advances in Developing Human Resources*, *1*(3), 80–95. doi:10.1177/152342239900100309

Marsick, V. J., & Watkins, K. E. (1990). *Informal and incidental learning in the workplace*. London, UK: Routledge.

Marsick, V. J., & Watkins, K. E. (2001). Informal and incidental learning. *New Directions for Adult and Continuing Education*, *2001*(89), 25–34. doi:10.1002/ace.5

Marsick, V. J., Watkins, K. E., Scully-Russ, E., & Nicolaides, A. (2017). *Rethinking informal and incidental learning in terms of complexity and the social context. Journal of Adult Learning, Knowledge and Innovation*. doi:10.1556/2059.01.2016.003

Martinez Witte, M., & Wolf, S. E. (2003). Infusing Mentoring and Technology Within Graduate Courses: Reflections in Practice. *Mentoring & Tutoring: Partnership in Learning, 11*(1).

Masikunis, G., Panayiotidis, A., & Burke, L. (2009). Changing the nature of lectures using a personal response system. *Innovations in Education and Teaching International*, *46*(2), 199–212. doi:10.1080/14703290902843935

Masrom, M., Nadzari, A. S., Mahmood, N. H., Zakaria, W. N., & Ali, N. R. (2016). Mobile learning in Malaysia education institutions. *Issues in Information Systems*, *17*(4), 152–157.

Masters, J., de Kramer, R. M., O'Dwyer, L. M., Dash, S., & Russell, M. (2012). The effects of online teacher professional development on fourth grade students' knowledge and practices in English language arts. *Journal of Technology and Teacher Education*, *20*(1), 21–46.

Masters, K., & Ellaway, R. (2008). e-Learning in medical education Guide 32 Part 2: Technology, management, and design. *Medical Teacher, 30*(5), 474–489. doi:10.1080/01421590802108349 PMID:18576186

Masters, K., Ellaway, R. H., Topps, D., Archibald, D., & Hogue, R. J. (2016). Mobile technologies in medical education: AMEE Guide No. 105. *Medical Teacher, 38*(6), 537–549. doi:10.3109/0142159X.2016.1141190 PMID:27010681

May, A. (2016, July 13). #AllLivesMatter hashtag is racist, critics say. *USA Today*. Retrieved from https://www.usatoday.com/story/news/nation-now/2016/07/13/why-saying-all-lives-matter-opposite-black-lives-matter/87025190/

Maycock, K. W., & Keating, J. G. (2017). The impact of an automated learning component against a traditional lecturing environment. *Journal of Computer Assisted Learning, 33*(6), 597–605.

Mayer, R. E., Stull, A., DeLeeuw, K., Almeroth, K., Bimber, B., Chun, D., ... Zhang, H. (2009). Clickers in college classrooms: Fostering learning with questioning methods in large lecture classes. *Contemporary Educational Psychology, 34*(1), 51–57. doi:10.1016/j.cedpsych.2008.04.002

Mazur, E. (2017). Diverse Disabilities and Dating Online. In *Identity, Sexuality, and Relationships among Emerging Adults in the Digital Age* (pp. 150–167). IGI Global. doi:10.4018/978-1-5225-1856-3.ch010

McCann. (2016). Mobile Learning in Workforce Development: Cultivating Creativity on Action Learning Teams through Higher-Order mLearning. In Handbook of Research on Mobile Learning in Contemporary Classrooms. Ed. Dominic Mentor. IGI Global. doi:10.4018/978-1-5225-0251-7.ch007

McCarthy, J. (2007). *What is artificial intelligence?* Retrieved from http://www-formal.stanford.edu/jmc/whatisai/whatisai.html

McDermott, I. E. (2016, November). Fun with virtual and augmented reality. *Online Searcher,* 27-29.

McEvers, K. (2016a, March 7). 'The New Yorker' charts the rise of Black Lives Matter – with Jelani Cobb. *NPR* [Radio podcast]. Retrieved from http://www.npr.org/sections/codeswitch/2016/03/2016/03/07/469545405/the-new-yorker-charts-the-rise-of-black-lives-matter?live=1

McEvers, K. (2016b, March 9). Jelani Cobb on his epic 'New Yorker' piece on Black Lives Matter. *NPR* [Radio podcast]. Retrieved from http://www.npr.org/sections/codeswitch/2016/03/09/469627740/jelani-cobb-on-his-epic-new-yorker-piece-on-black-lives-matter

McGrath, J., Mesiwala, A., Barrie, M., Boulger, C., Thompson, L., Mitzman, J., ... King, A. (2017). Novel Emergency Medicine Curriculum Utilizing Self-Directed Learning and the Flipped Classroom Method: Obstetric and Gynecologic Emergencies Small Group Module. *Journal of Education and Teaching in Emergency Medicine, 2*(4).

McLoughlin, C., & Lee, M. (2007). Social software and participatory learning: Pedagogical choices with technology affordances in the Web 2.0 era. ICT: Providing choices for learners and learning. Proceedings ASCILITE Singapore 2007.

McLuhan, M. (1964). *Understanding media: The extensions of man.* New York, NY: McGraw Hill.

McNally, D. (2010). *Global slump: The economics and politics of crisis and resistance.* Oakland, CA: PM Press.

McQuiggan, S., McQuiggan, J., Sabourin, J., & Kosturko, L. (2015). *Mobile learning: A handbook for developers, educators, and learners.* John Wiley & Sons. doi:10.1002/9781118938942

Meier, E. (2016). Beyond a Digital Status Quo: Re-conceptualizing Online Learning Opportunities. *Occasional Paper Series, 2015*(34), 2.

Meinel, C., Totschnig, M., & Willems, C. (2013). openHPI: Evolution of a MOOC platform from LMS to SOA. In *Proceedings of the 5th International Conference on Computer Supported Education (CSEDU), INSTICC, Aachen, Germany (Vol. 5)*. Academic Press.

Mentor, D. (2011). We Are New York Social Media for Adult Education. In C. Ho & M. Lin (Eds.), *Proceedings of E-Learn 2011--World Conference on E-Learning in Corporate, Government, Healthcare, and Higher Education* (pp. 1499-1504). Honolulu, HI: Association for the Advancement of Computing in Education (AACE). Retrieved December 24, 2017 from https://www.learntechlib.org/p/38929/

Mentor, D. (2014). *mClass: How to adapt to ever changing mlearning environments*. Presentation delivered at the 21st Century Academic Forum Conference at Harvard University The 21st century academic forum conference at Harvard, Cambridge, MA. doi: http://www.21caf.org/uploads/1/3/5/2/13527682/21st_century_academic_forum_conference_book.pdf

Mentor, D. (2014, February 14). Symposium Presentations. *UNESCO Mobile Learning Week 2014*. Retrieved March 12, 2017, from http://www.unesco.org/new/en/unesco/themes/icts/m4ed/unesco-mobile-learning-week-2014/

Mentor, D. (2014a). *M-Class: Mobile learning for teaching and learning*. Presented at the UNESCO Mobile Learning Week 2014, UNESCO Headquarters, Paris, France.

Mentor, D. (2014b). *Creating e- and m-learning Foundations to Engage Young Adults in a Workforce Development Program*. Presentation delivered at the ICELW 2014: The International Conference on E-Learning in the Workplace at Columbia University, New York, NY.

Mentor, D. (2016). EMxC3= e&mLearning Cultivating Connected Communities: Sustainable Workforce Talent Development. Handbook of Research on Mobile Learning in Contemporary Classrooms, 240-259.

Mentor, D. (2016). EMxC3= e&mLearning Cultivating Connected Communities: Sustainable Workforce Talent Development. In Handbook of Research on Mobile Learning in Contemporary Classrooms (pp. 240-259). Hershey, PA: IGI Global.

Mentor, D. (2016a). The Commuter's Learning Journey: Field Observations Informing Mobile Learning Initiatives. In Handbook of Research on Mobile Learning in Contemporary Classrooms (pp. 315-335). IGI Global.

Mentor, D. (2016b). The Commuter's Learning Journey: Field Observations Informing Mobile Learning Initiatives. In Handbook of Research on Mobile Learning in Contemporary Classrooms (pp. 315-335). IGI Global.

Mentor, D. (2016c). EMxC3= e&mLearning Cultivating Connected Communities: Sustainable Workforce Talent Development. Handbook of Research on Mobile Learning in Contemporary Classrooms, 240-259.

Mentor, D. (2017, April 25). Cultivating Digital TLC – Teaching and Learning Communities. *ModSim World 2017*. Retrieved July 23, 2017, from http://www.modsimworld.org/2017/documents/Working-Agenda-17-April.pdf

Mentor, D. (2017, April 27). Cultivating Digital TLC Teaching and Learning Communities. In *ModSim 2017 International Congress on Modelling and Simulation - Modeling and Simulation in the Age of Data*. Retrieved December 24, 2017, from http://modsimworld.org/papers/2017/Cultivating_Digital_TLC_Teaching_and_Learning_Communities.pdf

Mentor, D. (2018a). *Creating Teaching and Learning Communities for MOOCS*. Presented at the UNESCO Mobile Learning Week's Skills for a connected world 2018, UNESCO Headquarters, Paris, France.

Mentor, D. (2018b). *Including Digital Proficiency for Teaching and Learning benchmarks in a workforce development program*. Presented at the UNESCO Mobile Learning Week's Skills for a connected world 2018, UNESCO Headquarters, Paris, France.

Mentor, D. J. (2011). *Exploring social connectedness via mobile phone texting* (Ph.D. thesis). Teachers College, Columbia University. Retrieved December 25, 2017 from https://www.learntechlib.org/p/117314/

Mentor, D., & Ahmad, N. (2010, September 22). *Teaching America's First Course on Mobile Phone Learning*. Retrieved December 09, 2017, from https://er.educause.edu/articles/2010/9/teaching-americas-first-course-on-mobile-phone-learning

Mentor, D. (2015). From App Attack to Goal-Oriented Tablet Use. In H. An, S. Alon, & D. Fuentes (Eds.), *Tablets in K-12 Education: Integrated Experiences and Implications* (pp. 1–21). Hershey, PA: IGI Global. doi:10.4018/978-1-4666-6300-8.ch001

Mentor, D. (2015). Mobile phone use enhances social connectedness. In Z. Yan (Ed.), *Encyclopedia of mobile phone behavior* (pp. 673–693). Hershey, PA: IGI Global; doi:10.4018/978-1-4666-8239-9.ch056

Mentor, D. (2016a). EMxC3 = e&mLearning Cultivating Connected Communities: Sustainable Workforce Talent Development. In D. Mentor (Ed.), *Handbook of Research on Mobile Learning in Contemporary Classrooms* (pp. 240–259). Hershey, PA: IGI Global. doi:10.4018/978-1-5225-0251-7.ch012

Mentor, D. (2016a). *Handbook of Research on Mobile Learning in Contemporary Classrooms*. Hershey, PA: IGI Global. doi:10.4018/978-1-5225-0251-7

Mentor, D. (2016b). Empowering Urban Students through Regular and Collated Student Evaluation of Teachers. Presented at the *International Conference on Urban Education (ICUE 2016) Biennial Meeting San Juan*, Puerto Rico.

Mentor, D. (2018). Micro to macro social connectedness through mobile phone engagement. In D. B. A. Mehdi Khosrow-Pour (Ed.), *Encyclopedia of information science and technology* (4th ed.; pp. 6184–6194). Hershey, PA: IGI Global. doi:10.4018/978-1-5225-2255-3.ch537

Mentor, D., & Ahmad, N. (2010). Teaching America's First Course on Mobile Phone Learning. *EDUCAUSE Quarterly*, 33.

Mercer, S., & Ryan, S. (2010). A mindset for EFL: Learners' beliefs about the role of natural talent. *ELT Journal*, *64*(4), 436–444. doi:10.1093/elt/ccp083

Merchant, Z., Goetz, E. T., Cifuentes, L., Keeney-Kennicutt, W., & Davis, T. J. (2014). Effectiveness of virtual reality-based instruction on students' learning outcomes in K-12 and higher education: A meta-analysis. *Computers & Education*, *70*, 29 40. doi:10.1016/j.compedu.2013.07.033

Merriam, S. B., Caffarella, R. S., & Baumgartner, L. M. (2012). *Learning in adulthood: A comprehensive guide*. John Wiley & Sons.

Merriam, S., Caffarella, R., & Bamgartner, L. (2007). *Learning in Adulthood*. San Francisco, CA: Jossey-Bass.

Mezirow, J. (2000). Learning as Transformation: Critical Perspectives on a Theory in Progress. The Jossey-Bass Higher and Adult Education Series. Jossey-Bass Publishers.

Milkman, R., Luce, S., & Lewis, P. (2013). *Changing the subject: A bottom-up account of Occupy Wall Street in New York City*. New York, NY: Murphy Institute of the City University of New York. Retrieved from https://media.sps.cuny.edu/filestore/1/5/7/1_a05051d2117901d/1571_92f562221b8041e.pdf

Miller, H., & Griffiths, M. (2005). E-mentoring. In D. L. DuBois & M. J. Karcher (Eds.), *Handbook of Youth Mentoring*. Thousand Oaks, CA: Sage Publications. doi:10.4135/9781412976664.n20

Mi, M., Wu, W., Qiu, M., Zhang, Y., Wu, L., & Li, J. (2016). Use of Mobile Devices to Access Resources Among Health Professions Students: A Systematic Review. *Medical Reference Services Quarterly*, *35*(1), 64–82. doi:10.1080/02763869.2016.1117290 PMID:26794197

Minic, B., & Olenik, C. (2017, December 27). *A Demand-Driven Training Toolkit for Youth Employment*. Retrieved from http://www.makingcents.com/single-post/2017/12/27/A-Demand-Driven-Training-Toolkit-for-Youth-Employment

Mirani, L. (2010, October 2). Sorry, Malcolm Gladwell, the revolution may well be tweeted. *The Guardian*. Retrieved from https://www.theguardian.com/commentisfree/cifamerica/2010/oct/02/malcolm-gladwell-social-networking-kashmir

Mirazchiqyski, P. (2016). The digital divide: The role of socioeconomic status across countries. *Solsko Polje*, *27*(3/4), 23–52.

Mirrlees, T., & Alvi, S. (2016). *The MOOC: Rhetoric*. Political Economy and the Value of Technological Citizenship.

MIT Technology Review. (2017, October 10). *The way strangers meet via dating websites is changing society in unexpected ways, say researchers*. Retrieved December 09, 2017, from https://www.technologyreview.com/s/609091/first-evidence-that-online-dating-is-changing-the-nature-of-society/

Mohyeldin, A. (2011, January 20). Suicide sparked Tunisia revolution. *Al Jazeera English, Opinion*. Retrieved from http://english.aljazeera.net/indepth/opinion/2010/12/20101231161958792947.html

Montgomery, K. (n.d.). *Silicon Valley Now Selling Trade School Diplomas Called "Nanodegrees"*. Retrieved December 10, 2017, from http://valleywag.gawker.com/silicon-valley-now-selling-trade-school-diplomas-called-1638663780

Moore, K. (1999). Political protest and institutional change: The anti-Vietnam war movement and American science. In M. Giugni, D. McAdam, & C. Tilly (Eds.), *How social movements matter* (pp. 97–118). Minneapolis, MN: University of Minnesota Press.

Moraru, I. (2015, October 1). *Insight byte*. New York, NY: Accenture.

Moreno, R., & Mayer, R. E. (2000). A coherence effect in multimedia learning: The case for minimizing irrelevant sounds in the design of multimedia instructional messages. *Journal of Educational Psychology*, *92*(1), 117–125. doi:10.1037/0022-0663.92.1.117

Morino, M. (1997). *The impact of technology on youth in the 21st century*. Morino Institute.

Morino, M. (2001). *From access to outcomes: raising the aspirations for technology initiatives in low-income communities*. Morino Institute.

Morozov, E. (2010, February 20). The digital dictatorship. *The Wall Street Journal*. Retrieved from https://www.wsj.com/articles/SB10001424052748703983004575073911147404540

Morozov, E. (2009). Iran: Downside to the "Twitter revolution.". *Dissent*, *56*(4), 10–14. doi:10.1353/dss.0.0092

Moss, J., & Crowley, M. (2011). Effective learning in science: The use of personal response systems with a wide range of audiences. *Computers & Education*, *56*(1), 36–43. doi:10.1016/j.compedu.2010.03.021

Motsching-Pitrik, R., & Holzinger, A. (2002). Student-Centered Teaching Meets New Media: Concept and Case Study. *Journal of Educational Technology & Society*, *5*(4).

Mousa, A. A., & El-Salam, M. A. (2016). Employing QR Code as an Effective Educational Tool for Quick Access to Sources of Kindergarten Concepts. *World Academy of Science, Engineering and Technology, International Journal of Social, Behavioral, Educational, Economic. Business and Industrial Engineering*, *10*(7), 2358–2361.

Mouza, C. (2011). Promoting urban teachers' understanding of technology, content, and pedagogy in the context of case development. *Journal of Research on Technology in Education*, *44*(1), 1–29. doi:10.1080/15391523.2011.10782577

Mouza, C., & Barrett-Greenly, T. (2015). Bridging the app gap: An examination of a professional development initiative on mobile learning in urban schools. *Computers & Education*, *88*, 1–14. doi:10.1016/j.compedu.2015.04.009

Muhammad, K. G. (2014, December 1). The revolution will be live-tweeted: Why #BlackLivesMatter is the new model for civil rights. *The Guardian*. Retrieved from https://www.theguardian.com/commentisfree/2014/dec/01/black-lives-matter-civil-rights-movement-ferguson

Muhammad, A. J., Mitova, M. A., & Wooldridge, D. G. (2016). Utilizing Technology To Enhance Learning Environments: The Net Gen Student. *Journal of Family and Consumer Sciences*, *108*(2), 61–63. doi:10.14307/JFCS108.2.61

Mulcahy, K. L. J. (2017). *The effects of resources on the performance of competitive high school marching bands* (Doctoral dissertation). Boston University.

Müller, N., & Dikke, D. (2012). Towards Mobile Personalized Learning Management Systems. In *IADIS International Conference Mobile Learning* (pp. 115-122). Academic Press.

Müller, D. (2012). *Design Characteristics of Virtual Learning Environments: A Theoretical Integration and Empirical Test of Technology Acceptance and IS Success Research*. Springer Science & Business Media.

Murray, J. H. (2017). *Hamlet on the holodeck: The future of narrative in cyberspace*. MIT press.

Nabi, R. L., & Krcmar, M. (2016). It takes two: The effect of child characteristics on U.S. parents' motivations for allowing electronic media use. *Journal of Children and Media*, *10*(3), 285–303. doi:10.1080/17482798.2016.1162185

Nagda, B. A., Gregerman, S. R., Jonides, J., von Hippel, W., & Lerner, J. S. (1998). Undergraduate Student-Faculty Research Partnerships Affect Student Retention. *The Review of Higher Education*, *22*(1).

Nagin, D. (2016, October 16). Police, violence, and data: The BlackLivesMatter movement. *Heinz College News, Carnegie Mellon University*. Retrieved from http://www.heinz.cmu.edu/news/news-detail/index.aspx?nid=3860

Naismith, L., Sharples, M., Vavoula, G., & Lonsdale, P. (2004). *Literature review in mobile technologies and learning*. Academic Press.

Naismith, L., Lonsdale, P., Vavoula, G., & Sharples, M. (2004). *Literature Review in Mobile Technologies and Learning (Futurelab Series Report 11)*. Bristol: Futurelab.

Naismith, L., Lonsdale, P., Vavoula, G., & Sharples, M. (2004). *Literature review in mobile technologies and Learning*. NESTA Futurelab Series.

National Academy of Science, National Academy of Engineering, and Institute of Medicine. (2011). Expanding Underrepresented Minority Participation: America's Science and Technology Talent at the Crossroads. Washington, DC: The National Academies Press.

National Academy of Sciences. (2007). *Understanding Interventions that Encourage Minorities to Pursue Research Careers. Summary of a Workshop*. Washington, DC: Author.

National Institutes of Health. (2014). *Physician-Scientist Workforce Working Group Report*. Retrieved from http://acd.od.nih.gov/reports/PSW_Report_ACD_06042014.pdf

National Research Council. (2005). *Assessment of NIH Minority Research Training Programs: Phase 3*. Washington, DC: National Academies Press.

NEA Education Policy and Practice Department. (2017). *Technology in school: The ongoing challenge of access, adequacy, and equity*. Retrieved from http://www.nea.org/assets/docs/PB19_Technology08.pdf

Nelson, M. K., & Hauck, R. V. (2008). Clicking to learn: A case study of embedding radio-frequency based clickers in an introductory management information systems course. *Journal of Information Systems Education*, *19*(1), 55–64.

Nicolaides, A., & Marsick, V. J. (2016). Understanding Adult Learning in the Midst of Complex Social "Liquid Modernity". *New Directions for Adult and Continuing Education*, *2016*(149), 9–20. doi:10.1002/ace.20172

Niesyto, B., Buckingham, D., & Fisherkeller, J. E. (2003). Video culture: Crossing borders with young people's video productions. *Television & New Media*, *4*(4), 461–482. doi:10.1177/1527476403255813

Nieto, S. (2002). *Language, culture and teaching critical perspectives for a new century*. Mahwah, NJ: Lawrence Erlbaum Associates, Inc.

Niles-Hofmann, L. (2016, July 6). *Data-Driven Learning Design: How to Decode Learner Digital Body Language*. Retrieved January 24, 2018, from http://www.loriniles.com/ebook

Njie-Carr, V. P., Ludeman, E., Lee, M. C., Dordunoo, D., Trocky, N. M., & Jenkins, L. S. (2017). An integrative review of flipped classroom teaching models in nursing education. *Journal of Professional Nursing*, *33*(2), 133–144. doi:10.1016/j.profnurs.2016.07.001 PMID:28363388

Noe, R. A., Clarke, A. D. M., & Klein, H. J. (2014). Learning in the twenty-first-century workplace. *Annual Review of Organizational Psychology and Organizational Behavior*, *1*(1), 245–275. doi:10.1146/annurev-orgpsych-031413-091321

Noll, C., von Jan, U., Raap, U., & Albrecht, U. V. (2017). Mobile Augmented Reality as a feature for self-oriented, blended learning in medicine: Randomized controlled trial. *JMIR mHealth and uHealth*, *5*(9), e139. doi:10.2196/mhealth.7943 PMID:28912113

Noodle Staff. (2015). *41 most innovative K–12 schools in America*. Retrieved from.

Nordin, N. M., Embi, M. A., Norman, H., & Panah, E. (2017). A historical review of mobile learning research in Malaysia and its implications for Malaysia and the Asia-Pacific region. In *Mobile learning in higher education in the Asia-Pacific region* (pp. 137–150). Singapore: Springer. doi:10.1007/978-981-10-4944-6_7

Novak, G. M., Patterson, E. T., Gavrin, A. D., Christian, W., & Forinash, K. (1999). Just-in-time teaching. *American Journal of Physics*, *67*(10), 937–938.

O'Bannon, B. W., Waters, S., Lubke, J., Cady, J., & Rearden, K. (2017). Teachers and Students Poised to Use Mobile Phones in the Classroom. *Computers in the Schools*, *34*(3), 125–141. doi:10.1080/07380569.2017.1347454

O'Connor, G., & Schneider, A. (2017, April 3). *How Russian Twitter bots pumped out fake news during the 2016 election* [Radio podcast]. Retrieved from http://www.npr.org/sections/alltechconsidered/2017/04/03/522503844/how-russian-twitter-bots-pumped-out-fake-news-during-the-2016-election

O'Hara, K. E. (2014). From filtering to Freire: Critical use of the internet in urban classrooms. *Journal of Arts and Humanities*, *3*(6), 9.

O'Neal, L., Gibson, P., & Cotten, S. (2017). Elementary school teachers' beliefs about the role of technology in 21st-century teaching and learning. *Computers In The Schools*, 1–15. doi:10.1080/07380569.2017.1347443

O'Neil, J., & Marsick, V. J. (2007). *Understanding action learning*. New York: AMACOM.

O'Reilly, T. (n.d.). *What is Web 2.0: Design Patterns and Business Models for the Next Generation of Software*. Retrieved from: http://www.oreillynet.com/lpt/a/6228

O'Sullivan, M. K., & Dallas, K. B. (2017). A collaborative approach to implementing 21st century skills in a high school senior research class. *Education Libraries*, *33*(1), 3–9. doi:10.26443/el.v33i1.284

Obrenović, Ž. (2011). Design-based research: what we learn when we engage in design of interactive systems. *Interactions, 18*(5), 56-59.

Occupy Wall Street. (2011a). About Occupy Wall Street. *Occupy Wall Street website*. Retrieved from http://occupywallst.org/about/

Occupy Wall Street. (2011b). You can't evict an idea whose time has come. *Occupy Wall Street website*. Retrieved from http://occupywallst.org/article/you-cant-evict-idea-whose-time-has-come/

Odom, K. L., Roberts, L. M., Johnson, R. L., & Cooper, L. A. (2007). Exploring Obstacles to and Opportunities for Professional Success Among Ethnic Minority Medical Students. *Academic Medicine*, *82*(2), 146–153. doi:10.1097/ACM.0b013e31802d8f2c PMID:17264692

Okan, Z. (2003). Edutainment: Is learning at risk? *British Journal of Educational Technology*, *34*(3), 255–264. doi:10.1111/1467-8535.00325

Olteanu, A., Weber, I., & Gatica-Perez, D. (2016, March). *Characterizing the demographics behind the #BlackLivesMatter movement.* AAAI Spring Symposia on Observational Studies Through Social Media and Other Human-Generated Content, Stanford, CA. Retrieved from http://arxiv.org/abs/1512.05671

Online Courses Have Three Critical Components (and Learning Management Systems Are Not One of Them). (2017). *Distance Learning*, *14*(2), 64-63.

OpenNet. (2010). Global internet filtering map. *OpenNet*. Retrieved from http://map.opennet.net/filtering- pol.html

Ortagus, J. C. (2017). From the periphery to prominence: An examination of the changing profile of online students in American higher education. *The Internet and Higher Education*, *32*(Supplement C), 47–57. doi:10.1016/j.iheduc.2016.09.002

Ortega, J., & Hergovich, P. (2017). *The Strength of Absent Ties: Social Integration via Online Dating*. arXiv preprint arXiv:1709.10478

Ospina, M. B., Bond, K., Karkhaneh, M., Buscemi, N., Dryden, D. M., Barnes, V., & Shannahoff Khalsa, D. (2008). Clinical trials of meditation practices in health care: Characteristics and quality. *Journal of Alternative and Complementary Medicine (New York, N.Y.)*, *14*(10), 1199–1213. PubMed

Oubenaissa, L., Giardina, M., & Battacharya, M. (2002). Designing a framework for the implementation of situated online, collaborative, problem-based activity: Operating within a local and multi-cultural learning context. *International Journal on E-Learning*, *1*(3), 41–46.

Owen, T. (2017, August 12). Car slams into Black Lives Matter protesters. *Vice News*. Retrieved from https://news.vice.com/story/armed-white-supremacists-march-in-charlottesville-state-of-emergency-declared

Owen, E., Song, H., & Kidd, T. (2007). Re-examining the sociocultural factors affecting technology use in mathematics classroom practices. *International Journal of Web-Based Learning and Teaching Technologies*, *4*(2), 37–45.

Pandolfo, N. (2012). As some schools plunge into technology, poor schools are left behind. *The Hechinger Report, 24*.

Papadakis, M. (2017). *Experiences of first-generation students at one ivy league campus* (EdD). Teachers College Columbia University, New York, NY. Retrieved from https://search.proquest.com/docview/1906740375?accountid=13314 (1906740375)

Pask, G. (1976). *Conversation theory: Applications in education and epistemology*. Elsevier Publishing Company.

Pass, F., & Renkl, A. (2003). *Cognitive Load Theory and Instructional Design: Recent Developments*. Academic Press.

Pathak, B. (2016). Improving the Educational Effectiveness of MOOCs by Using Web-Based Information and Communication Technology. In *Revolutionizing Education through Web-Based Instruction* (pp. 186–201). IGI Global. doi:10.4018/978-1-4666-9932-8.ch011

Patterson, C., Stephens, M., Chiang, V., Price, A. M., Work, F., & Snelgrove-Clarke, E. (2017). The significance of personal learning environments (PLEs) in nursing education: Extending current conceptualizations. *Nurse Education Today*, *48*, 99–105. doi:10.1016/j.nedt.2016.09.010 PMID:27744138

Paul, K. (2017, October 21). *How online dating affects divorce rates*. Retrieved December 09, 2017, from https://www.marketwatch.com/story/what-online-dating-could-do-to-divorce-rates-2017-10-16

Paulsen, M. F. (2002). Online Education Systems: Discussion and definition of terms. *NKI Distance Education, 202*.

Pedró, F. (2006). *The new millennium learners: Challenging our views on ICT and learning (No. 9228)*. Inter-American Development Bank.

Pegrum, M. (2014). *Mobile learning: Languages, literacies and cultures*. Springer. doi:10.1057/9781137309815

Pelster, B., Haims, J., Stempel, J., & Vyver, B. (2016). Learning: Employees take charge. Global Human Capital Trends 2016. *The new organization: Different by design*. Retrieved from https://www2.deloitte.com/us/en/pages/human-capital/articles/introduction-human-capital-trends-2016.html

Penney, J., & Dadas, C. (2013). (Re)tweeting in the service of protest: Digital composition and circulation in the Occupy Wall Street movement. *New Media & Society,* 1–17. doi:10.1177/1461444813479593

Perna, L. W., Ruby, A., Baruch, R. F., Wang, N., Scull, J., Ahmad, S., & Evans, C. (2014). *Educational Researcher*. DOI 10.3102/001389X14562423

Peters, M. A. (2016). The postcolonial university. *Linguistic and Philosophical Investigations*, *15*, 77.

Pfeiffer, T., Feiner, S. K., & Mayol-Cuevas, W. W. (2016). 6.4 Group 4: Eyewear Computing for Skill Augmentation and Task Guidance. *Eyewear Computing–Augmenting the Human with Head-Mounted Wearable Assistants*, *23*(3), 199.

Pfund, C., Pribbenow, C. M., Branchaw, J., Lauffer, S. M., & Handelsmann, J. (2006). The Merits of Training Mentors. *Sciencemag.org, 311*, 473-474.

Picciano, A. G., & Seaman, J. (2009). *K-12 online learning*. The Sloan Consortium.

Pickering, J. D., Henningsohn, L., DeRuiter, M. C., de Jong, P. G., & Reinders, M. E. (2017). Twelve tips for developing and delivering a massive open online course in medical education. *Medical Teacher*, 1–6. PMID:28485629

Pittman, T., & Gaines, T. (2015). Technology integration in third, fourth and fifth grade classrooms in a Florida school district. *Educational Technology Research and Development*, *63*(4), 539–554. doi:10.100711423-015-9391-8

Plass, J. L., Moreno, R., & Brünken, R. (2010). *Cognitive load theory*. Cambridge University Press. doi:10.1017/CBO9780511844744

Plaza, I., Demarzo, M. M. P., Herrera-Mercadal, P., & García-Campayo, J. (2013). Mindfulness based mobile applications: Literature review and analysis of current features. *JMIR mHealth and uHealth*, *1*(2), e24. doi:10.2196/mhealth.2733 PMID:25099314

Polanyi, M. (1966). *The tacit dimension*. London, UK: Routledge & Kegan Paul.

Ponce, L. B., Méndez, J. A. J., & Peñalvo, F. J. G. (2014, November). A systematic review of using mobile devices in medical education. In *Computers in Education (SIIE), 2014 International Symposium on* (pp. 205-210). IEEE. 10.1109/SIIE.2014.7017731

Pöntinen, S., Dillon, P., & Väisänen, P. (2017). Student teachers' discourse about digital technologies and transitions between formal and informal learning contexts. *Education and Information Technologies*, *22*(1), 317–335. doi:10.100710639-015-9450-0

Popenici, S. (2015). Deceptive Promises: The Meaning of MOOCs. *Macro-Level Learning Through Massive Open Online Courses (MOOCs): Strategies and Predictions for the Future*, 158-167.

Porfilio, B. J., Roychoudhury, D., & Gardner, L. (2013). Ending the "war against youth": Social media and hip-hop culture as sites of resistance, transformation and (re)conceptualization. *The Journal for Critical Education Policy Studies*, *11*(4), 85–105. Retrieved from https://www.researchgate.net/publication/306107562

Prensky, M. (2001). Digital natives, digital immigrants part 1. *On the Horizon*, *9*(5), 1–6. doi:10.1108/10748120110424816

Prestridge, S., & Tondeur, J. (2015). Exploring elements that support teachers engagement in online professional development. *Education in Science*, *5*(3), 199–219. doi:10.3390/educsci5030199

Preszler, R. W., Dawe, A., Shuster, C. B., & Shuster, M. (2007). Assessment of the effects of student response systems on student learning and attitudes over a broad range of biology courses. *CBE Life Sciences Education*, *6*(1), 29–41. doi:10.1187/cbe.06-09-0190 PMID:17339392

Quinn, C. (2005). *Engaging Learning: Designing e-Learning Simulation Games*. San Francisco, CA: Pfeiffer.

Quinn, C. N. (2011). *Designing mLearning: tapping into the mobile revolution for organizational performance*. Wiley.

Rai, L., & Chunrao, D. (2016). Influencing factors of success and failure in MOOC and general analysis of learner behavior. *International Journal of Information and Education Technology (IJIET)*, *6*(4), 262–268. doi:10.7763/IJIET.2016.V6.697

Ralph, T. W. (1949). Basic principles of curriculum and instruction. Syllabus for Education.

Rana, N. P., & Dwivedi, Y. K. (2015). Using clickers in a large business class: Examining use behavior and satisfaction. *Journal of Marketing Education*, *38*(1), 47–64. doi:10.1177/0273475315590660

Rana, N. P., Dwivedi, Y. K., & Al-Khowaiter, W. A. A. (2016). A review of literature on the use of clickers in the business and management discipline. *International Journal of Management Education*, *14*(2), 74–91. doi:10.1016/j.ijme.2016.02.002

Retrieved November 19, 2017

Richards, R. (2012). Exploring Formative Assessment with a Project Using Mobile Phones. *Ubiquitous Learning. International Journal (Toronto, Ont.)*, *4*(2).

Rideout, V. J., Foehr, U. G., & Roberts, D. F. (2010). *Generation m2: Media in the lives of 8- to 18-year-olds*. Kaiser Family Foundation. Retrieved 5/20/17 from http://www.kff.org/entmedia/upload/8010.pdf

Rikala, J., & Kankaanranta, M. (2012, October). The Use of Quick Response Codes in the Classroom. In mLearn (pp. 148-155). Academic Press.

Rikala, J., & Kankaanranta, M. (2013). Mobile learning. A Review of Current Research. *Reports of the Department of Mathematical Information Technology Series E. Educational Technology*, 1–65.

Robbins, P., & Aydede, M. (2009). A short primer on situated cognition. In P. Robbins & M. Aydede (Eds.), *The Cambridge Handbook of Situated Cognition*. Cambridge, MA: Cambridge University Press.

Robinson, L. (2009). A taste for the necessary: A bourdieuian approach to digital inequality. *Information Communication and Society*, *12*(4), 488–507. doi:10.1080/13691180902857678

Robinson, L. (2011). Information-channel preferences and information-opportunity structures. *Information Communication and Society*, *14*(4), 472–494. doi:10.1080/1369118X.2011.562224

Roblyer, M. D., & Wiencke, W. R. (2003). Design and use of a rubric to access and encourage interactive qualities in distance courses. *American Journal of Distance Education*, *17*(2), 77–98. doi:10.1207/S15389286AJDE1702_2

Rochadiat, A. M., Tong, S. T., & Novak, J. M. (2017). Online dating and courtship among Muslim American women: Negotiating technology, religious identity, and culture. *New Media & Society*.

Rogers, E. M. (2010). *Diffusion of innovations*. Simon and Schuster.

Rogers, S. (2016). Bridging the 21st century digital divide. *TechTrends*, *60*(3), 197–199. doi:10.100711528-016-0057-0

Roo, J. S., & Hachet, M. (2017, October). One Reality: Augmenting How the Physical World is Experienced by combining Multiple Mixed Reality Modalities. In *Proceedings of the 30th Annual ACM Symposium on User Interface Software and Technology* (pp. 787-795). ACM. 10.1145/3126594.3126638

Rosenfeld, M. J. (2017). Marriage, Choice, and Couplehood in the Age of the Internet. Academic Press.

Rosewarne, L. (2016). *Intimacy on the internet: Media representations of online connections* (Vol. 86). Routledge.

Rosin, H. (2013). The touch-screen generation: Young children-- even toddlers-- are spending more and more time with digital technology. Thousands of apps appealing to kids just out of diapers are now released every year. Should parents recoil or rejoice? An examination of what tablets and smartphones are doing to young children's brains. *Atlantic (Boston, Mass.)*, (3), 56.

Ross, J. (2015, August 19). How Black Lives Matter moved from a hashtag to a real political force. *The Washington Post*. Retrieved from https://www.washingtonpost.com/news/the-fix/wp/2015/08/19/how-black-lives-matter-moved-from-a-hashtag-to-a-real-political-force/?utm_term=.26d7d0519fdc

Ross, S. M., Morrison, G. R., & Lowther, D. L. (2010). Educational technology research past and present: Balancing rigor and relevance to impact school learning. *Contemporary Educational Technology*, *1*, 19.

Russell-Bennett, R., Drennan, J., Raciti, M., Bowring, N., & Burns, K. (2017). *Using service design to understand digital preferences for improving applications for tertiary education in under-represented markets*. Academic Press.

Ryan, Y. (2011, January 20). The tragic life of a street vendor. *Al Jazeera*. Retrieved from http://www.aljazeera.com/indepth/features/2011/01/201111684242518839.html

Sadiki, L. (2010, December 27). Tunisia: The battle of Sidi Bouzid. *Al Jazeera English, Opinion*. Retrieved from http://english.aljazeera.net/indepth/opinion/2010/12/20101227142811755739.html

Şahin, F., & Mentor, D. (2016). Using Mobile Phones for Assessment in Contemporary Classrooms. Handbook of Research on Mobile Learning in Contemporary Classrooms, 116-138.

Sahin, F., & Mentor, D. (2017). Creating Teaching and Learning Accountabilities Through Data Analytic Feedback Loops. In *ModSim 2017 International Congress on Modelling and Simulation - Modeling and Simulation in the Age of Data*. Retrieved December 24, 2017 from http://modsimworld.org/papers/2017/Creating_Teaching_and_Learning_Accountabilities_Through_Data_Analytic_Feedback_Loops.pdf

Sahin, I. (2006). Detailed review of Rogers' diffusion of innovations theory and educational technology-related studies based on Rogers' theory. *TOJET: The Turkish Online Journal of Educational Technology, 5*(2).

Şahiner, M. (2016). Öğrenci Merkezli Eğitimde Yeni Bir Yöntem: Ters Yüz Edilmiş Sınıflar. *Turkiye Klinikleri Journal of Medical Education-Special Topics*, *1*(1), 24–32.

Saleeb, N., & Dafoulas, G. (2013). *Effects of Virtual World Environments in Student Satisfaction: An Examination of the Role of. Governance, Communication, and Innovation in a Knowledge Intensive Society*. Academic Press.

Sambunjak, D., Straus, S. E., & Marusic, A. (2006). Mentoring in Academic Medicine: A Systematic Review. *Journal of the American Medical Association, 296*(9), 1103. doi:10.1001/jama.296.9.1103 PMID:16954490

Saravani, S. J. (2010). Standards informing design of library service delivery to mobile devices and nomadic learners. In *Vala2010 abstracts* (pp. 1345–1415). Melbourne: VALA2010 Conference.

Savino, D. M. (2014). The impact of MOOCs on human resource training and development. *Journal of Higher Education Theory and Practice, 14*(3), 59.

Sayek, İ. (2016). Tıp Eğitiminin Gelişimi ve Değişim: 21. Yüzyılda Tıp Eğitimi. In Tıp Eğiticinin El Kitabı (pp. 3-11). Ankara: Güneş Tıp Kitabevleri.

Scardamalia, M. (2002). Collective cognitive responsibility for the advancement of knowledge. In B. Smith (Ed.), *Liberal education in a knowledge society* (pp. 67–98). Chicago: Open Court.

Schank, R. (1997). *Virtual Learning: A Revolutionary Approach to Building a Highly Skilled Workforce*. New York, NY: McGraw Hill.

Schank, R., & Neaman, A. (2001). Motivation and failure in educational simulation design. In K. D. Forbus & P. Feltovich (Eds.), *Smart machines in education: The coming revolution in educational technology* (pp. 37–69). Cambridge, MA: The MIT Press.

Schechter, D. (2011). Occupy: Dissecting Occupy Wall Street. New York, NY: NewsDissector.org & Coldtype.net.

Schneider, N. (2011, October 12). From Occupy Wall Street to occupy everywhere. *The Nation*. Retrieved from https://www.thenation.com/article/occupy-wall-street-occupy-everywhere/

Schön, D. (1983). *The reflective practitioner: How professionals think in action*. New York, NY: Basic Books.

Schön, D. (1989). *Educating the reflective practitioner: Toward a new design for teaching and learning*. San Francisco, CA: Jossey-Bass.

Schradie, J. (2011). The digital production gap: The digital divide and web 2.0 collide. *Poetics, 39*(2), 145–168. doi:10.1016/j.poetic.2011.02.003

Schroeder, H. (2016). Four essential tips for professional development success. *eLearn Magazine*. Retrieved from http://elearnmag.acm.org/featured.cfm?aid=3018842

Schwartz, J., Collins, L., Stockton, H., Wagner, D., & Walsh, B. (2017). *Rewriting the Rules for the Digital Age*. 2017 Deloitte Global Human Capital Trends.

Scott, J. E., & Walczak, S. (2009). Cognitive engagement with a multimedia ERP training tool: Assessing computer self-efficacy and technology acceptance. *Information & Management, 46*(4), 221–232. doi:10.1016/j.im.2008.10.003

Selwyn, N. (2016). *Education and technology: Key issues and debates*. Bloomsbury Publishing.

Servon, L. J., & Nelson, M. K. (2001). Community technology centers and the urban technology gap. *International Journal of Urban and Regional Research, 25*(2), 419–426. doi:10.1111/1468-2427.00320

Seymour, E., Hunter, A.-B., Laursen, S. L., & Deantoni, T. (2004). Establishing the Benefits of Research Experiences for Undergraduates in the Sciences: First Findings from a Three-Year Study. *Science Education, 88*(4), 493–534. doi:10.1002ce.10131

Sezer, B., Onan, A., & Elçin, M. (2016). Sürekli Tıp Eğitiminde Bilişim Teknolojileri. *Turkiye Klinikleri Journal of Medical Education-Special Topics*, *1*(3), 1–6.

Shaffer, D. M., & Collura, M. J. (2009). Evaluating the effectiveness of a personal response system in the classroom. *Teaching of Psychology*, *36*(4), 273–277. doi:10.1080/00986280903175749

Shah, D. (2016, December 25). *By the numbers: MOOCS in 2016*. Retrieved from https://www.class-central.com/report/mooc-stats-2016/

Shank, D. B., & Cotten, S. R. (2014). Does technology empower urban youth? The relationship of technology use to self-efficacy. *Computers & Education*, *70*, 184–193. doi:10.1016/j.compedu.2013.08.018

Shao, F., Tian, W., Lin, W., Jiang, G., & Dai, Q. (2016). Toward a blind deep quality evaluator for stereoscopic images based on monocular and binocular interactions. *IEEE Transactions on Image Processing*, *25*(5), 2059–2074. doi:10.1109/TIP.2016.2538462 PMID:26960225

Shapiro, A. M., Sims-Knight, J., O'Rielly, G. V., Capaldo, P., Pedlow, T., Gordon, K., & Monteiro, K. (2017). Clickers can promote fact retention but impede conceptual understanding: The effect of the interaction between clicker use and pedagogy on learning. *Computers & Education*, *111*, 44–59. doi:10.1016/j.compedu.2017.03.017

Sharecare. (2016). *Sharecare Acquires Virtual Reality Company to Transform Patient Engagement and Medical Education - Sharecare*. Retrieved May 17, 2017, from https://about.sharecare.com/press-releases/sharecare-acquires-virtual-reality-company-transform-patient-engagement-medical-education

Sharehu, A. L., & Emmanuel, A. E. (2015). Readiness of Teachers and Pupils for Use of Mobile Devices as Support for Effective Pedagogy in Nigeria: Could Location be a Major Determinant? *Online Journal of Distance Learning Administration*, *18*(3), 1–11.

Sharples, M., Taylor, J., & Vavoula, G. (2007). A Theory of Learning for the Mobile Age. In R. Andrews & C. Haythornthwaite (Eds.), *The Sage Handbook of Elearning Research* (pp. 221–247). London: Sage. doi:10.4135/9781848607859.n10

Shedletsky, L. (2006). Internet training for older adult learners: An intergenerational mentoring approach. *The LLI Review*, *1*, 34–43.

Sheninger, E. C., & Murray, T. C. (2017). *Learning Transformed: 8 Keys to Designing Tomorrow's Schools, Today*. ASCD.

Sherman, W. R., & Craig, A. B. (2002). *Understanding virtual reality: Interface, application, and design*. Elsevier.

Sheth, A. (2010). Computing for human experience: Semantics-empowered sensors, services, and social computing on the ubiquitous web. *IEEE Internet Computing*, *14*(1), 88–91. doi:10.1109/MIC.2010.4

Sheth, A., Anantharam, P., & Henson, C. (2016). Semantic, cognitive, and perceptual computing: Paradigms that shape human experience. *Computer*, *49*(3), 64–72. doi:10.1109/MC.2016.75

Shirky, C. (2008). *Here Comes Everybody: The Power of Organizing Without Organizations*. Penguin Press.

Siau, J., Sheng, H., & Nah, F. F. H. (2006). Use of a classroom response system to enhance classroom interactivity. *IEEE Transactions on Education*, *49*(3), 398–403. doi:10.1109/TE.2006.879802

Siddiqui, S. (2016, August 9). Russell Simmons talks Black Lives Matter. *Politics for humans podcast. Guardian U.S.* [Radio podcast]. Retrieved from https://www.theguardian.com/politics/audio/2016/aug/09/russell-simmons-black-lives-matter-clinton-trump-politics-for-humans

Simpson, V., & Oliver, M. (2007). Electronic voting systems for lectures then and now: A comparison of research and practice. *Australasian Journal of Educational Technology*, *23*(2), 187–208. doi:10.14742/ajet.1264

Sinclair, H. (2017, August 15). Black Lives Matter calls for U.S. to ban Confederate symbols after Charlottesville violence. *Newsweek*. Retrieved from http://www.newsweek.com/black-lives-matter-ban-confederate-symbols-charlottesville-violence-651106

Slaouti, D., & Barton, A. (2007). Opportunities for practice and development: Newly qualified teachers and the use of information and communications technologies in teaching foreign languages in English secondary school contexts. *Journal of In-service Education*, *33*(4), 405–424. doi:10.1080/13674580701687807

Smedley, B. D., Stith, A. Y., Colburn, L., & Evans, C. H. (2001). *The Right Thing to Do, The Smart Thing to Do: Enhancing Diversity in the Health Professions*. Washington, DC: National Academy Press.

Smith, A., & Anderson, M. (2016). *5 facts about online dating*. Retrieved December 09, 2017, from http://www.pewresearch.org/fact-tank/2016/02/29/5-facts-about-online-dating/

Smith, A., & Anderson, M. (2016, February 29). *5 facts about online dating*. Retrieved December 09, 2017, from http://www.pewresearch.org/fact-tank/2016/02/29/5-facts-about-online-dating/

Smith, S. (n.d.). *SVP HR Operations*. The Telecommunications Company. Retrieved from http://about.att.com/category/all_news.html

Solorzano, D., & Yosso, T. (2002). Critical race methodology: Counter-storytelling as an analytical framework for education research. *Qualitative Inquiry*, *8*(1), 23–44. doi:10.1177/107780040200800103

Song, S., & Owens, E. (2011). Rethinking technology disparities and instructional practices within urban schools: Recommendations for school leadership and teacher training. *Journal of Technology Integration in The Classroom*, *3*(2), 23–36.

Sprague, E. W., & Dahl, D. W. (2010). Learning to click: An evaluation of the personal response system clicker technology in introductory marketing courses. *Journal of Marketing Education*, *32*(1), 93–103. doi:10.1177/0273475309344806

Staples, A., Pugach, M. C., & Himes, D. J. (2005). Rethinking the technology integration challenge: Cases from three urban elementary schools. *Journal of Research on Technology in Education*, *37*(3), 285–311. doi:10.1080/15391523.2005.10782438

Starr, E., Ganco, M., & Campbell, B. (2017). *Strategic Human Capital Management in the Context of Cross-Industry and Within-Industry Mobility Frictions*. Robert H. Smith School Research Paper No. RHS 2753720. http://dx.doi.org/10/2139/ssrn.2753720

Stiglitz, J. E. (2011, May). Of the 1%, by the 1%, for the 1%. *Vanity Fair*. Retrieved from http://www.vanityfair.com/news/2011/05/top-one-percent-201105

Stowell, J. R. (2015). Use of clickers vs mobile devices for classroom polling. *Computers & Education*, *82*, 329–334. doi:10.1016/j.compedu.2014.12.008

Stowell, J. R., & Nelson, J. M. (2007). Benefits of electronic audience response systems on student participation, learning, and emotion. *Teaching of Psychology*, *34*(4), 253–258. doi:10.1080/00986280701700391

Stuart, S., Brown, M., & Draper, S. (2004). Using an electronic voting system in logic lectures: One practitioner's application. *Journal of Computer Assisted Learning*, *20*(2), 95–102. doi:10.1111/j.1365-2729.2004.00075.x

Sturdy, S. (2007). Scientific method for medical practitioners: The case method of teaching pathology in early twentieth-century Edinburgh. *Bulletin of the History of Medicine*, *81*(4), 760–792. doi:10.1353/bhm.2007.0093 PMID:18084106

Suarez-Orozco, C. S.-O., & Todorova, I. (2008). Learning a new land. Cambridge, MA: The Belknap Press of Harvard University Press.

Subramony, D. (2011). Socio-cultural issues in educational technology integration. *Colleagues, 6*(1), Article 10. Retrieved from http://scholarworks.gvsu.edu/colleagues/vol6/iss1/10

Subramony, D. P. (2014). *Reframing the digital divide within a 'Flat World' Context: The McJulien Lecture.* Paper presented at the Association for Educational Communications and Technology (AECT) International Convention 2014, Jacksonville, FL.

Subramony, D. P. (2007). Understanding the complex dimensions of the digital divide: Lessons learned in the Alaskan arctic. *The Journal of Negro Education*, *76*(1), 57–67.

Sun, J. C. Y. (2014). Influence of polling technologies on student engagement: An analysis of student motivation, academic performance, and brainwave data. *Computers & Education*, *72*, 80–89. doi:10.1016/j.compedu.2013.10.010

Sunstein, C. R. (2017). *Republic: Divided democracy in the age of social media*. Princeton, NJ: Princeton University Press. doi:10.1515/9781400884711

SuperData. (Ed.). (2016). *Virtual reality's hardware hurdles hurt forecast*. Retrieved May 16, 2017, from https://www.superdataresearch.com/virtual-realitys-hardware-hurdles-hurt-forecast/

Sweeney, W. B. (2012). Teaching surgery to medical students. *Clinics in Colon and Rectal Surgery*, *25*(03), 127–133. doi:10.1055-0032-1322525 PMID:23997667

Swinnerton, B. J., Morris, N. P., Hotchkiss, S., & Pickering, J. D. (2017). The integration of an anatomy massive open online course (MOOC) into a medical anatomy curriculum. *Anatomical Sciences Education*, *10*(1), 53–67. doi:10.1002/ase.1625 PMID:27315160

Tamim, R. M., Bernard, R. M., Borokhovski, E., Abrami, P. C., & Schmid, R. F. (2011). What forty years of research says about the impact of technology on learning a second-order meta-analysis and validation study. *Review of Educational Research*, *81*(1), 4–28. doi:10.3102/0034654310393361

Tapscott, D. (1998). *Growing Up Digital: The Rise of the Net Generation*. New York: McGraw-Hill.

Täuscher, K., & Kietzmann, J. (2017). Learning from Failures in the Sharing Economy. *MIS Quarterly Executive*, *16*(4).

Taylor, E. W., & Cranton, P. (2012). *The handbook of transformative learning: Theory, research, and practice*. John Wiley & Sons.

Teng, C. E., & Joo, T. M. (2017). Analyzing the Usage of Social Media: A Study on Elderly in Malaysia. *World Academy of Science, Engineering and Technology, International Journal of Social, Behavioral, Educational, Economic. Business and Industrial Engineering*, *11*(3), 713–719.

Tettegah, S. Y., & Mayo, C. (2005). Urban education and technology in the digital Age. *Urban Education*, *40*(4), 363–367. doi:10.1177/0042085905276374

The Body VR. (2015). *Want to stay up to date with our announcements? Sign up below*. Retrieved May 23, 2017, from http://thebodyvr.com/

The Economist. (2017, January 12). *Established education providers v new contenders*. Retrieved December 10, 2017, from https://www.economist.com/news/special-report/21714173-alternative-providers-education-must-solve-problems-cost-and

The Leadership Alliance. (2017). *Summer Research - Early Identification Program (SR-EIP).* Retrieved from http://theleadershipalliance.org/programs/summer-research

The Nellie Mae Education Foundation. (2011). *Integrating technology with student-centered learning.* Quincy, MA: Author.

The White House. (2016). *My Brother's Keeper 2016 progress report. Two years of expanding opportunity and creating pathways to success*. Washington, DC: Author. Retrieved from wh.gov/mbk

Thistlethwaite, J. E., Davies, D., Ekeocha, S., Kidd, J. M., MacDougall, C., Matthews, P., & Clay, D. (2012). The effectiveness of case-based learning in health professional education. A BEME systematic review: BEME Guide No. 23. *Medical Teacher*, *34*(6), e421–e444. doi:10.3109/0142159X.2012.680939 PMID:22578051

Tloaele, M., Hofman, A., Naidoo, A., & Winnips, K. (2014). Using clickers to facilitate interactive engagement activities in a lecture room for improved performance by students. *Innovations in Education and Teaching International*, *51*(5), 497–509. doi:10.1080/14703297.2013.796725

Tondeur, J., van Braak, J., Ertmer, P. A., & Ottenbreit-Leftwich, A. (2017). Understanding the relationship between teachers' pedagogical beliefs and technology use in education: A systematic review of qualitative evidence. *Educational Technology Research and Development*, *65*(3), 555–575. doi:10.100711423-016-9481-2

Tough, A. (1967). *Learning without a teacher*. Toronto: Ontario Institue for Studies in Education.

Toven-Lindsey, B. A. (2017). *Digital Literacy and Career Capital: How College Experiences are Preparing Students for the Transition to Work* (Doctoral dissertation). UCLA.

Townes, C. (2015, October 22). Obama explains the problem with "All Lives Matter." *ThinkProgress*. Retrieved from https://thinkprogress.org/obama-explains-the-problem-with-all-lives-matter-780912d54888

Traxler, J., & Kukulska-Hulme, A. (2016). Introduction to the next generation of mobile learning. *Mobile learning: The next generation*, 1-10.

Traxler, J. (2007). Defining, Discussing and Evaluating Mobile Learning: The moving finger writes and having writ..... *The International Review of Research in Open and Distributed Learning*, *8*(2). doi:10.19173/irrodl.v8i2.346

Tredinnick, L. (2006). Web 2.0 and Business: A pointer to the intranets of the future? *Business Information Review*, *23*(4), 228–234. doi:10.1177/0266382106072239

Trees, A. R., & Jackson, M. H. (2007). The learning environment in clicker classrooms: Student processes of learning and involvement in large university-level courses using student response systems. *Learning, Media and Technology*, *32*(1), 21–40. doi:10.1080/17439880601141179

Trentin, G. (1997). Telematics and on-line teacher training: The POLARIS project. *Journal of Computer Assisted Learning*, *13*(4), 261–270. doi:10.1046/j.1365-2729.1997.00029.x

Trew, J. L., & Nelsen, J. L. (2012). Getting the most out of audience response systems: Predicting student reactions. *Learning, Media and Technology*, *37*(4), 379–394. doi:10.1080/17439884.2011.621957

Tufekci, Z. (2010, September 27). What Gladwell gets wrong: The real problem is scale mismatch (Plus, weak and strong ties are complementary and supportive). *Technosociology*. Retrieved from http://technosociology.org/?p=178

Turkay, S., Hoffman, D., Kinzer, C. K., Chantes, P., & Vicari, C. (2014). Toward understanding the potential of games for learning: Learning theory, game design characteristics, and situating video games in classrooms. *Computers in the Schools*, *31*(1-2), 2–22. doi:10.1080/07380569.2014.890879

Turkle, S. (2006). Tethering. In C. A. Jones, B. Arning, & J. Farver (Eds.), *Sensorium: Embodied Experience, Technology, and Contemporary Art* (pp. 220–226). Cambridge, MA: The MIT Press.

Turkle, S. (2008). Always-on/always-on-you: The tethered self. In J. E. Katz (Ed.), *Handbook of Mobile Communication Studies* (pp. 121–137). Cambridge, MA: MIT Press. doi:10.7551/mitpress/9780262113120.003.0010

Tweed, S. R. (2013). Technology implementation: teacher age, experience, self-efficacy, and professional development as related to classroom technology integration. *Electronic Theses and Dissertations.* Paper 1109. Retrieved from http://dc.etsu.edu/etd/1109

U.S. Department of Commerce. (1999). *Falling through the net: Defining the digital divide.* Washington, DC: National Telecommunications and Information Administration.

U.S. Department of Education, Office of Elementary and Secondary Education, Office of State Support. (2015). *Improving basic programs operated by local educational agencies (Title I, Part A).* Author.

Udacity. (2014). *Nanodegree Programs.* Retrieved December 10, 2017, from https://www.udacity.com/nanodegree

United Nations *Educational, Scientific and Cultural Organization (UNESCO). (2012). Turning* on mobile learning: Global themes. UNESCO working paper series on mobile learning. Retrieved September 2017, from http://unesdoc.unesco.org/ images/0021/002164/216451E.pdf

United Nations Educational, Scientific and Cultural Organization (UNESCO). (2017). Retrieved from http://www.unesco.org/new/en/unesco/themes/icts/

United Nations World Youth Report. (2016). *The world youth report on youth civic engagement.* New York, NY: United Nations Department of Economic and Social Affairs.

Van der Stappen, E., & Zitter, I. (2016). Exploring Design Principles for Technology-Enhanced Workplace Learning. In Bled eConference (p. 14). Academic Press.

van Dijk, J. A. G. M. (2006). Digital divide research, achievements and shortcomings. *Poetics*, *34*(4-5), 221–235. doi:10.1016/j.poetic.2006.05.004

Van Raaij, E. M., & Schepers, J. J. (2008). The acceptance and use of a virtual learning environment in China. *Computers & Education*, *50*(3), 838–852. doi:10.1016/j.compedu.2006.09.001

Vasudevan, L. (2004). *Telling different stories differently: The possibilities of (counter)storytelling with African -American adolescent boys* (Doctoral Dissertation). Retrieved from http://repository.upenn.edu/dissertations

Vélez, V., Perez Huber, L., Benavides Lopez, C., de la Luz, A., & Solórzano, D. G. (2008). Battling for human rights and social justice: A Latina/o critical race media analysis of Latina/o student youth activism in the wake of 2006 anti-immigrant sentiment. *Social Justice (San Francisco, Calif.)*, *35*(1), 7–27. Retrieved from http://www.jstor.org/stable/29768477

Verger, A., Lubienski, C., & Steiner-Khamsi, G. (2016). The emergence and structuring of the global education industry: Towards an analytical framework. In A. Verger, C. Lubienski, & G. Steiner-Khamsi (Eds.), *World Yearbook of Education 2016: The Global Education Industry.* New York: Routledge.

Villanueva, K., & Vaidya, J. (2016). Transforming Learning with Mobile Games: Learning with mGames. In D. Mentor (Ed.), *Handbook of Research on Mobile Learning in Contemporary Classrooms* (pp. 260–278). Hershey, PA: IGI Global. doi:10.4018/978-1-5225-0251-7.ch013

Vinh, T. L., Bouzefrane, S., Farinone, J., Attar, A., & Kennedy, B. P. (2015). Middleware to integrate mobile devices, sensors, and cloud computing. *Procedia Computer Science*, *52*, 234–243. doi:10.1016/j.procs.2015.05.061

Virtual Reality Society. (2016). *Virtual Reality in Healthcare.* Retrieved May 16, 2017, from https://www.vrs.org.uk/virtual-reality-healthcare/

Vitsas, P. A. (2016). Commercial simulator applications in flight test training. *Journal of Aerospace Engineering*, *29*(4), 1–12. doi:10.1061/(ASCE)AS.1943-5525.0000589

Voelkel, S., & Bennett, D. (2014). New uses for a familiar technology: Introducing mobile phone polling in large classes. *Innovations in Education and Teaching International*, *51*(1), 46–58. doi:10.1080/14703297.2013.770267

Volery, T., & Lord, D. (2000). Critical success factors in online education. *International Journal of Educational Management*, *14*(5), 216–223. doi:10.1108/09513540010344731

Von Ins, C. (2016). *A Working Learner Is a Modern Learner*. Retrieved March 31, 2018, from https://www.td.org/insights/a-working-learner-is-a-modern-learner?_ga=2.139790707.1582201096.1522499837-1585515252.1522499837

Wahbeh, H., Elsas, S.-M., & Oken, B. S. (2008). Mind–body interventions: Applications in neurology. *Neurology*, *70*(24), 2321–2328. doi:10.1212/01.wnl.0000314667.16386.5e PMID:18541886

Walker, D. d., Lindner, J. R., Murphrey, T. P., & Dooley, K. (2016). Learning management system usage. *Quarterly Review Of Distance Education*, *17*(2), 41–50.

Walsh, E., Daems, W., & Steckel, J. (2017). Assistive Pointing Device Based on a Head-Mounted Camera. *IEEE Transactions on Human-Machine Systems*, *47*(4), 590–597. doi:10.1109/THMS.2017.2649884

Walther, J. B., Hoter, E., Ganayem, A., & Shonfeld, M. (2015). Computer-mediated communication and the reduction of prejudice: A controlled longitudinal field experiment among Jews and Arabs in Israel. *Computers in Human Behavior*, *52*, 550–558. doi:10.1016/j.chb.2014.08.004

Wamuyu, P. K. (2017). Bridging the digital divide among low income urban communities. Leveraging use of community technology centers. *Telematics and Informatics*, *34*(8), 1709–1720. doi:10.1016/j.tele.2017.08.004

Wang, C. Y., Wu, H. K., Lee, S. W. Y., Hwang, F. K., Chang, H.-Y., Wu, Y. T., ... Tsai, C. C. (2014). A review of research on technology-assisted school science laboratories. *Journal of Educational Technology & Society*, *17*(2), 307–320.

Watkins, K. E., Marsick, V. J., & Faller, P. G. (2012). Transformative learning in the workplace: Leading learning for self and organizational change. The handbook of transformative learning, 373-387.

Watkins, K. E., & Marsick, V. J. (2014). Adult education & human resource development: Overlapping and disparate fields. *New Horizons in Adult Education and Human Resource Development*, *26*(1), 42–54. doi:10.1002/nha3.20052

Watkins, K., & Marsick, V. (1993). *Sculpting the learning organization: Lessons in the art and science of systemic change*. San Francisco, CA: Jossey-Bass.

Watkins, K., & O'Neil, J. (2013). The Dimensions of the Learning Organization Questionnaire: A Nontechnical Manual. *Advances in Developing Human Resources*, *15*(2), 133–147. doi:10.1177/1523422313475854

Watson, K. (2017, March 30). Russian bots still interfering in U.S. politics after election, says expert witness. *CBS News*. Retrieved from https://www.cbsnews.com/news/russian-bots-still-interfering-in-u-s-politics-after-election-expert/

Watson, T. N., & Bogotch, I. (2015). Reframing parent involvement: What should urban school leaders do differently? *Leadership and Policy in Schools*, *14*(3), 257–278. doi:10.1080/15700763.2015.1024327

Watters, A. (2014). *The monsters of education technology. Creative Commons Attribution-ShareAlike 4.0 License*. Retrieved October 15,2016 from http://monsters.hackeducation.com/

Weinberg, F. J., & Lankau Melanie, J. (2011). Formal Mentoring Programs: A Mentor-Centric and Longitudinal Analysis. *Journal of Management*, *37*(6), 1527–1557. doi:10.1177/0149206309349310

Weiss, J. (2007). "Eyes on me regardless": Youth responses to high school surveillance. *Educational Foundations*, *21*(1-2), 47–69.

Wendling, M. (2015, March 12). "Blue Lives Matter" trends after officers shot. *BBC Trending*. Retrieved from http://www.bbc.com/news/blogs-trending-31853299

Weng, C., & Tang, Y. (2014). The relationship between technology leadership strategies and effectiveness of school administration: An empirical study. *Computers & Education, 76*, 91–107. doi:10.1016/j.compedu.2014.03.010

Wenger, E. (2010). Communities of practice and social learning systems: the career of a concept. *Social Learning Systems and Communities of Practice, 3*, 179-198.

West, W., & Vosloo, S. (2013). *UNESCO policy guidelines for mobile learning*. Paris: UNESCO. Retrieved from http://unesdoc.unesco.org/images/0021/002196/219641e.pdf

Wielgosz, J., Schuyler, B. S., Lutz, A., & Davidson, R. J. (2016). Long-term mindfulness training is associated with reliable differences in resting respiration rate. *Scientific Reports*, 6. PMID:27272738

Wildi-Yune, J., & Cordero, C. (2015). *Corporate Digital Learning: How to Get It "Right"*. Retrieved from https://www.imd.org/globalassets/publications/working-papers/docs/working-paper_corporate-digital-learning-final--05-05-15.pdf

Wilkin, S., Davies, H., & Eynon, R. (2017). Addressing digital inequalities amongst young people: Conflicting discourses and complex outcomes. *Oxford Review of Education, 43*(3), 332–347. doi:10.1080/03054985.2017.1305058

Williams, L. C. (2017, May 15). Where are they now? The Russian bots that disrupted the 2016 election. *ThinkProgress* [Radio podcast]. Retrieved from https://thinkprogress.org/russian-bots-where-are-they-now-e2674c19017b/

Wilson, B. G. (2017). Constructivism for active, authentic learning. In R. A. Reiser & J. V. Dempsey (Eds.), *Trends and Issues in Instructional Design and Technology* (pp. 61–67). Boston, MA: Pearson.

Wong, L.-H. (2013). Analysis of Students' After-School Mobile-Assisted Artifact Creation Process in a Seamless Language Learning Environment. *Journal of Educational Technology & Society, 16*(2), 198–211.

Woodill, G. (2013, September 03). *MobiLearnAsia: Asia's Growing Mobile Learning Conference*. Retrieved December 03, 2017, from https://gowithfloat.com/2013/09/mobilearnasia-asias-growing-mobile-learning-conference/

Wroblewski, L. (2012). *Mobile first: Préface de Jeffrey Zeldmann*. Editions Eyrolles.

Wu, P. H., Hwang, G. J., Tsai, C. C., Chen, Y. C., & Huang, Y. M. (2011). A pilot study on conducting mobile learning activities for clinical nursing courses based on the repertory grid approach. *Nurse Education Today, 31*(8), e8–e15. doi:10.1016/j.nedt.2010.12.001 PMID:21196068

Wu, P. H., Hwang, G. J., & Tsai, W. H. (2013). An expert system-based context-aware ubiquitous learning approach for conducting science learning activities. *Journal of Educational Technology & Society, 16*(4), 217–230.

Wyne, M. F. (2015). Merging mobile learning into traditional education, 2013–2016. *The International Conference on E-Learning in the Workplace*.

Wyner, J. (2014). *What excellent community colleges do*. Cambridge, MA: Harvard Education.

Wyn, J. (2009). *Touching the future: Building skills for life and work*. Melbourne, Australia: Australian Council for Educational Research Press.

Yang, H. (2017). *Exploring a Semi-Virtual Reality System Impacting Learning Curves of College Students*. Academic Press.

Year, Up. (2016). *2016 annual report: Betting big on year up*. Boston, MA: Author.

Yiu, L., & Parker, J. (2000). *Action Learning in Cyberspace: A Case Example of a Technology-Supported Laboratory Management Development Program*. Paper presented at the Transformative Learning Conference, New York, NY.

Yoo, Y. (2010). Computing in everyday life: A call for research on experiential computing. *Management Information Systems Quarterly*, *34*(2), 213–231. doi:10.2307/20721425

Yorks, L., & Kasl, E. (2002). Towards a theory and practice for whole person learning: Reconceptualizing experience and the role of affect. *Adult Education Quarterly*, *52*(3), 176–192. doi:10.1177/07417136020523002

Yosso & Solorzano. (2001). Critical Race Theory in Chicana/o Education. Paper presented at the National Association for Chicana and Chicano Studies Annual Conference, San Jose, CA.

Yourstone, S. A., Kraye, H. S., & Albaum, G. (2008). Classroom questioning with immediate electronic response: Do clickers improve learning? *Decision Sciences Journal of Innovative Education*, *6*(1), 75–88. doi:10.1111/j.1540-4609.2007.00166.x

Yousef, A. M. F., Chatti, M. A., Schroeder, U., & Wosnitza, M. (2014, July). What drives a successful MOOC? An empirical examination of criteria to assure design quality of MOOCs. In *Advanced Learning Technologies (ICALT), 2014 IEEE 14th International Conference on* (pp. 44-48). IEEE. 10.1109/ICALT.2014.23

Yuan, L., Powell, S., & Cetis, J. (2013). *MOOCs and open education: Implications for higher education*. Academic Press.

Yuen, S. C. Y., Yaoyuneyong, G., & Johnson, E. (2011). Augmented reality: An overview and five directions for AR in education. *Journal of Educational Technology Development and Exchange*, *4*(1), 11. doi:10.18785/jetde.0401.10

Zheng, L. (2015). In-service professional development in an online environment: What are South Australian English as an additional language or dialect teachers' views? *Professional Development in Education*, *41*(3), 527–545. doi:10.1080/19415257.2014.902860

Zheng, M., & Waller, M. P. (2017). ChemPreview: An augmented reality-based molecular interface. *Journal of Molecular Graphics & Modelling*, *73*, 18–23. doi:10.1016/j.jmgm.2017.01.019 PMID:28214437

Zhu, C. (2012). Student satisfaction, performance, and knowledge construction in online collaborative learning. *Journal of Educational Technology & Society*, *15*(1), 127–136.

Zielezinski, M. B. (2016). Using technology to support underserved students. *Education Digest*, *82*(3), 30–34.

Zuckerman, E. (2011, January 15). The first Twitter revolution? *Foreign Policy*. Retrieved from http://foreignpolicy.com/2011/01/15/the-first-twitter-revolution-2/

About the Contributors

Dominic Mentor (Ed.D.) is an Adjunct Asst. Prof. at Teachers College Columbia University and corporate trainer who creates self-directed, digitally infused autonomous learning systems. He initiated and co-designed the USA's first mLearning graduate class and a social media fellowship for the NYC Mayor's Office of Adult Education. He won the Optimas Gold Vision 2014 award from Workforce Magazine and the IELA 2015 blended eLearning award. He was keynote speaker at the 2016 International Conference on Globalization, Education, & Diversity, 2015 KDP Honors Society, the 2014 Educational Technology summit in South Africa and 2012 Technology in Education conference at William Patterson Univ. He conceived of and served as Editor and contributing author of the Handbook of Research on Mobile Learning in Contemporary Classrooms as well as chapters in Tablets in K-12 Education: Integrated Experiences and Implications, and the Encyclopedia of Mobile Phone Behavior. Other publications include articles in Educause Journal and Center for African Studies' Forum. Presentations include UNESCO (Paris, France), Harvard's 21st Century Academic Forum Conference (Boston, USA), Urban Education International Conference (2014, 2016) (NY) and eLearning in the Workplace 2014 conference.

* * *

Paul Acquaro holds a Doctorate of Education from Teachers College, Columbia University and has focused on the development and implementation of online learning tools. He currently works in higher education administration and is an adjunct instructor at New York University.

Renae Beaumont is a Clinical Psychologist and Research Associate at the Department of Child and Adolescent Psychiatry, Weill Cornell Medicine, New York Presbyterian Hospital. She is the developer of the Secret Agent Society Program, which is used in schools, hospials and clinics in 11 countries worldwide to teach children social-emotional resilience skills. Renae is passionate about using technology to improve mental health outcomes for children and youth. Over the past 12 years, she has published numerous research papers and book chapters and spoken at international conferences on using strengths-based approaches and technological innovations to upskill and empower children and the parents and professionals who support them.

Edward Bednar has worked in information technology for 28 years, much of that time in chief architect roles at Sabre, IBM, and LexisNexis. He is currently an associate director at Deloitte where he leads strategy and architecture in the firm's global CTO organization. In this role, he is responsible for enterprise architecture, platform architecture, and architecture delivery, including enterprise cloud hosting and related strategic infrastructure programs. Ed holds an M.A. from Columbia University and a B.B.A. from Baylor University. His academic and research interests are situated computing, affective computing, artificial intelligence, and computational linguistics.

S. Ayhan Caliskan is an Associate Professor of Medical Education at Ege Unversity Faculty of Medicine İzmir, Turkey. He graduated from same medical school in 1992 and received his doctorate in Public Health in 2003. He has been working at the Department of Medical Education since 1999. He received fellowship from FAIMER in 2010. His research areas include, assessment, e-learning, skills training, and simulated patients.

Regina Casale is a doctoral candidate at Teachers College at Columbia University. Regina has been a United States Institute of Peace Fellow, has been recognized by CARECEN and Suffolk County Human Rights Commissioner for her work in education and human rights, and received a Foreign Language Area Studies scholarship where she lived in the Napo, with a Quichua community in the Amazon. Her focus on teaching language through a human rights lens has been presented in several educational forums such as; Teachers College, IEARN in Egypt, AATSP in Costa Rica and San Antonio, Conferencia de Interculturalidad Andina and the Foro Mundial de Migracion in Ecuador. She has written several grants to sustain LIFE through My Eyes, a project which she originated in Ecuador.

Simon C. H. Chan (PhD, The Hong Kong Polytechnic University) is a teaching fellow at Department of Management and Marketing, The Hong Kong Polytechnic University. He has published research papers in Academy of Management Journal, Journal of Organizational Behavior, Human Relations, International Journal of Human Resources Management, Personnel Review. His main research interests focus mainly on leadership and organizational behavior.

Kadir Demir holds a Master of Computer Education and Instructional Technology (CEIT) from Dokuz Eylul University in 2014. He is a Ph.D. student in CEIT at Anadolu University. He works as a Research Assistant at Dokuz Eylul University. His research areas include mobile learning, graphic and animation, wearable technologies, computational thinking and digital literacies.

Ruth Gotian, EdD, MS is Assistant Dean for Mentoring, Executive Director of the Mentoring Academy, Chief Learning Officer in Anesthesiology and Assistant Professor of Education in Anesthesiology at Weill Cornell Medicine. She previously ran the Weill Cornell/Rockefeller/Sloan Kettering Tri-Institutional MD-PhD Program for over two decades. Her research and work is focused on what makes people successful. Dr. Gotian has a doctorate in Adult Learning and Leadership from Teachers College Columbia University, and both bachelors and masters degrees in Business Management from SUNY Stony Brook. Starting her career in finance, Dr. Gotian made the move back to higher education, a field she has been immersed in for over a quarter of a century.

Sharon Hinton is a Research Manager at the University of Queensland in Brisbane, Australia and a private consultant. She has a PhD in psychology from the University of Queensland, where she conducted a trial of a telehealth parenting program for parents of children with developmental disorders. Sharon has considerable expertise in the development, delivery and evaluation of e-learning programs and technologies across multiple vocational sectors including health, law, engineering and education. Sharon has a particular passion for employing patient-centred technology to enhance population health management.

Himanshu Joshi is a doctoral student in the Adult Learning and Leadership department at Teachers College, Columbia University. He has worked on programs to improve teaching in schools and leadership, teamwork and culture in organizations. He served as a research associate at Columbia Business School, researching Personal Leadership and Delivering Transformational learning through technology enabled applications, particularly mobile based platforms. He has also served as General Manager with XSEED Institute (iDiscoveri Leadership) where he focused on Learning interventions that facilitate Leadership Capability development for Middle and Senior Management, Vision & Alignment for top teams, Building Culture, Enabling change and embedding enabling mindsets for key business teams or cross organisation.

Ozan Karaca is working as Instructional Technologist, Instructional Production Specialist in Institution of Health Science in Ege University. He is working in this department, on the adaptation of postgraduate education with e-learning and the organization of basic scientific research education. Furthermore a Medical Education Ph.D. candidate in that Institute. He is studying for the Ph.D. thesis, on lifelong learning in health education. In addition, he earned his master and bachelor degrees in Computer and Instructional Technology.

Stephen Ko is a Teaching Fellow in Entrepreneurship and Strategic Management at the Department of Management and Marketing, The Hong Kong Polytechnic University. He teaches courses on Entrepreneurship and Research Methods. Currently, he serves as a member of the Board of Directors of the Global Business and Technology Association, and as an academic advisor of the Asia Case Research Centre of the University of Hong Kong.

Victoria J. Marsick is a professor of Adult and Organizational Learning in the Department of Organization & Leadership, Teachers College, Columbia University. Since the publication of her first book, Learning in the Workplace, she has brought a rich understanding of the nature of individual, team and organizational learning to scholars and practitioners of adult and continuing education and Human Resource Development. Marsick is Academic Program Director of Adult Learning & Leadership programs. She has held served in the past as head of the innovative Adult Education Guided Independent Study (AEGIS) program at Teachers College, as the department chair, and as a co-director of research centers with a focus on adult and organizational learning.

Adam Neaman is Head of Learning Design and Delivery at Citadel Global Equities in the Greater New York City Area. Previously Adam worked on Learning Design Leadership, Everwise (Silicon Valley startup focused on mentoring and leadership development) as well as at McKinsey & Company. Adam is a results-driven learning design leader with 20 years experience in world-class global corporations and vendor organizations. Passionate about research-grounded design and integrating learning into work.

Kate Sofronoff is an Associate Professor in the School of Psychology at the University of Queensland. Kate has worked in autism research for the past 20 years with a focus on interventions for children and adolescents and their parents. She was involved in the seminal trials of cognitive behaviour therapy for children with high functioning autism spectrum conditions and anxiety and has continued to work with colleagues to extend this work to other social emotional programs with a particular focus on the Secret Agent Society multi-media program. Kate is currently working on projects through the Australian Autism Co-operative Research Collaboration and has recently collaborated on eTherapy and online programs to assist academics working with students on the spectrum and parents of children with developmental disorders.

Dino Sossi has produced media for AOL, CBS newsmagazine "60 Minutes," CNN, the New York Times, the United Nations and Viacom's vh1. His documentary films have screened at festivals in New York and Los Angeles, academic institutions such as Berkeley, Cambridge, Columbia, Harvard, Oxford and Pennsylvania, as well as the United Nations headquarters in New York. Dino's work has been broadcast on CBC, CTV, Discovery USA, Globe & Mail, IFC, Life, MTV Canada, MuchMore-Music, One, Pridevision and PrimeTV. His digital storytelling has been exhibited at the San Francisco Museum of Modern Art. Dino taught and was a digital media consultant at Columbia. He also taught digital storytelling at Adelphi University. Dino has performed storytelling at the Moth StorySLAM in New York. Dino teaches at Columbia University's School of International and Public Affairs (SIPA) and the School of Professional Studies at New York University (NYU). Dino earned his Doctorate, Master of Arts and Master of Education in Instructional Technology and Media at Columbia University. He earned first class honors for his Master of Philosophy thesis at the University of Cambridge where he also performed with the Footlights comedy troupe. Dino previously studied at Columbia Law School's School Law Institute and the Oxford Internet Institute Summer Doctoral Programme. He was previously a fellow at the Berkman Center for Internet and Society at Harvard University. For more information, please visit www.DinoSossi.com or www.TheHomeFilm.com.

Angelia K. Swinton is entering her third year as a doctoral student in Curriculum and Instruction with a focus in Urban Education at Texas A & M University. She has been a K-12 educator for 25 years. Angelia is a mom to teenagers, Gabrielle and Jordan. Her current research interests include: factors that contribute to academic or behavioral struggles in school; socioeconomic, racial/ethnic, and geographic disparities in children's health and well-being; and educational equity among vulnerable populations.

Lauren Ashley Williams uses her interests and passions in urban education, multicultural education, teacher empathy, and the cultural beliefs of pre-service and in-service teachers to drive her research and continued education as a doctoral student at Texas A&M University. Her passions revolve around students of poverty and the teacher selection and retention of effective teachers and administrators for those students.

Index

O

P

R

S

T

U

V

W

Y

www.ingramcontent.com/pod-product-compliance
Ingram Content Group UK Ltd.
Pitfield, Milton Keynes, MK11 3LW, UK
UKHW050951290926
15470UKWH00005B/1764

9 781522 541110